Table of Contents

Applications and Challenges of Reconfigurable Intelligent Surfaces in 6G

Nalini M.
Sri Sairam Engineering College, India

Malathy Sathyamoorthy
KPR Institute of Engineering and Technology, India

Rajesh Kumar Dhanaraj
Symbiosis International University, India

IGI Global
Scientific Publishing
Publishing Tomorrow's Research Today

Vice President of Editorial	Melissa Wagner
Managing Editor of Acquisitions	Mikaela Felty
Managing Editor of Book Development	Jocelynn Hessler
Production Manager	Mike Brehm
Cover Design	Phillip Shickler

Published in the United States of America by
IGI Global Scientific Publishing
701 East Chocolate Avenue
Hershey, PA, 17033, USA
Tel: 717-533-8845
Fax: 717-533-7115
Website: https://www.igi-global.com E-mail: cust@igi-global.com

Library of Congress Cataloging-in-Publication Data

LCCN: 2024055467 (CIP Data Pending)
ISBN13: 9798369380994
Isbn13Softcover: 9798369381007
EISBN13: 9798369381014

British Cataloguing in Publication Data
A Cataloguing in Publication record for this book is available from the British Library.

Detailed Table of Contents

A new technology that can improve wireless communication systems, especially in next-generation networks like 6G, is Reconfigurable Intelligent Surfaces (RIS). A wide range of passive, adjustable components make up the surfaces of RIS, which can be configured to dynamically reflect or refract incoming electromagnetic waves. The wireless propagation environment can be actively controlled by RIS by modifying the phase, amplitude, and polarization of these waves. This results in stronger, more direct transmission channels, less interference, and fewer signal blockages. RIS allows for low latency, large data speeds, and energy-efficient communication in 6G applications. RIS uses less energy and produces less noise than traditional relay and antenna systems because it modifies existing signals rather than creating new ones.

A novel technology that has emerged is the so-called reconfigurable intelligent surface. Its purpose is to change the properties of electromagnetic waves to improve wireless communication networks. In this article, the authors analyse the use cases, embedding methods, and research problems of RIS. Special attention is paid to integrating RIS into the communication system, issues of system architecture, its implementation features, and the functions afforded by RIS in the 5G and 6G. It mentioned some important challenges such as problems with scalability, energy efficiency, and material limitations to appreciate the future Development and the adoption of RMS in important areas.

As such, backscatter communication with RIS's support is the suitable technology for using in 6G IoT networks, according to the efficiency, reliability, security, and low energy consumption. The analysis indicates that the integration of the RIS with backscatter communication enables achieving additional enhancements of up to 2.5 dB in the channel gain and up to 17% in the data rate as well as up to 3 dBm/Hz in energy efficiency. In particular, this book chapter will be devoted to an introduction to RIS-assisted backscatter communication, as well as to the elaboration of certain technical principles of this system, the definition of its possibilities in the context of improving large-scale IoT networks for the sixth generation. This chapter will also contain information concerning these problems, directions of further research, and potential applications of this technology.

Reconfigurable Intelligent Surfaces and Artificial Intelligence combined offer a major breakthrough in wireless communication technology. These are a promising technology that has the potential to significantly improve network performance and signal propagation in wireless communication. RIS is capable of optimizing wireless channels, resulting in increased signal strength, decreased interference, and expanded coverage by dynamically modifying the electromagnetic parameters of the surrounding environment. This study provides an extensive overview of the current state of AI-powered RIS-assisted networks, and investigates the functioning processes, deployment approaches, and possible applications of RIS technology. The limitations of hardware design, computational complexity, interoperability, and security issues are highlighted as major research problems. This review paper explore as a foundational reference for research challenges and providing insights into the transformative potential of AI-RIS and guiding future innovations in this rapidly evolving field.

They have made the conversation much better because so many people use them. One example is the intelligent surface (RIS), which can be changed to help wireless devices work better. Some studies say that ML-optimized (RIS)--assisted wireless communication systems might be a good choice since they help the signal stay strong in the wireless frequency band. That being said, we can't say which method or ways of doing things are the best because there are so many of them. More than one study has looked at these two ideas as ways to improve radio networks. As part of the study, they want to make a list of all of them. The project's goal is to find the best ways to use machine learning to make a system that works with RIS. There are different groups for each type of speed boost, database, machine learning, and hard code.

Reconfigurable intelligent surfaces (RIS) and wirelessly piloted unmanned aerial vehicles (UAVs) are a revolutionary step in wireless communication and energy transfer. This chapter focuses on the integration of RIS technology and wirelessly piloted UAVs, including various design challenges and potential applications. RIS made of programmable metamaterials can dynamically control electromagnetic waves, thereby improving signal strength, coverage, and efficiency. When combined with wireless energy transfer (WPT) technologies, UAVs can operate for long periods of time without requiring frequent battery replacement or landing.The major challenges faced in this integration include optimizing RIS configurations for effective WPT, ensuring uninterrupted communication in dynamic environments, & maintaining energy efficiency. In addition, this chapter investigates the implications of RIS-assisted WPT for UAVs in various fields such as disaster management, agricultural monitoring, & logistics.

This study looks into a new method for instantaneous Doppler positioning using Iridium signals, which are interpreted as satellite signals of opportunity (SOP). It begins by describing the structure and features of the Iridium SOP-based positioning system. The method for instantaneous Doppler positioning is then described, followed by a discussion of how measurement inaccuracies and satellite orbital errors affect positioning accuracy. Furthermore, the impact of constellation shape on accuracy, particularly in terms of positioning between receiver locations, is evaluated. Simulations and real-world trials with IRIDIUM NEXT signals demonstrate that static receivers can reach 22-meter (1σ) two-dimensional positioning accuracy with height assistance and unimpeded sky views.

Reconfigurable Intelligent Surfaces (RIS) are emerging as a groundbreaking technology in the realm of wireless communications, with significant implications for enhancing physical layer security. This chapter delves into the integration of RIS with advanced security techniques, exploring how this innovative technology can be harnessed to address the growing challenges of securing wireless networks. this chapter delves into the future trends and advancements in RIS technology, including next-generation RIS architectures and their potential integration with emerging technologies like 6G. It explores how RIS could pave the way for innovative security protocols and play a pivotal role in advancing secure wireless network infrastructures. RIS technology enables the dynamic and intelligent modification of radio environments through programmable surfaces, which can adjust and optimize signal paths to improve both communication efficiency and security. This chapter provides valuable insights into the current applications and future prospects of RIS in enhancing wireless network security.

Reconfigurable Intelligent Surfaces (RIS) are innovative, engineered structures composed of a large array of passive or semi-passive reflecting elements, known as meta-atoms, capable of controlling the behaviour of electromagnetic waves. These elements can be dynamically reconfigured to manipulate the amplitude, phase, polarization, and direction of the reflected signals. RIS systems are typically controlled via software or electronic components that adjust how each meta-atom interacts with incoming signals, thus enabling real-time control of the radio propagation environment. Unlike conventional communication elements like active antennas, which directly transmit or receive signals, RIS modifies the characteristics of an impinging signal without active power amplification.

Reconfigurable Intelligent Surfaces are emerging as a transformative technology in wireless communication, particularly in the context of massive Multiple-Input Multiple-Output (mMIMO) systems and Non-Orthogonal Multiple Access (NOMA) networks. This chapter provides an in-depth exploration of how RIS can enhance the performance of mMIMO and NOMA networks, focusing on both practical applications and research challenges. RIS technology enables dynamic control over the wireless environment by adjusting signal reflections and enhancing signal propagation, which can significantly improve the efficiency and effectiveness of mMIMO and NOMA systems. For mMIMO, RIS can optimize spatial beamforming and mitigate interference, leading to enhanced capacity and coverage. In NOMA networks, RIS can This chapter offers a comprehensive overview of the potential of RIS to revolutionize mMIMO and NOMA networks, while also addressing the critical research challenges that must be overcome to fully realize its benefits.

Rapidly growing wireless communication technologies have promoted the use of reconfigurable intelligent surfaces (RIS) in smart cities. This chapter explains how RIS technology can help transform urban environments, improving connectivity, efficiency, and sustainability. Reconfigurable intelligent surface technology has the ability to control electromagnetic waves in a flexible and dynamic way. This ability can solve many problems of modern cities, such as network congestion, energy shortages, and the integration of various IoT devices. This chapter thoroughly analyzes the main challenges associated with the implementation of RIS, such as the design and optimization of RIS systems, real-time reconfiguration algorithms, and compatibility with existing communication infrastructure. It also discusses future research and development directions of RIS technology and new advances that can further enhance its capabilities.

Reconfigurable Intelligent Surfaces (RIS) are emerging as a pivotal technology in next-generation wireless networks, particularly in enhancing Device-to-Device (D2D) communications. This chapter explores the applications, challenges, and future directions of RIS in D2D contexts. By leveraging programmable meta-materials, RIS can significantly improve spectrum efficiency, enhance reliability, and optimize energy consumption in D2D networks. However, challenges such as beamforming optimization, channel estimation, and practical deployment remain. Moreover, the integration of machine learning (ML) and artificial intelligence (AI) is crucial for optimizing RIS configurations and addressing dynamic network conditions. As we approach the era of 6G, RIS-assisted D2D communications promise to revolutionize how devices interact, enabling ultra-reliable and low-latency communication for various applications, including IoT, smart cities, and autonomous systems.

RISs could improve radio broadcasts by adding virtual channels, for example. This would make the sound better and less noisy. If you change how electromagnetic waves move and behave, you can make things that gather energy, take pictures, and find things better. People can stay safe and private with RISs because they can let or stop certain bands of electromagnetic energy. This article gives a brief history of RIS buildings. It also talks about how the buildings at RIS are planned and put together. Why and how RIS technology can be used, like how it can make Wi Fi links faster, save energy, improve tracking and picture quality, and make monitoring and security stronger. Finally, we discuss some businesspeople issues and attempt to guess how RIS technology will be utilized. We think much progress will be made in this area because RISs can improve many tools and apps.

The deployment of Reconfigurable Intelligent Surfaces (RIS) in 6G networks promises improved performance but introduces challenges in optimization due to dynamic environments and complex configurations. This chapter proposes a Hybrid Nesterov Accelerated Gradient-Golden Eagle Optimization (HNAG-GEO) algorithm designed to efficiently manage RIS configurations. By integrating fast convergence from Nesterov Accelerated Gradient and the exploration strengths of Golden Eagle Optimization, HNAG-GEO addresses key issues like fluctuating user demands and varying channel conditions. Simulation results highlight significant improvements: a 40% increase in throughput (7.85 Gbps), 30% reduction in latency (0.85 ms), enhanced coverage (4.85 km^2), and superior spectral efficiency (18.50 bps/Hz). Although computation time is slightly higher at 45 seconds, the overall gains justify this trade-off. The proposed approach outperforms traditional and AI-based algorithms, making it a robust solution for optimizing RIS in 6G networks.

Preface

The future of wireless communication is evolving rapidly, with 6G networks on the horizon promising to revolutionize connectivity on a global scale. One of the most exciting advancements in this space is the development of Reconfigurable Intelligent Surfaces (RIS). RIS technology, which utilizes engineered surfaces to manipulate electromagnetic waves, holds the potential to reshape the way we design and deploy communication systems, offering new levels of efficiency, security, and scalability.

This book is dedicated to exploring the multifaceted role of RIS in 6G networks, examining its applications, challenges, and transformative potential. Through a series of chapters written by leading experts in the field, we offer both a deep dive into the underlying technology of RIS and a broad perspective on its various uses in the next generation of wireless networks.

Chapter Overview

- **Chapter 1** introduces the concept of RIS, explaining how these programmable surfaces can actively control wireless propagation by modifying the phase, amplitude, and polarization of incoming signals. This chapter sets the foundation for understanding how RIS will power the evolution of 6G, offering energy-efficient, low-latency communication.
- **Chapter 2** expands on the hardware implementation of RIS, outlining its integration into communication systems and the associated research challenges, including scalability, energy efficiency, and material limitations. This chapter provides critical insights for the successful deployment of RIS technology in real-world environments.
- **Chapter 3** takes a closer look at RIS applications in secure wireless communication, focusing on how RIS can be used to enhance system and channel models, optimize signal processing techniques like precoding and channel estimation, and ultimately improve the security of wireless networks.

- **Chapter 4** discusses the design of RIS-enhanced smart radio environments, analyzing how machine learning can optimize network performance. This chapter also examines how RIS can improve multi-antenna systems, including beamforming patterns and resource management.
- **Chapter 5** delves into the intersection of RIS and backscatter communication, highlighting its potential for enabling low-energy, high-reliability communication in large-scale IoT networks, which will play a crucial role in the development of 6G.
- **Chapter 6** explores the combination of Artificial Intelligence (AI) with RIS, reviewing how AI-powered systems can optimize RIS-assisted networks, addressing challenges such as computational complexity, interoperability, and security.
- **Chapter 7** takes a practical approach, discussing the use of machine learning to design RIS systems that can adapt to dynamic wireless environments, optimizing signal strength and reducing interference for more efficient communication.
- **Chapter 8** focuses on the integration of RIS with unmanned aerial vehicles (UAVs), discussing the potential for wirelessly powered UAVs to operate more efficiently and for longer periods, with applications in fields such as disaster management, agriculture, and logistics.
- **Chapter 9** shifts to RIS in satellite communications, exploring how this technology can be used to improve positioning systems and satellite network efficiency, enabling more accurate and reliable data transmission.
- **Chapter 10** examines RIS's role in enhancing physical layer security, offering insights into how programmable surfaces can mitigate security vulnerabilities in wireless networks and protect sensitive data.
- **Chapter 11** focuses on RIS-assisted MIMO (multiple-input, multiple-output) and NOMA (non-orthogonal multiple access) networks, providing a detailed analysis of how RIS can improve network capacity, coverage, and interference management.
- **Chapter 12** explores the applications of RIS in smart cities, discussing how the technology can address urban challenges such as network congestion and energy inefficiency, while also enabling the integration of IoT devices in these environments.
- **Chapter 13** discusses RIS applications in Device-to-Device (D2D) communications, examining how RIS can improve spectrum efficiency, energy consumption, and reliability in these networks, with a particular focus on future 6G systems.

- **Chapter 14** presents an exploration of self-powered RIS, offering solutions for energy-efficient deployment and optimization of RIS systems, and how these advancements can enhance overall system performance.
- **Chapter 15** introduces bio-inspired optimization techniques for enhancing RIS performance, proposing new algorithms to address challenges such as fluctuating user demands and changing channel conditions in 6G networks.
- **Chapter 16** proposes a Hybrid Nesterov Accelerated Gradient-Golden Eagle Optimization (HNAG-GEO) algorithm designed to efficiently manage RIS configurations. By integrating fast convergence from Nesterov Accelerated Gradient and the exploration strengths of Golden Eagle Optimization, HNAG-GEO addresses key issues like fluctuating user demands and varying channel conditions. Simulation results highlight significant improvements: a 40% increase in throughput (7.85 Gbps), 30% reduction in latency (0.85 ms), enhanced coverage (4.85 km^2), and superior spectral efficiency (18.50 bps/Hz). Although computation time is slightly higher at 45 seconds, the overall gains justify this trade-off. The proposed approach outperforms traditional and AI-based algorithms, making it a robust solution for optimizing RIS in 6G networks.

This collection of chapters presents a diverse range of perspectives on RIS and its role in the future of wireless communication. From its integration into network architectures to its ability to enhance security, optimize energy use, and enable new IoT applications, RIS stands at the forefront of the next wave of innovation in 6G networks. Whether you're a researcher, engineer, or industry professional, this book will provide you with the knowledge and insights necessary to understand and contribute to the development of RIS technology and its applications in next-generation communication systems.

Chapter 1
Reconfigurable Intelligent System (RIS): Introduction

Janaki Ramal
https://orcid.org/0009-0006-9954-8296
Saveetha Engineering College, India

S. Malathi
SRM Valliammai Engineering College, India

ABSTRACT

A new technology that can improve wireless communication systems, especially in next-generation networks like 6G, is Reconfigurable Intelligent Surfaces (RIS). A wide range of passive, adjustable components make up the surfaces of RIS, which can be configured to dynamically reflect or refract incoming electromagnetic waves. The wireless propagation environment can be actively controlled by RIS by modifying the phase, amplitude, and polarization of these waves. This results in stronger, more direct transmission channels, less interference, and fewer signal blockages. RIS allows for low latency, large data speeds, and energy-efficient communication in 6G applications. RIS uses less energy and produces less noise than traditional relay and antenna systems because it modifies existing signals rather than creating new ones.

DOI: 10.4018/979-8-3693-8099-4.ch001

INTRODUCTION

Introduction to Reconfigurable Intelligent Surfaces (RIS)

The transition from 5G to 6G: Give an overview of 6G, the upcoming wireless communication standard. Highlight the most significant anticipated improvements in coverage, latency, data speeds, and intelligence after 5G.

The need for 6G should be discussed, with an emphasis on the growing demands for high-speed communication, widespread connection, and the Internet of Things. Underline that in order to support novel use cases like holographic communication, extended reality (XR), and digital twins, networks must be low-latency, high-capacity,and energy-efficient.Give a succinct introduction to RIS technology and its potential application in 6G. Describe how reconfiguring the wireless environment with RIS can increase energy efficiency, extend network coverage, and improve signal propagation.

Overview of Reconfigurable Intelligent Surfaces (RIS)

What Does RIS Stand For?

Remarkably, artificial structures (meta-surfaces) with the ability to regulate electromagnetic waves are known as reconfigurable optical surfaces, or RIS (Liaskos et al., 2018). What are the tunable features of RIS elements that allow for the manipulation of incoming signals' phase, amplitude, and polarization?

Fundamental Elements and Components of RIS: Talk about the meta-materials and intelligent reflectarrays that make up the physical framework of RIS (5GAmericas, 2020). Point out how the energy-efficient signal processing made possible by RIS technology is made possible by inexpensive, passive or semi-passive components.

Describe the fundamental principles behind the RIS's ability to reflect, refract, and divert wireless signals. Stress how RIS may be programmed and tuned to allow it to dynamically optimise the wireless environment.

Analogous to Conventional Technologies (like Relays and MIMO): Analyze RIS in comparison to massive MIMO and conventional active relay systems. RIS has a number of benefits, including the capacity to rearrange wireless channels without the need for active signal creation and a lower power usage.

Key Advantages of RIS in 6G Networks

Increasing Network Coverage and Capacity: Explain how RIS can increase the reach of 6G networks, particularly in difficult situations like cities, interior environments, and places with a lot of obstacles (such as walls and buildings). RIS can extend coverage into places with poor signal reception and successfully bypass obstructions.

Green Communication and Energy Efficiency: Talk about how RIS can save energy in comparison to conventional infrastructure. Describe how RIS helps to promote sustainable and environmentally friendly communication in 6G networks by functioning passively or semi-passively and requiring little energy to reflect or divert signals (Liaskos et al., 2018).

Enhanced Spectral Efficiency: Describe how using the same frequency band for various communications through RIS might improve spectral efficiency (5G PPP Technology Board, 2021). Higher capacity and throughput are thus possible without the need for increased spectrumallocation.

The low cost and ease of deployment of RIS panels make them a viable option for upgrading 6G networks. They are lightweight, inexpensive, and simple to install on existing infrastructure, such as buildings, walls, and poles.

Enhanced Security and Privacy: Explain how RIS can make 6G networks' physical security more secure. RIS can prevent eavesdropping and improve wireless communication privacy by managing the electromagnetic wave's propagation.

RIS FOR MILLIMETER-WAVE (mmWave) AND TERAHERTZ (THz) COMMUNICATION

Overview of THz and mmWave Bands in 6G

Give a summary of the THz (0.1-10 THz) and mmWave (30-300 GHz) bands, which are essential for reaching the ultra-high data rates that 6G promises. Describe the advantages of these frequency bands, such their wide bandwidth and fast data transfer, as well as the drawbacks (like signal attenuation and short propagation lengths).

Overcoming Propagation Difficulties: Talk about how RIS can lessen the difficulties associated with signal propagation in these high-frequency bands. Dense urban surroundings are the perfect fit for RIS since it can improve non-line-of-sight (NLoS) communication and reroute signals around obstructions.

Beamforming and Directional Communication: Examine how RIS's beamforming feature works to guide signals toward particular users or devices (Alexandropoulos, 2021). By directing the signal energy in a particular direction, lowering interference, and enhancing connectivity, this helps maximize the effectiveness of mmWave and THz communication (Renzo et al., 2019).

Role of RIS in Network Optimization

The ability to regulate the wireless environment in real time thanks to RIS allows for the optimization of signal routes and adaptation to changing network conditions, such as user mobility and environmental changes (Gao et al., 2020). This is known as dynamic control of the wireless environment.

Resource Allocation and Traffic Management: Talk about how RIS can help optimize network resources by dynamically modifying the way signals propagate in order to lessen interference, balance network loads, and efficiently manage traffic.

Explain how RIS contributes to interference cancellation through intelligent regulation of signal.Absorption and reflection in the context of interference management. As a result, there is less interference between symbols and the network performs better overall.

Integration with AI and Machine Learning: Describe the ways in which RIS can be used in conjunction with AI/ML algorithms to anticipate and address network requirements. AI can be used to optimize the RIS setup in real-time, reducing latency and improving network performance (Gao et al., 2020).

Key Applications of RIS in 6G

Technical Restrictions: Talk about the technical restrictions of RIS, such as signal attenuation, intricate control systems, and the requirement for exact metasurface tuning for best results (Huang et al., 2019).

Power Supply and Maintenance: Discuss the difficulties of providing power to and maintaining RIS in practical settings. Although RIS can function in a passive manner, certain configurations necessitate low power for active tuning, which poses design issues during implementation (Wu & Zhang, 2019).

Integrate with Current Infrastructure: Examine the challenges associated with incorporating RIS into the current generation of 5G networks and other wireless infrastructure. New standards and design considerations might be necessary to provide compatibility and interoperability with current communication methods.

Environmental Factors: Talk about how the weather, such as rain, fog, and temperature variations, affects the performance of RIS, particularly in the THz and mmWave bands that are vulnerable to air absorption (Alexandropoulos, 2021).

Future Research Directions for RIS in 6G

Developments in Intelligent Reflect Arrays and Meta-Materials: Talk about the current research being done to create more sophisticated meta-materials that can govern electromagnetic waves with more flexibility (Alexandropoulos, 2021).

AI-Driven RIS Optimization: Examine how AI/ML algorithms may be used to dynamically optimize RIS setups, allowing for more intelligent and independent network management.

RIS in Quantum Communication: Take into consideration the possible applications of RIS in cryptography and quantum communication, where exact manipulation of electromagnetic waves may be necessary to provide safe communication paths.

Hybrid RIS and Active Surfaces: Explain the idea of hybrid RIS systems, which incorporate both active and passive parts (5GAmericas, 2020).

Overview of Reconfigurable Intelligent Surfaces (RIS) Design

With the help of revolutionary technology like Reconfigurable Intelligent Surfaces (RIS), wireless communication systems in the 6G future will be more reliable, scalable, and efficient (Gao et al., 2020). RIS is designed to regulate and modify electromagnetic waves. RIS provides a way to transform signal transmission, enhancing both performance and reliability, as the need for wireless connectivity grows due to new applications such as the Internet of Things (IoT), autonomous systems, and real-time immersive experiences (Liaskos et al., 2018).

Reconfigurable intelligent surfaces (RISs), commonly referred to as smart surfaces, represent a relatively new technological innovation with the capacity to completely transform wireless communications. These surfaces can be used to control electromagnetic waves in any desired way since they are composed of a large number of small, individually adjustable parts (Wu & Zhang, 2019). By rearranging these components' dimensions and positions, reconfigurable intelligence electromagnetic waves can be reflected, absorbed, or transmitted by surfaces to carry out a multitude of distinct tasks (Molisch et al., 2017).

An rising number of devices in modern civilization rely on wireless connections to interact with one another and access the internet, making wireless communications an essential component of modern life (Basar et al., 2017). In the upcoming years, there will likely be an increase in demand for wireless data, necessitating the development of new technologies. The capacity and coverage of wireless networks have been successfully increased using traditional wireless communication technologies, such as multiple-input multiple-output (MIMO) systems, but their performance has hit a limit (Marzetta, Larsson, Yang, & Ngo, 2016).

THE FUTURE OF MOBILE COMMUNICATIONS APPEARS BRIGHT, WITH POTENTIAL NEW USE CASES AND RIGOROUS REQUIREMENTS OF 5G AND BEYOND CELLULAR NETWORKS

But between the transmitter and the receiver, the propagation medium has been thought of as a randomly acting agent. As wireless communications enter the digital era, the quality of the received signal is deteriorating because of the random interactions between the sent radio waves and the surrounding objects. A thorough review of the literature on reconfigurable intelligent surfaces (RISs) and assisted application areas is presented in this presentation (Chen, Diao, Chen, & Liu, 2020). By resolving the detrimental aspects of environmental wireless propagation, the RIS allows network operators to adjust the characteristics of radio waves' scattering, reflection, and refraction (Marzetta, Larsson, Yang, & Ngo, 2016). Additionally, amplitude, phase, frequency, and potentially other aspects of the wavefront can be efficiently controlled by the RIS.

The general belief is that while 6G will bring about a true revolution by enabling Internet-of-everything/Internet-of-sensing applications, 5G can be considered an evolution of 4G in terms of physical-layer technology utilized (Basar & Yildirim, 2021). A recent white paper from a significant industry group predicts that holography, digital replication, and fully immersive virtual reality would be among the main uses of 6G from a communication engineering standpoint. These science-fictional apps will need the user to peak data speeds of one gigabit per second and 10 Gbps, which are 50 and 10 times fasterthan5G, in turn, in addition to significantly improved energy and spectrum efficiency (Renzo et al., 2019). To improve these important key performance indicators, we will inevitably need to turn to more radical communication paradigms, even in the face of tremendous efforts and advancements in the physical layer. Like 5G's long-ago standardization, new candidate technologies are emerging to meet the demanding needs of 6G wireless networks in the future, including Wireless power transmission, AI-powered communications, TeraHertz (THz)/visible light communications, interconnected extraterrestrial networks, enormous cell-free index modulation, MIMO systems, and advanced waveforms and intelligent radio settings reconfigurable intelligent surfaces enable/assist Under the revolutionary paradigm of communication through RISs, network operators can regulate the end-to-end wireless propagation channel among end terminals. In order to enhance the functionality of target communication systems—particularly those experiencing significant attenuation and degradation—software-controlled RIS can adjust their impinging signals over the air (Chen, Diao, Chen, & Liu, 2020).

In the past Throughout the past several years, interest in unique methods of communication that take advantage of implicit unpredictability of the propagation surroundings. Within this index modulation that takes advantage of the indices of components of the intended communication systems to more details, seemed like a fascinating an alternative to the well-known MIMO modes and waveforms that comply with criteria By utilizing the wireless channel to improve various communication objectives, including error performance, capacity, secrecy, outage, and energy efficiency, in addition to information conveyance as in index modulation schemes, RIS-empowered communications go beyond this paradigm. Stated differently, intelligent software-defined reconfigurable intelligent surfaces might be thought of as an advancement over software-defined networks of wireless communication.

A specially created artificial surface made of electromagnetic (EM) material with integrated electronics, electronically controlled, and unnatural wireless communication capabilities is called a reconfigurable intelligent surface (RIS) (Huang et al., 2020). In the scenario that is most frequently thought of, a RIS's many tiny, inexpensive, and nearly passive components can easily alter the incident signal over the air to increase signal coverage and/or caliber. This technique is not the same as current large-scale MIMO transceivers and operational beamformers along with the collaborative relaying and communication via backscatter paradigms. The most unique characteristic of RISs is their inexpensive, almost passive unit cell panel, each having the ability to apply a phase that is controlled change to the field of impact (for example, by two-state reflection coefficients in an early RIS architecture RISs do not require radio frequency (RF) signal processing, amplification, filtering, mixing, or up/down conversion when utilized as pure tunable reflectors. Considering this, huge MIMO and active relaying systems can be supplemented or replaced by RIS-enabled communication (Marzetta, Larsson, Yang, & Ngo, 2016).

Over the past three years, there have been notable advancements in the field of RIS-powered communications, and the underlying RIS technology has reached a point of maturity where both business and academia are investigating its promising future use-cases (Niu & Li, 2019). In particular, researchers have concentrated on combining active and passive beam formation issues within the framework of RIS wireless systems. Channel modeling in various circumstances and performance evaluations, security at the physical layer multiple non-orthogonal NOMA access techniques incoherent modulating automobile networks hybrid designs that are passive and sensitive, engaged and cooperative designs that are reflective and transmissive in-depth education techniques for reflection modulating actual prototyping experiments in addition to with a plethora of additional intriguing application areas (Basar & Poor, 2021).

Essential Elements of RIS

Meta Surfaces

Sub-wavelength-sized components intended to regulate wave propagation make up the core of RIS technology. It is possible to electronically modify these components to change the signals' amplitude, phase, and even polarization.

Phase Shifters: To improve signal quality and directionality, RIS is outfitted with phase shifters that enable controlled wave reflection and refraction.

Control Mechanisms: Based on user needs and network conditions, RIS is frequently linked to a distributed or centralized controller that dynamically modifies the surface configuration.

Reasons for Using RIS in 6G Networks

1. **Handling the Limitations of 5G**

Even while 5G networks offer much better connectivity than 4G networks, they still have issues with energy usage, infrastructure expenses, and coverage restrictions, particularly in densely populated areas. These problems should be resolved with the rollout of 6G networks, with RIS playing a crucial role by addressing particular pain points:

Coverage Limitations:

Signal blockages brought on by structures, moving cars, and other obstructions frequently result in "dead zones" in densely populated regions. RIS can get around this by changing signal pathways on the fly (Liaskos et al., 2018).

Energy Efficiency:

The energy usage of traditional network components, such as relays and base stations, becomes unsustainable as 6G aspires to handle trillions of devices. With its reduced energy consumption for signal reflection, RIS provides a low-power substitute.The Requirement of Environmentally Friendly Communications One of the main issues with wireless networks is energy usage (Gao et al., 2020). In this sense, unlike typical active elements, RIS greatly advances the development of green communication technologies by reflecting signals without requiring extra power for signal amplification (Renzo et al., 2019).

Extensive Device Networking:

Billions of IoT devices will require 6G networks to support them while consuming the least amount of power and spectrum. By enhancing the signal propagation environment without requiring extra infrastructure, RIS helps to enable huge connection.

Significance of RIS:

Reconfigurable Intelligent Surfaces (RIS) hold great relevance in 6G technology as they have the ability to completely transform wireless communication infrastructures. The following explains why RIS is seen as essential to the creation of 6G networks.

Improved Signal Transmission and Management:

By cleverly regulating the electromagnetic waves' refraction, scattering, and reflection, RIS makes it possible to manipulate the wireless propagation environment. Better signal strength, coverage, and dependability are made possible by this control, particularly in complicated contexts like metropolitan regions where barriers can impede signal quality (Marzetta, Larsson, Yang, & Ngo, 2016).

Efficiency in Energy Use:

In comparison to conventional active components like relays and repeaters, RIS uses a very little amount of power. For 6G networks to be viable, this energy efficiency is essential, especially in situations where a dense deployment of communication infrastructure is needed (Basar et al., 2019). Signal propagation can be improved using RIS without appreciably increasing the energy footprint.

Enhanced Spectral Efficiency:

By raising the signal-to-noise ratio (SNR) and facilitating more effective radio spectrum usage, RIS increases spectral efficiency. This is crucial for 6G networks, which are designed to facilitate enormous data throughput and connection in situations with constrained spectrum.

Lower Expenses for Infrastructure:

RIS lessens the need for new, costly infrastructure, such as cell towers, because it may be installed on walls or other existing structures. This increases network capacity and coverage while lowering deployment costs (Strinati et al., 2021).

Enhanced Coverage and Connectivity:

When it comes to weak signal regions, including indoor spaces or "urban canyons" (cities with dense populations sandwiched between tall buildings), RIS can be of assistance. Reconfiguring wave reflections allows RIS to better cover the network by directing signals to users who are located in remote areas (Basar et al., 2017).

Assistance with Advanced Communication Services:

Beamforming: Advanced beamforming, which reduces interference and improves communication quality by directing signals more accurately towards users, is made possible by RIS (Molisch et al., 2017). For services like enhanced mobile broadband (eMBB) and ultra-reliable low-latency communication (URLLC), this is essential.

Massive Connectivity: RIS is made to handle the enormous connectivity requirements of 6G, which will involve the simultaneous and dependable connecting of billions of devices, including IoT sensors and smartphones.

Adaptability and Flexibility:

When obstacles or users move around in the wireless environment, RIS can adjust dynamically to meet these changes. Maintaining high-quality communication in situations such as self-driving cars, smart cities, and real-time remote operations requires this flexibility.

Privacy and Security:

Controlling signal propagation is one way that RIS may improve network security. It can lessen the chance of eavesdropping and make it more difficult for unauthorized users to intercept conversations by directing signals in particular directions.

Figure 1. Intelligent Reflecting Surface (IRS)

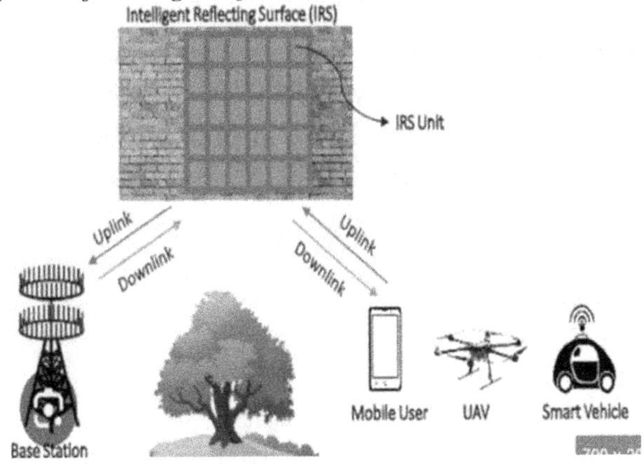

RIS's Technical Advantages

Spectral Efficiency:

Spectral efficiency, or the effectiveness with which the available spectrum is exploited for data transmission, is one of the most important measures in wireless communications. By reducing interference, raising the signal-to-noise ratio (SNR), and managing wave propagation, RIS improves spectral efficiency.

Efficiency in Energy Use:

Conventional communication systems depend on energy-intensive active components like relays and repeaters. On the other hand, RIS is a passive technology that achieves excellent communication quality with low power consumption. Because of its energy-efficient design, RIS is a desirable choice for long-term, extensive 6G deployments.

Improved Reliability and Coverage:

RIS enables more reliable communication by mitigating the effects of obstacles and interference (Marzetta, Larsson, Yang, & Ngo, 2016). For instance, in urban areas where buildings obstruct signals, RIS can reflect and direct waves around these obstacles, ensuring a more stable connection for users in otherwise poor-coverage areas (Marzetta, Larsson, Yang, & Ngo, 2016).

Low Cost and Easy Deployment:

Since RIS can be installed on existing infrastructure like building facades, streetlights, or walls, it can be deployed cost-effectively without requiring new, expensive cell towers or base stations. This capability reduces operational and capital expenditure for network operators, while still improving coverage and capacity (Huang et al., 2020).

Operation Modes

Reflecting RISs

RIS-enabled smart wireless settings were first thought to be limited to passive RISs with very negligible power consumption unit elements.Moreover, by packing an increasing number of software-controlled antenna elements onto a two-dimensional (2D) surface of finite size, active RIS systems naturally evolve from conventional massive MIMO systems. This results in an active RIS that functions as a large intelligent surface (LIS), with element spacing decreasing as the number of elements increases (Marzetta, Larsson, Yang, & Ngo, 2016). A condensed integration of several small active reconfigurable processing networks in RIS elements able to create a workable continuous antenna aperture.The One can transmit and receive data using an active RIS framework signals using the hologram along the whole surface idea. The array of discrete photonic antennas is yet another useful application of active RISs. It include converters, active optical-electric detectors,as well as modulators for broadcast, reception, and RF or optical signal conversion (Alexandropoulos, 2021).

Figure 2. Reflecting off a passive RIS vs an active RIS

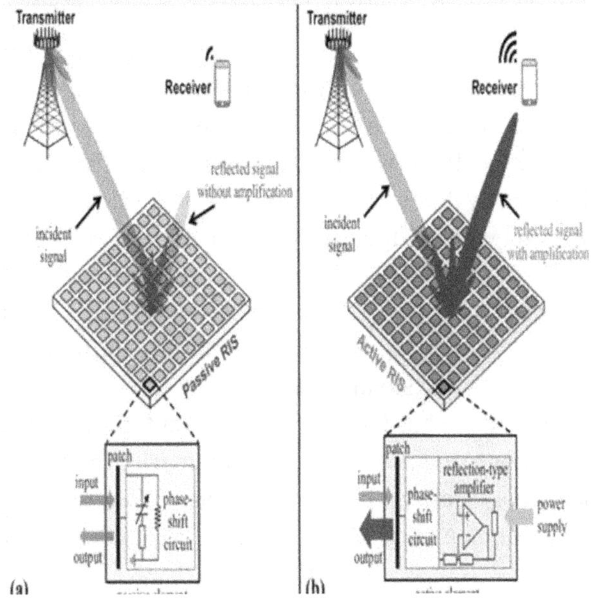

Passive RISs

Passive RISs can be configured to modify an impinging electromagnetic field in a programmable manner. They function similarly to a passive metal mirror or wave collector (Ho-Van, Sofotasios, Alexandropoulos, & Freear, 2015). When compared to an active RIS, a passive RIS is typically made up of inexpensive, nearly passive components that don't need special power sources (Huang et al., 2020). Energy harvesting can power its embedded sensors and circuitry modules, a strategy that could potentially make they are actually energy-neutral (Molisch et al., 2017). Whatever their particular applications, what constitutes a passive RIS technology that is appealing in terms of energy usage from their ability to manipulate radio waves encroaching upon them and transmitting the received signal without using an RF chain or power amplifier, and even without using an advanced signal executing.Lastly, because of their incredibly cheap hardware costs and power consumption, passive RIS structures can be seamlessly included into wireless communication environments. They can be used in laptop cases, factory and room ceilings, building facades, and even human clothing (Alexandropoulos, Papadogiannis, & Sofotasios, 2011).

RISs That Are Discrete

Several discrete low-power unit cells make up a discrete holographic multiple input multiple outputsurface (HMIMOS) and metamaterials that are software-tunable (Basar et al., 2019). The methods to electronically alter the unit's electromagnetic characteristics cells range from readily available electrical parts liquid crystals, microelectromechanical systems electromechanical switches, systems, and other Metamaterials that can be reconfigured (Molisch et al., 2017). This arrangement is very distinct from the traditional MIMO antenna configuration.

Continuous RISs: A contiguous RIS can therefore produce a spatially continuous receiver aperture by integrating an almost limitless number of elements into a constrained surface area (Wu & Zhang, 2019). Contiguous RISs differ significantly from discrete RISs in terms of hardware and implementation, as will be discussed in the fabrication techniques subsection's sequel.characteristics of reflection (5GAmericas, 2020). As previously stated,.he active discrete surface is an additional variety regarding photonic antenna arrays (Basar & Poor, 2021).

REFERENCES

Alexandropoulos, G. C.George C. Alexandropoulos. (2021). Low complexity full duplex MIMO systems: Analog canceler architectures, beamforming design, and future directions. *ITU Journal : ICT Discoveries*, 2(2), 1–19. DOI: 10.52953/GFFF9448

Alexandropoulos, G. C., Lerosey, G., Debbah, M., & Fink, M. (2020). Reconfigurable intelligent surfaces and metamaterials: The potential of wave propagation control for 6G wireless communications. *IEEE ComSoc TCCN Newslett*, 6(1), 25–37.

Alexandropoulos, G. C., Papadogiannis, A., & Sofotasios, P. C. (2011). A comparative study of relaying schemes with decode-and-forward over Nakagami-m fading channels. Journal of Computational Networks and Communications, 2011(1), 560528.1–560528.14.

Alexandropoulos, G. C., Shlezinger, N., & Hougne, P. D. (2021). Reconfigurable intelligent surfaces for rich scattering wireless communications: Recent experiments, challenges, and opportunities. *IEEE Communications Magazine*, 59(6), 28–34. DOI: 10.1109/MCOM.001.2001117

Basar, E., & Poor, H. V. (2021). Present and future of reconfigurable intelligent surface-empowered communications. *IEEE Signal Processing Magazine*, 38(6), 146–152. DOI: 10.1109/MSP.2021.3106230

Basar, E., Renzo, M. D., Rosny, J. D., Debbah, M., Alouini, M.-S., & Zhang, R. (2019). Wireless communications through reconfigurable intelligent surfaces. *IEEE Access : Practical Innovations, Open Solutions*, 7, 116753–116773. DOI: 10.1109/ACCESS.2019.2935192

Basar, E., Wen, M., Mesleh, R., Renzo, M. D., Xiao, Y., & Haas, H. (2017). Index modulation techniques for next generation wireless networks. *IEEE Access : Practical Innovations, Open Solutions*, 5, 16693–16746. DOI: 10.1109/ACCESS.2017.2737528

Basar, E., & Yildirim, I. (2021). Reconfigurable intelligent surfaces for future wireless networks: A channel modeling perspective. *IEEE Wireless Communications*, 28(3), 108–114. DOI: 10.1109/MWC.001.2000338

Björnson, E., & Sanguinetti, L. (2020). Power scaling laws and near-field behaviors of massive MIMO and intelligent reflecting surfaces. *IEEE Open Journal of the Communications Society*, 1, 1306–1324. DOI: 10.1109/OJCOMS.2020.3020925

Calvanese Strinati, E., Alexandropoulos, G. C., Wymeersch, H., Denis, B., Sciancalepore, V., D'Errico, R., Clemente, A., Phan-Huy, D. T., Carvalho, E. D., & Popovski, P. (2021). Reconfigurable, intelligent, and sustainable wireless environments for 6G smart connectivity. *IEEE Communications Magazine*, 59(10), 99–105. DOI: 10.1109/MCOM.001.2100070

Chen, L., Diao, X., Chen, J., & Liu, R. (2020). QoS assurance in IAB network. In *Proceedings of the 2020 IEEE IWCMC* (pp. 1860–1865). Limassol, Cyprus.

Di, B., Zhang, H., Song, L., Li, Y., Han, Z., & Poor, H. V. (2020). Hybrid beamforming for reconfigurable intelligent surface based multi-user communications: Achievable rates with limited discrete phase shifts. *IEEE Journal on Selected Areas in Communications*, 38(8), 1809–1822. DOI: 10.1109/JSAC.2020.3000813

Gao, Y., Chen, J., Liu, Z., Zhang, B., Ke, Y., & Liu, R. (2020). Machine learning-based energy saving scheme in wireless access networks. In *Proceedings of the 2020 IEEE IWCMC* (pp. 1573–1578). Limassol, Cyprus. DOI: 10.1109/IWCMC48107.2020.9148536

Ho-Van, K., Sofotasios, P. C., Alexandropoulos, G. C., & Freear, S. (2015). Bit error rate of underlay decode-and-forward cognitive networks with best relay selection. *Journal of Communications and Networks (Seoul)*, 17(2), 162–171. DOI: 10.1109/JCN.2015.000030

Huang, C., Hu, S., Alexandropoulos, G. C., Zappone, A., Yuen, C., Zhang, R., Renzo, M. D., & Debbah, M. (2020). Holographic MIMO surfaces for 6G wireless networks: Opportunities, challenges, and trends. *IEEE Wireless Communications*, 27(5), 118–125. DOI: 10.1109/MWC.001.1900534

Huang, C., Zappone, A., Alexandropoulos, G. C., Debbah, M., & Yuen, C. (2019). Reconfigurable intelligent surfaces for energy efficiency in wireless communication. *IEEE Transactions on Wireless Communications*, 18(8), 4157–4170. DOI: 10.1109/TWC.2019.2922609

Kaina, N., Dupré, M., Lerosey, G., & Fink, M. (2014). Shaping complex microwave fields in reverberating media with binary tunable metasurfaces. *Scientific Reports*, 4(1), 6693. DOI: 10.1038/srep06693 PMID: 25331498

Liaskos, C., Nie, S., Tsioliaridou, A., Pitsillides, A., Ioannidis, S., & Akyildiz, I. F. (2018). A new wireless communication paradigm through software-controlled metasurfaces. *IEEE Communications Magazine*, 56(9), 162–169. DOI: 10.1109/MCOM.2018.1700659

Marzetta, T. L., Larsson, E. G., Yang, H., & Ngo, H. Q. (2016). *Fundamentals of Massive MIMO*. Cambridge University Press. DOI: 10.1017/CBO9781316799895

Molisch, A. F., Ratnam, V. V., Han, S., Li, Z., Nguyen, S. L. H., Li, L., & Haneda, K. (2017). Hybrid beamforming for massive MIMO: A survey. *IEEE Communications Magazine*, 55(9), 134–141. DOI: 10.1109/MCOM.2017.1600400

Niu, J.-P., & Li, G. Y. (2019). An overview on backscatter communications. *Journal of Communications and Information Networks*, 4(2), 1–14. DOI: 10.23919/JCIN.2019.8917868

Renzo, M. D., Debbah, M., Phan-Huy, D. T., Zappone, A., Alouini, M.-S., Yuen, C., Sciancalepore, V., Alexandropoulos, G. C., Hoydis, J., & Gacaninet, H.. (2019). Smart radio environments empowered by reconfigurable AI meta-surfaces: An idea whose time has come. *EURASIP Journal on Wireless Communications and Networking*, 2019(1), 1–20. DOI: 10.1186/s13638-019-1438-9

Samsung. (2012). Samsung's 6G white paper lays out the company's vision for the next generation of communications technology. Retrieved from https://news.samsung.com/global/samsungs-6gwhite-paper-lays-out-the-companysvision-for-the-next-generationof-communications-Technology

Strinati, E. C., Alexandropoulos, G. C., Sciancalepore, V., Renzo, M. D., Wymeersch, H., Phan-Huy, D. T., Crozzoli, M., D'Errico, R., Carvalho, E. D., Popovski, P., ... (2021). Wireless environment as a service enabled by reconfigurable intelligent surfaces: The RISE-6G perspective. In Proceedings of the 2021 Joint EuCNC & 6G Summit (pp. 562–567). Porto, Portugal.

Tang, W., Chen, M. Z., Chen, X., Dai, J. Y., Han, Y., Renzo, M. D., Zeng, Y., Jin, S., Cheng, Q., & Cui, T. J. (2021). Wireless communications with reconfigurable intelligent surface: Path loss modeling and experimental measurement. *IEEE Transactions on Wireless Communications*, 20(1), 421–439. DOI: 10.1109/TWC.2020.3024887

Wu, Q., & Zhang, R. (2019). Intelligent reflecting surface enhanced wireless network via joint active and passive beamforming. *IEEE Transactions on Wireless Communications*, 18(11), 5394–5409. DOI: 10.1109/TWC.2019.2936025

Chapter 2
Hardware Implementation of Reconfigurable Intelligent Surfaces:
Applications and Research Challenges

M. Nalini
https://orcid.org/0000-0001-6457-7990
Sri Sai Ram Engineering College, India

Sakthi Sabapathy Raja V. K.
https://orcid.org/0009-0000-9504-533X
Sri Sai Ram Engineering College, India

Sri Sai Ram Engineering College, India

Rajesh Kumar
Dhanaraj Symbiosis International University, India

ABSTRACT

A novel technology that has emerged is the so-called reconfigurable intelligent surface. Its purpose is to change the properties of electromagnetic waves to improve wireless communication networks. In this article, the authors analyse the use cases, embedding methods, and research problems of RIS. Special attention is paid to integrating RIS into the communication system, issues of system architecture, its implementation features, and the functions afforded by RIS in the 5G and 6G. It mentioned some important challenges such as problems with scalability, energy

DOI: 10.4018/979-8-3693-8099-4.ch002

efficiency, and material limitations to appreciate the future Development and the adoption of RMS in important areas.

I. TABLE OF CONTENTS

1. INTRODUCTION

The rise of Reconfigurable Intelligent Surfaces (RIS) has sparked excitement in the field of wireless communication. RIS extends the control of wireless channels by varying the way reflected signals behave, and thus can change the propagation environment on the fly. Consequently, RIS technology has become crucial in the design of next-generation networks like 5G and 6G, which offer better coverage, higher throughput, and lower power consumption among other capabilities. The article looks at different purposes and strategies for RIS management, problems that arise, and prospects for further development.

Figure 1. RIS in communication

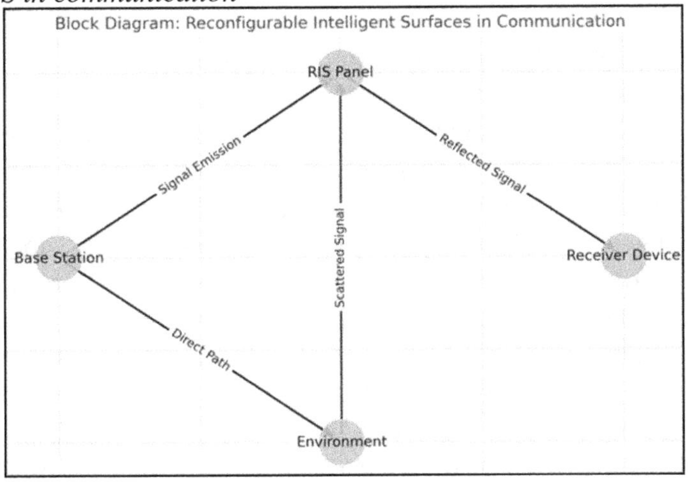

2. OVERVIEW OF RECONFIGURABLE INTELLIGENT SURFACES

Definition and Significance of RIS in Wireless Communications

Reconfigurable Intelligent Surfaces (RIS) represent an essential innovation in wireless communication. They are composed of engineered surfaces consisting of numerous tiny controllable devices capable of bending electromagnetic waves. RIS dictate the laws of reflection, refraction, or absorption of these waves by modifying their phase, amplitude, or polarization, hence reconfiguring the wireless space (Wu, Q., & Zhang, R.,2019).

It is the RIS importance with regards to wireless networks that enables better signal strength, higher data rates, and energy efficiency. Expanding over the traditional active elements (for instance, repeaters, or relays), RIS is mostly passive and has negligible power requirements. They have been the subject of research as they enable the creation of 'smart' radio environments, where signals can be focused on particular users, enhancing coverage and connection quality. To meet the demands of 5G and the emerging 6G world that requires higher efficiencies and flexibilities, RIS technology has presented an alternative strategy which is cheap, effective and scalable by designing the wireless environment.

Historical Development and Advances in Metasurfaces

The evolution of reflective intelligent surfaces parallels the development of metamaterials which modified surface waves in propagation. The investigation into the manipulated surfaces started with research areas such as electromagnetics and optics with the early fixating on various types of materials, especially martensitic materials. These are materials which have aid in housing a structured fashion a geometrical array of subunit885 called artificial devices or 'meta' atoms which can alter properties found in normal materials such as attenuation without refraction.

Metasurfaces were considered to be the evolution of 3D metamaterials the first-generation deployment of 2D changing surfaces. As the different types of shapes became available, metasurfaces evolved from passive to active, where every meta-atom can be programmed. The technology shift towards programmable metasurfaces allowed researchers to modulate the effects of the surface leading to the development of RIS. These surfaces formed the basis of electronic control alteration in the dispersion of signals directing them only to intended users due to the real-time needs of the network, making it suitable for contemporary communication systems.

Key Technologies that Enabled RIS: Metamaterials and Programmable Surfaces

1. **Metamaterials**

The core technology behind RIS is based on metamaterials, which are specially designed-structures that interact with electromagnetic waves in unique ways. Each meta-atom is crafted with distinct characteristics (like size, shape, and material) that influence its interaction with particular wave frequencies. Metamaterials facilitate RIS by offering the essential components needed to manage the propagation of electromagnetic waves at the surface level.

2. **Programmable Surfaces**

The shift from static to dynamic wave manipulation in reconfigurable intelligent surfaces (RIS) has been achieved through the use of programmable metasurfaces. These surfaces feature electronically controllable elements that allow each section to modify its response to incoming waves. Typically, each meta-atom can be programmed using integrated circuits or tunable components, such as PIN diodes and varactors, which facilitate real-time adjustments to the surface's phase and amplitude. This capability for real-time adaptability enables programmable surfaces to react to changes in the environment or user needs, providing a flexible and responsive enhancement to networks.

3. **Tunable Materials**

Recent advancements in tunable materials like graphene and phase-change materials have significantly boosted the development of reconfigurable intelligent surfaces (RIS). For instance, graphene possesses unique electrical properties that can be modified by applying an electric field, making it particularly suitable for RIS applications that demand high-speed and precise control. Additionally, phase-change materials can switch between different states, each with unique electromagnetic properties, providing an effective way to incorporate reconfigurable functionalities in RIS.

4. **Microelectronics and Control Algorithms**

Microelectronics provide the control necessary to manipulate each meta-atom on an RIS dynamically. Coupled with advanced control algorithms, these electronics coordinate the elements to achieve the desired signal transformation, enabling complex operations like beamforming and signal redirection. Machine learning and artificial intelligence are increasingly used to optimize RIS control in real-time, adapting to complex wireless environments and maximizing performance.

3. RIS IN WIRELESS COMMUNICATIONS

Integration of RIS with Existing Communication Networks

Integrating Reconfigurable Intelligent Surfaces (RIS) into existing wireless networks introduces a fundamentally new approach to signal transmission and reception. Unlike traditional network infrastructure, which relies on active components like base stations, relays, and antennas to transmit and amplify signals, RIS passively shapes the wireless environment by controlling how electromagnetic waves propagate. RIS devices, typically installed on walls, ceilings, or other surfaces, work by dynamically adjusting each unit's phase, amplitude, and polarization to control the direction and strength of reflected signals.

To integrate RIS into communication networks, researchers and engineers have developed hybrid systems where RIS functions alongside conventional components. For example, RIS can be placed near cellular towers to redirect signals around obstacles, enhancing the coverage area without needing additional active equipment. In dense urban environments, RIS helps mitigate signal blockages caused by buildings or other structures. Moreover, RIS can assist in controlling interference by steering signals in specific directions, benefiting both network providers and end-users through a more reliable and predictable network environment (Liang, Y.-C., et al.,2019).

Role in Creating "Smart" Radio Environments

One of the most transformative potentials of RIS is its role in establishing smart radio environments. In these intelligent environments, RIS actively manages the wireless channel to optimize signal paths, reduce interference, and improve connectivity. Smart radio environments leverage RIS to dynamically alter the electromagnetic landscape in response to changes, such as user mobility or varying environmental conditions.

Smart radio environments can anticipate and adjust to network demands by adapting RIS settings in real time. This adaptability helps networks meet quality-of-service requirements in high-density areas, manage resources efficiently, and maintain consistent connectivity for moving users, like those in vehicles or on public transportation. The creation of such environments is especially relevant for next-generation networks, such as 5G and 6G, which demand higher data rates, lower latency, and seamless coverage for diverse applications ranging from augmented reality (AR) to the Internet of Things (IoT) (Di Renzo, M., et al.,2020).

Use Cases of RIS in Wireless Communications

1. **Improving Signal Quality**
   - **Signal Focusing**: RIS can focus signals toward specific users, improving signal-to-noise ratios (SNR) and reducing the power required by end devices. This capability is particularly beneficial in scenarios with weak signals or noisy environments, as RIS can boost clarity and reliability.
   - **Reducing Multi-Path Fading**: By controlling the reflection angles and phases of signals, RIS can reduce multi-path fading, a common problem where signals take multiple paths and interfere with each other. The controlled reflections allow for clearer signal paths, enhancing overall communication quality (Björnson, E., Sanguinetti, L., & Debbah, M.,2021).
2. **Enhancing Data Throughput**
   - **Capacity Increase**: By efficiently managing signal reflections and directing them toward intended receivers, RIS can significantly increase the data throughput in crowded or high-density areas. This effect is

achieved by creating multiple focused beams to serve different users simultaneously, known as spatial multiplexing.

  - **Optimising Beamforming**: RIS can collaborate with massive MIMO (Multiple-Input Multiple-Output) systems to direct beams with higher precision. This collaboration allows operators to use bandwidth more effectively and serve more users without compromising data rates.

3. **Extending Coverage**
   - **Non-Line-of-Sight (NLOS) Environments**: In environments where direct line-of-sight (LOS) communication is obstructed, such as inside buildings or in urban canyons, RIS can reflect signals around obstacles to maintain connectivity. RIS provides an efficient alternative to repeaters or additional base stations, which can be more expensive and require additional power.
   - **Supporting Rural and Remote Areas**: RIS offers a low-cost and energy-efficient solution for extending network coverage in remote or rural areas. By strategically placing RIS in critical locations, such as hills or tall structures, signals from distant base stations can be redirected to cover areas without limited access.

Case Study Examples

- **Urban Connectivity Enhancement**: In a dense urban area, RIS panels installed on building facades can improve cellular coverage by redirecting signals around obstacles, creating an uninterrupted path to users. This setup significantly enhances coverage without adding more base stations, thereby reducing infrastructure costs.
- **Indoor Network Optimization**: In large indoor environments, like airports or shopping malls, RIS technology can help direct strong signals to specific areas, avoiding dead zones or interference issues common with traditional Wi-Fi or cellular networks. By carefully configuring RIS to create clear paths to mobile devices, users experience improved connectivity throughout the area.
- **High-Speed Rail and Transportation**: Moving vehicles face frequent signal loss and interference due to high speeds. RIS panels installed along rail tracks or highways can reflect signals toward moving vehicles, helping maintain high-quality connectivity. This application is critical for connected vehicle systems, including autonomous driving and on-board passenger entertainment.

4. RIS ARCHITECTURES AND HARDWARE IMPLEMENTATIONS

Basic RIS Hardware Structure

The fundamental architecture of Reconfigurable Intelligent Surfaces (RIS) typically includes three main components:

1. **Antenna Arrays**:

RIS is built on a surface composed of multiple antenna elements, often arranged in an array structure. Each element, also referred to as a "meta-atom," is designed to control electromagnetic waves. The array design allows the surface to manipulate how waves reflect or scatter by adjusting the properties (like phase and amplitude) of each element. These arrays are generally arranged in a grid pattern to provide high coverage and precision in wave manipulation.

2. **Control Circuitry**:

To enable real-time reconfigurability, RIS requires control circuits that adjust the electromagnetic properties of each element on demand. These circuits are often integrated at the back of the surface, allowing for compact and low-power operation. The control circuitry, governed by algorithms or even AI, fine-tunes the RIS's response to incoming signals, enabling functions like beamforming, redirection, and focusing.

3. **Tunable Elements**:

Tunable or programmable components allow each element in the array to alter its electromagnetic response, such as phase shifting, reflecting, or absorbing signals. This tuning is achieved through materials that can change properties under different stimuli—electrical, magnetic, or thermal. Common tunable elements include PIN diodes, varactors, and tunable dielectric materials, which provide control over the RIS's response and enable real-time adaptability.

Implementation Approaches: Passive vs. Active RIS

There are two main approaches to implementing RIS, depending on the desired functionality and power requirements:

1. **Passive RIS**:

Passive RIS relies on materials and components that control waves without actively generating or amplifying signals. Each element reflects incoming signals by modifying their phase or polarisation but does not add additional power to the signal. Passive RIS is advantageous in terms of power efficiency and cost, as it minimizes energy consumption and reduces interference. However, it has limited control over signal strength, as it cannot amplify or regenerate the signal.

2. **Active RIS**:

Active RIS integrates active components, such as amplifiers, within each element to boost or modulate signals. This type of RIS can actively influence both the direction and strength of signals, allowing for greater control and flexibility. Active RIS typically consumes more power and is more complex to implement but is effective in scenarios where signal strength is a critical factor, such as long-range or high-density urban applications.

Comparison of RIS Hardware Configurations and Materials

RIS hardware configurations vary depending on the type of materials and technologies used. Here are a few common materials and configurations:

1. **Liquid Crystals**:

Liquid crystal-based RIS leverage the ability of liquid crystal molecules to change orientation in response to an electric field, altering their electromagnetic response. Liquid crystals are advantageous due to their fine-tuning ability, low power consumption, and flexibility. However, they have slower response times, which can limit their use in applications requiring rapid adjustments.

2. **Tunable Metasurfaces**:

Metasurfaces made from tunable materials, such as graphene or phase-change materials, offer rapid reconfigurability and a high degree of control. Graphene, for instance, is known for its conductivity and can quickly respond to electrical changes. These metasurfaces are suitable for fast and precise signal manipulation and are often used in dynamic wireless environments.

3. **PIN Diodes and Varactors**:

These semiconductor devices are commonly used in RIS for electronic tuning. PIN diodes switch between reflective and transmissive states, while varactors adjust the capacitance of each element, allowing for phase control. PIN diodes are fast and reliable, while varactors enable finer control over the phase but may require additional circuitry.

Challenges in Miniaturizing and Integrating RIS with Current Hardware

1. **Power Efficiency**:

Although passive RIS consumes minimal power, achieving sufficient control over signals without compromising power efficiency remains challenging, especially in active RIS. Power management is crucial, particularly for large-scale RIS arrays, as inefficient power consumption can lead to overheating and increased operational costs.

2. **Miniaturization and Design Complexity**:

Designing RIS for compact applications involves miniaturizing each element while preserving its tunability and performance. As the size of each unit decreases, the complexity of control circuits increases, posing integration and manufacturing challenges. Miniaturized RIS arrays also need to maintain a balance between high resolution (more elements per area) and processing speed to avoid performance degradation.

3. **Material Stability and Durability**:

RIS elements, particularly those using advanced materials like liquid crystals or phase-change materials, must withstand varying environmental conditions, including temperature and humidity changes. Ensuring stability over long operational periods, especially outdoors, is a challenge that needs addressing to enable reliable, long-term RIS deployment.

4. **Signal Interference and Noise Control**:

As RIS manipulates electromagnetic waves, managing interference among adjacent elements and between multiple RIS panels is essential. Proper shielding and advanced algorithms are required to mitigate signal distortion and noise, especially in dense urban environments or high-frequency bands.

5. **Seamless Network Integration**:

Integrating RIS into existing networks without disrupting current operations is another challenge. RIS panels must communicate effectively with network controllers and adapt to existing protocols. Ensuring seamless handover and coordination between RIS and other network elements, such as base stations and antennas, is crucial for efficient network performance.

5. BEAMFORMING WITH RIS

How RIS Technology Supports Beamforming in Communication Networks

Reconfigurable Intelligent Surfaces (RIS) offer a new approach to beamforming by controlling how signals are reflected across a surface of individually tunable elements. Each element of the RIS can adjust the phase and amplitude of incoming signals, which allows RIS to steer, focus, or scatter signals in desired directions without requiring traditional active antennas or complex electronics. This "passive" approach to beamforming, known as intelligent beamforming, leverages RIS's ability to dynamically adjust its configuration based on real-time needs in the environment.

In a typical communication network, RIS-enabled beamforming provides support in the following ways:

1. **Dynamic Control of Signal Paths:** By configuring each RIS element to reflect signals toward specific areas, networks can establish direct and highly controlled signal paths. This adaptability allows networks to overcome obstacles, minimize losses, and ensure that users receive strong signals regardless of their location.
2. **Optimization of Wireless Channel Conditions:** RIS-based beamforming adapts to changing environmental factors like user movement, interference sources, or multipath fading. For instance, RIS can create favourable wireless channels by adjusting reflections to minimize interference or amplify desired signals, which helps maintain connectivity quality.
3. **Energy Efficiency and Cost Savings:** Unlike traditional active beamforming, RIS achieves beamforming without adding additional power to the signal. This makes RIS a low-cost, energy-efficient solution for scenarios where passive reflection can meet network needs, such as improving connectivity indoors or around obstacles in dense urban areas.

Table 1. Aspects and beamforming

Aspect	Traditional Beamforming	RIS-Enabled Beamforming
Hardware Requirements	Requires active antennas and amplifiers	Utilizes passive, tunable surfaces
Power Consumption	Consumes significant power due to active signal generation and processing	Consumes minimal power, as it reflects rather than transmits
Complexity	High complexity; often requires multiple antennas and beamforming algorithms	Lower complexity; adjusts reflection patterns without signal amplification
Flexibility	Limited to specific directions; depends on array design	Highly flexible; can redirect and reshape signals in real-time
Cost	High due to complex circuitry and power needs	Lower cost as it doesn't rely on active electronics
Scalability	Limited by cost and power needs for each additional antenna	Scales easily by adding more RIS panels

RIS-based beamforming differs fundamentally from traditional methods in that it manipulates the environment rather than relying on active transmission. Traditional beamforming typically requires multi-antenna arrays (e.g., MIMO) to direct signals, while RIS changes the electromagnetic properties of the surface itself to guide signals toward intended locations. This difference makes RIS particularly suited for applications requiring flexible and energy-efficient solutions (Basar, E. 2020).

Applications in Signal Direction Control, Interference Management, and Precision Communication

1. **Signal Direction Control:**

RIS is highly effective for *steering signals* in specific directions, which is particularly useful in urban environments, high-rise buildings, or other areas with obstructions. For example, RIS panels installed on walls or buildings can reflect signals toward areas with low coverage, such as blind spots, ensuring that users receive strong, directed signals. This level of control is crucial for maintaining quality in areas where direct line-of-sight (LOS) is unavailable.

2. **Interference Management:**

By carefully controlling how each element of RIS reflects signals, networks can direct unwanted signals away from areas where they may cause interference. This "reflective nulling" approach allows RIS to reduce interference between users and base stations, particularly in dense, multi-user environments. By adjusting reflections to mitigate overlap in the paths of nearby signals, RIS minimizes interference, improving overall network performance and user experience.

3. **Precision Communication:**

RIS enables a new level of precision in beamforming by providing more granular control over signal direction and intensity. This makes RIS ideal for applications requiring precise coverage and high-quality transmission, such as augmented reality (AR), virtual reality (VR), and autonomous vehicle networks, where signal accuracy and reliability are essential. For instance, in an AR environment, RIS can focus beams on users while avoiding reflective interference from other surfaces, ensuring seamless connectivity.

Figure 2. RIS Enchanced signal strength vs frequency

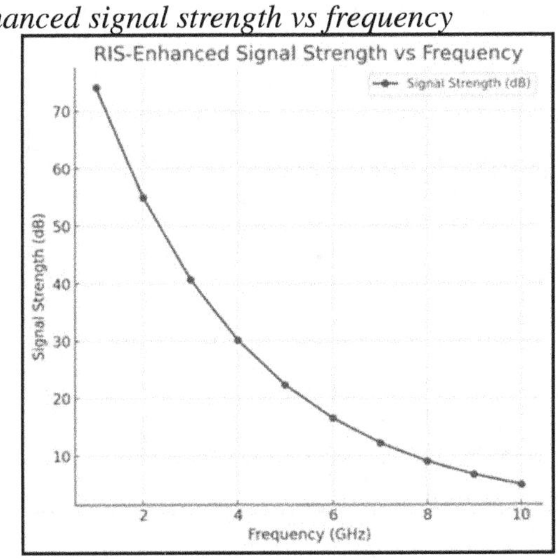

Case Examples of RIS-Enabled Beamforming Applications

- **Smart Cities:** RIS panels installed on buildings in smart cities can help direct signals to specific areas, such as bus stops, public parks, or office lobbies, ensuring consistent connectivity in high-traffic areas while managing interference from overlapping signals.
- **Indoor Office Networks:** In offices or large indoor areas, RIS can help optimize Wi-Fi and cellular coverage by directing signals to areas with high user density, like conference rooms, while avoiding signal spillover into low-use zones.

- **Vehicle-to-Infrastructure Communication:** In autonomous vehicle networks, RIS can assist in reliable, low-latency communication by reflecting signals around obstacles or directly toward moving vehicles. This targeted reflection ensures vehicles receive continuous, strong connectivity as they navigate through complex environments, such as urban intersections or tunnels.
- **Rural and Remote Area Connectivity:** RIS panels can be strategically positioned in rural areas to reflect and extend signals from distant base stations, bringing reliable connectivity to underserved locations. This approach offers a cost-effective alternative to building more towers in sparsely populated areas.

Figure 3. Efficiency of RIS operations vs frequency

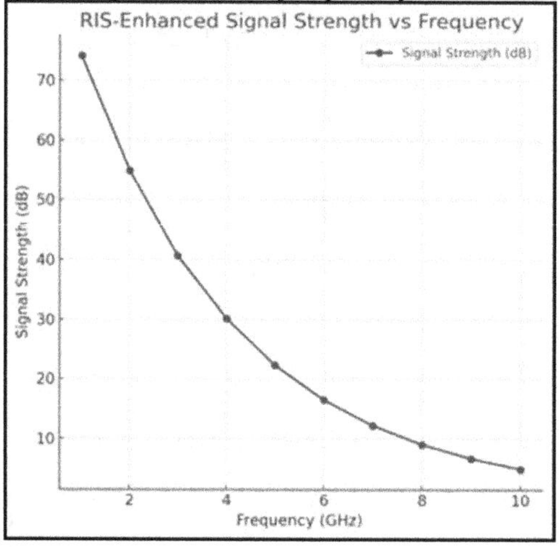

6. METAMATERIALS AND META-ATOM DESIGN

Overview of Metamaterials and the Role of Meta-atoms in RIS

Metamaterials are artificial materials engineered to have properties not typically found in naturally occurring materials. These materials derive their unique characteristics from their internal structure rather than their composition. Metamaterials can manipulate electromagnetic waves in unconventional ways, allowing for functionalities like negative refraction, superlensing, and cloaking. They are typically structured on a scale smaller than the wavelength of the light or signals they manipulate, which gives rise to their extraordinary behaviours.

In the context of **Reconfigurable Intelligent Surfaces (RIS)**, metamaterials play a pivotal role in enabling advanced signal manipulation. The fundamental unit of a metamaterial is the **meta-atom**, which is a small, engineered structure that can respond to electromagnetic fields. Each meta-atom can be independently designed to control aspects such as phase, amplitude, and polarization of incident signals. By configuring an array of these meta-atoms, RIS can achieve precise control over the reflected and transmitted waves, facilitating enhanced beamforming, interference management, and signal quality improvement.

Design Considerations for Achieving Tunability and Responsiveness

When designing meta-atoms for use in RIS, several key considerations must be addressed to ensure optimal performance:

1. **Tunable Electromagnetic Response**:

Meta-atoms must be capable of altering their electromagnetic response based on external stimuli, such as electric fields, magnetic fields, or temperature changes. This tunability is essential for achieving the dynamic reconfigurability that characterizes RIS. Design strategies may include integrating phase-shifting components like PIN diodes, varactors, or MEMS (Micro-Electro-Mechanical Systems) to facilitate rapid changes in the meta-atom's behaviour.

2. **Scalability and Integration**:

The design of meta-atoms should allow for scalability, meaning that they can be effectively arranged in large arrays without performance loss. Furthermore, the ability to integrate meta-atoms with existing electronic and communication systems is crucial for practical deployment. This requires consideration of manufacturing processes that can produce dense arrays of meta-atoms reliably and economically.

3. **Size and Geometry**:

The physical dimensions and shapes of meta-atoms must be optimized to ensure that they interact effectively with the relevant frequencies of operation. The geometry influences the resonance characteristics, which dictate how the meta-atoms respond to incident waves. Careful design is necessary to achieve the desired operational bandwidth while maintaining a compact size for integration into RIS.

4. **Loss Minimization**:

Minimizing energy losses within meta-atoms is essential for maintaining signal integrity. Material choices and structural designs should be aimed at reducing resistive losses and enhancing the quality factor (Q-factor) of the meta-atom. This ensures that most of the incident signal is effectively manipulated rather than dissipated as heat.

Advances in Materials Used for Metasurfaces

Recent advancements in materials have significantly enhanced the capabilities of metasurfaces, broadening their applicability and functionality. Key materials include:

1. **Graphene**:

Graphene, a single layer of carbon atoms arranged in a two-dimensional lattice, exhibits exceptional electrical and optical properties. Its high carrier mobility allows for rapid tuning of its electromagnetic response by applying external electric fields. Graphene-based meta-atoms can be used to achieve ultra-fast reconfigurability, making them ideal for applications in high-frequency communication and advanced sensing.

2. **Phase-Changing Materials (PCMs)**:

PCMs can change their phase between solid and liquid states, which alters their optical properties. Materials like vanadium dioxide (VO_2) and germanium telluride (GeTe) can be used in meta-atom designs to enable significant changes in reflectance and transmission based on temperature or optical pumping. This property allows for substantial tunability and reconfigurability in RIS applications.

3. **Liquid Crystals**:

Liquid crystals provide a versatile platform for tunable meta-atoms due to their ability to change alignment in response to applied electric fields. This change in alignment alters their refractive index, enabling dynamic control over the phase and amplitude of reflected signals. Liquid crystal-based metasurfaces offer a balance between tunability and operational speed, making them suitable for various communication applications.

4. **Dielectric Metamaterials**:

Recent advancements in dielectric materials have led to the development of high-performance metasurfaces that do not rely on metallic components. These materials can achieve low loss while providing substantial control over electromagnetic waves. Innovations in dielectric metamaterials are paving the way for lightweight, efficient, and environmentally stable RIS designs.

5. **Topological Metamaterials**:

This emerging class of materials exploits the principles of topology to achieve robust wave manipulation. Topological metamaterials can support unique electromagnetic wave states immune to defects and disorders. Integrating topological principles into RIS design opens up new avenues for creating highly stable and efficient communication environments.

7. SIGNAL PROCESSING TECHNIQUES FOR RIS

Signal Processing Needs for RIS-Enabled Systems

Signal processing is integral to the effective operation of RIS-enabled systems, as it governs how the signals are manipulated to achieve desired communication outcomes. Key needs include:

1. **Real-Time Signal Processing**: RIS requires fast processing capabilities to adjust phase shifts and amplitudes dynamically based on real-time channel conditions. This necessitates algorithms capable of quick calculations to ensure minimal latency and maximum responsiveness.
2. **Channel Estimation**: Accurate estimation of the wireless channel is crucial for RIS. Techniques such as pilot signals and machine learning algorithms can be employed to assess channel state information (CSI), allowing for optimal tuning of the RIS elements.
3. **Feedback Mechanisms**: Efficient feedback loops are needed to continuously update the RIS configuration based on changing environmental conditions and user demands. This can involve both user feedback and network performance metrics.

4. **Interference Management**: Signal processing techniques must be employed to minimize interference from neighbouring signals. This involves employing algorithms that dynamically adjust the RIS to enhance desired signals while suppressing interference.
5. **Multi-User Processing**: In scenarios involving multiple users, RIS must be capable of processing signals for different users simultaneously. Advanced signal processing techniques, including multi-user detection and scheduling, can be integrated to facilitate this (Björnson, E., Sanguinetti, L., & Debbah, M.,2021).

Optimization Algorithms for Controlling Phase Shifts and Amplitude

Optimizing the control of phase shifts and amplitudes in RIS is essential for enhancing signal quality and achieving efficient communication. Key algorithms include:

1. **Gradient Descent Algorithms**: These algorithms iteratively adjust phase and amplitude settings by calculating gradients based on performance metrics like signal-to-noise ratio (SNR). This approach is effective for converging to optimal solutions but may require multiple iterations.
2. **Genetic Algorithms**: These evolutionary algorithms can be used to explore a wide solution space for optimal phase configurations. By mimicking natural selection processes, genetic algorithms can effectively manage complex optimization problems in RIS configurations.
3. **Particle Swarm Optimization (PSO)**: PSO algorithms leverage a swarm of particles to explore potential configurations. Each particle represents a possible solution, and they adjust their positions based on their own experience and that of their neighbours, making PSO particularly effective for multi-objective optimization in RIS applications.
4. **Reinforcement Learning**: Techniques from reinforcement learning can be employed to enable RIS to learn optimal phase and amplitude settings over time based on feedback from the environment. This approach allows for adaptive optimization based on historical data and current conditions.
5. **Convex Optimization**: For scenarios with well-defined performance metrics, convex optimization techniques can provide global optimal solutions. These techniques are particularly valuable for straightforward cases where the objective function is convex.

Machine Learning and Artificial Intelligence Approaches for RIS Optimization

The integration of machine learning (ML) and artificial intelligence (AI) into RIS optimization is a rapidly evolving field, offering innovative solutions for complex challenges:

1. **Deep Learning Models**: Neural networks can be trained to predict optimal phase configurations based on historical channel data. This approach enables the RIS to adapt to varying conditions without needing exhaustive optimization in real time.
2. **Supervised Learning for Channel Estimation**: ML models can be employed for channel estimation tasks by training on labelled datasets. This leads to improved accuracy in predicting channel conditions, allowing for better RIS configurations.
3. **Unsupervised Learning for Anomaly Detection**: Unsupervised learning techniques can help identify patterns in network performance, enabling the detection of anomalies or shifts in user behaviour, which can then be addressed by adjusting RIS settings.
4. **Federated Learning**: In scenarios with multiple RIS units, federated learning allows them to collaboratively learn optimal configurations without sharing sensitive data. This approach enhances privacy while improving overall network performance.
5. **Optimization under Uncertainty**: AI techniques can handle uncertainty in channel conditions by employing probabilistic models that take into account varying levels of noise and interference, leading to more robust RIS configurations.

8. ENERGY EFFICIENCY AND POWER CONSUMPTION

Evaluation of RIS for Energy-Efficient Wireless Communication

Reconfigurable Intelligent Surfaces represent a paradigm shift in energy efficiency for wireless communication. The evaluation of RIS for energy efficiency can be based on several metrics:

1. **Reduced Transmission Power**: By reflecting signals toward users, RIS can reduce the power required from base stations. This decreases the overall energy consumption in the network, making RIS a compelling alternative to traditional active repeaters.

2. **Dynamic Resource Allocation**: RIS can dynamically allocate resources based on real-time demand, reducing unnecessary power usage during low-traffic periods. This adaptability enhances overall system energy efficiency.
3. **Low-power Operation**: The passive nature of RIS means that they consume significantly less power compared to traditional active components, which often require constant power to amplify signals. This inherently makes RIS a more sustainable option.

Power-Saving Potential Compared to Traditional Repeaters or Amplifiers

The power-saving potential of RIS compared to traditional communication systems is significant:

1. **Elimination of Active Components**: Traditional repeaters and amplifiers consume high levels of power for active transmission. In contrast, RIS primarily relies on passive reflection, dramatically lowering energy consumption.
2. **Spatial Efficiency**: RIS can provide coverage over larger areas with fewer components, further reducing the power required per unit of coverage. This spatial efficiency translates into energy savings in densely populated environments.
3. **Lower Backhaul Requirements**: By improving signal strength and quality through reflective techniques, RIS can decrease the load on backhaul links, reducing the energy required for data transmission back to the core network.

Methods to Optimize RIS Energy Consumption, Including Adaptive Control Strategies

To maximize the energy efficiency of RIS, various optimization methods can be employed:

1. **Adaptive Control Strategies**: Implementing algorithms that adaptively control the phase shifts and amplitudes based on current network conditions can significantly reduce energy consumption. These algorithms can ensure that only the necessary configurations are applied, avoiding over-processing.
2. **Energy Harvesting Techniques**: Integrating energy harvesting technologies, such as solar panels or RF energy harvesting, can provide a self-sustaining energy source for RIS, minimizing reliance on external power supplies.
3. **Sleep Mode Functionality**: By implementing sleep modes or low-power states during periods of low activity, RIS can conserve energy without compromising performance during peak usage times.

4. **Optimal Deployment Strategies**: Strategic placement and configuration of RIS can enhance coverage while minimizing energy expenditure. This involves using simulations to model various deployment scenarios and selecting configurations that provide maximum coverage with minimal power use.
5. **Network Coordination**: Coordinating with other network components, such as base stations and user devices, can optimize the overall energy consumption of the communication system. By sharing channel information and collaboratively managing power distribution, RIS can contribute to a more energy-efficient network (Huang, C., et al. 2019).

9. SECURITY CHALLENGES IN RIS

RIS Vulnerabilities and Implications for Network Security

Reconfigurable Intelligent Surfaces (RIS) introduce new security challenges in wireless communication networks. Key vulnerabilities include:

1. **Increased Attack Surface**: The introduction of RIS expands the attack surface, as they become potential targets for adversaries aiming to exploit their functionalities. Each surface unit can be manipulated if security measures are not robust.
2. **Data Integrity Risks**: The ability of RIS to reconfigure means that unauthorized manipulation could lead to compromised data integrity. An attacker could redirect signals or alter communications, leading to misinformation.
3. **Dependence on Network Infrastructure**: RIS operate within the broader context of existing network infrastructure. Vulnerabilities in associated systems, such as backhaul connections and control signalling, can impact the security of the entire network.
4. **Channel Estimation Attacks**: Adversaries may attempt to spoof or mislead channel estimation processes, affecting the optimization of phase shifts and potentially leading to performance degradation or interception of signals.

Potential Threats: Eavesdropping, Physical Attacks on RIS Hardware

1. **Eavesdropping**: RIS may inadvertently facilitate eavesdropping if not properly secured. As they reflect signals, unauthorized users could capture communication without detection. This risk is particularly significant in environments where sensitive data is transmitted.

2. **Physical Attacks**: RIS hardware can be subject to physical attacks, including tampering or damage to the surfaces. Such attacks could lead to the loss of control over the RIS and its functionalities, making it vulnerable to exploitation.
3. **Denial of Service (DoS) Attacks**: Attackers may target RIS with DoS attacks, overwhelming them with traffic or causing them to malfunction, which could disrupt service availability for legitimate users.

Solutions for Secure Deployment, Including Encryption and Monitoring Techniques

1. **Robust Encryption Protocols**: Implementing strong encryption techniques for data transmitted through RIS can mitigate the risk of eavesdropping and unauthorized access. End-to-end encryption ensures that only authorized users can decrypt and access the transmitted data.
2. **Intrusion Detection Systems (IDS)**: Deploying IDS can help monitor network traffic and identify suspicious activities. By analyzing patterns and behaviours, these systems can alert network administrators to potential security breaches.
3. **Secure Control Channels**: Establishing secure communication channels for RIS control signals is vital. Using techniques like secure multi-party computation can help ensure that only authenticated parties can control the RIS configuration.
4. **Physical Security Measures**: Securing the physical locations of RIS installations through surveillance, access controls, and tamper-resistant hardware can help prevent physical attacks.
5. **Regular Security Audits**: Conducting periodic security assessments and vulnerability testing can help identify and mitigate potential threats to RIS systems (Renzo, M. D., et al. 2020).

10. RIS IN 5G AND 6G TECHNOLOGIES

Role of RIS in Achieving the High Standards Set by 5G/6G

1. **Enhanced Spectral Efficiency**: RIS significantly improves spectral efficiency by optimizing the propagation of wireless signals. By intelligently directing and enhancing signals, RIS can increase the amount of data transmitted within the same bandwidth.
2. **Network Slicing**: In 5G and future 6G networks, network slicing enables the creation of virtual networks tailored to specific applications. RIS can be dynamically reconfigured to serve different slices efficiently, meeting varying quality of service (QoS) requirements.

3. **Support for Massive MIMO**: The integration of RIS with massive multiple-input multiple-output (MIMO) technologies can further enhance network capacity and performance, allowing for simultaneous connections to a higher number of users (Wu, Q., et al. 2020).

Enhancements in Network Capacity, Latency Reduction, and User Experience

1. **Increased Network Capacity**: RIS can dynamically manage signal propagation, improving capacity in dense urban environments where traditional infrastructure struggles. This capability is vital for accommodating the increasing demand for mobile data.
2. **Latency Reduction**: By optimizing signal paths and minimizing interference, RIS can significantly reduce latency, enhancing the responsiveness of applications, particularly in critical sectors like autonomous driving and telemedicine.
3. **Improved User Experience**: By providing enhanced coverage and data rates, RIS can greatly improve the overall user experience, making services more reliable and efficient, especially in challenging environments.

Implementation Case Studies: Smart Cities, Internet of Things (IoT) Applications

1. **Smart Cities**: In smart city deployments, RIS can manage the communication needs of various IoT devices, ensuring reliable data transmission for traffic management, environmental monitoring, and public safety applications.
2. **IoT Applications**: RIS can support diverse IoT applications by enhancing connectivity and data rates for smart homes, industrial automation, and health monitoring systems, enabling real-time data collection and analysis.

11. PRACTICAL IMPLEMENTATION CHALLENGES

Deployment Challenges, Including Integration with Existing Infrastructure

1. **Infrastructure Compatibility**: Integrating RIS with existing network infrastructure poses challenges, particularly when legacy systems are involved. Ensuring compatibility and seamless operation requires careful planning and potential upgrades.

2. **Cost Considerations**: To ensure viable implementation strategies, the financial implications of deploying RIS technology, including the costs of installation, maintenance, and upgrades, must be addressed.
3. **Training and Skill Development**: There is a need for skilled personnel who can design, deploy, and maintain RIS systems. Training programs may be necessary to bridge the knowledge gap in this emerging technology.

Environmental Impact and Regulatory Considerations

1. **Environmental Assessments**: Deploying RIS requires thorough environmental assessments to understand the potential impact on local ecosystems and ensure compliance with regulations.
2. **Regulatory Compliance**: Navigating the regulatory landscape can be complex, particularly in terms of frequency spectrum allocation, safety standards, and environmental regulations. Collaboration with regulatory bodies is essential for successful deployment.

Scalability Issues and Solutions for Large-Scale Networks

1. **Scalability of Deployment**: As the number of RIS installations increases, managing and scaling the network becomes a challenge. Solutions may include the use of centralized control systems that can efficiently manage large numbers of RIS units.
2. **Load Balancing**: Effective load balancing algorithms will be necessary to optimize resource allocation across multiple RIS, ensuring equitable distribution of communication resources.
3. **Interference Management**: As more RIS units are deployed, managing interference between them becomes critical. Advanced signal processing and machine learning algorithms can be employed to mitigate interference dynamically.

Real-World Case Studies of RIS Implementations

6. **Deployment in Urban Areas**: Several pilot projects in urban areas have showcased the potential of RIS in enhancing connectivity in high-density environments, demonstrating significant improvements in user experience and data throughput.
7. **Industry Collaborations**: Collaborations between academia and industry have led to successful RIS implementations in specialized applications, such as smart agriculture, where reliable data transmission is crucial for monitoring crop conditions.

8. **Testing in 5G Networks**: Initial tests in 5G networks have highlighted the benefits of integrating RIS, particularly in scenarios with high user density, showcasing improved signal quality and coverage.

12. FUTURE DIRECTIONS AND RESEARCH OPPORTUNITIES

Areas for Innovation: New Materials, Adaptive Designs, Real-Time Optimization

1. **New Materials Development**: Research into novel materials is essential for enhancing the performance of RIS. Materials such as liquid crystals, 2D materials (e.g., graphene), and phase-change materials offer promising avenues for creating more efficient and responsive surfaces. The exploration of hybrid materials combining different properties could lead to RIS that can adapt to a wider range of environmental conditions.
2. **Adaptive Designs**: Innovations in adaptive design will allow RIS to adjust dynamically to changing signal environments. This includes the development of surfaces that can alter their physical properties or configurations in response to real-time feedback from the communication environment. For instance, incorporating smart materials that can change their characteristics based on surrounding electromagnetic fields could improve signal fidelity.
3. **Real-Time Optimization Algorithms**: The advancement of signal processing techniques and optimization algorithms is crucial for the effective operation of RIS. Research into machine learning and artificial intelligence (AI) can facilitate real-time optimization of RIS configurations, enabling them to make autonomous adjustments based on the current communication context, user demand, and network conditions.

Exploration of RIS Beyond Traditional Communications, Such as Healthcare, Industry, and Space

1. **Healthcare Applications**: The integration of RIS in healthcare settings can enhance the performance of medical devices, particularly in wireless health monitoring systems. By improving signal strength and coverage in hospitals and clinics, RIS can facilitate better real-time data transmission for patient monitoring and telemedicine applications.
2. **Industrial Applications**: In industrial environments, RIS can support the growing number of IoT devices, ensuring reliable connectivity in smart factories. The ability to enhance signal quality in challenging environments, such as factories

with metal structures, can enable seamless data transmission for automation systems, predictive maintenance, and remote monitoring.

3. **Space Communications**: RIS can also play a significant role in space communications, where traditional infrastructure may be impractical. By deploying RIS in orbiting satellites or space stations, researchers can improve the quality and efficiency of communication signals transmitted over vast distances, aiding in data collection and transmission for space missions.

Multidisciplinary Approaches Combining Materials Science, AI, and Communications

1. **Collaborative Research**: Future advancements in RIS technology will benefit from interdisciplinary research that combines materials science, communications engineering, and artificial intelligence. Collaborative projects can lead to innovative solutions that leverage the strengths of each discipline, fostering the development of next-generation RIS technologies.
2. **Simulation and Modeling**: The use of advanced simulation tools and modelling techniques will be essential for designing and testing new RIS concepts. By creating virtual environments, researchers can predict the performance of various materials and designs before physical implementation, significantly reducing development time and costs.
3. **Standardization and Interoperability**: As RIS technology matures, establishing standards for design, implementation, and operation will be crucial. Research efforts should focus on creating frameworks that ensure interoperability between different RIS systems and existing communication networks, facilitating broader adoption and integration.

13. CONCLUSION

In summary, Reconfigurable Intelligent Surfaces (RIS) present a transformative opportunity for advancing wireless communication technology. By enabling more efficient signal propagation, enhancing network capacity, and improving user experiences, RIS has the potential to reshape the landscape of telecommunications.

Key implementation strategies involve addressing challenges such as integration with existing infrastructure, security vulnerabilities, and the need for real-time optimization. Moreover, the exploration of RIS applications beyond traditional communications into sectors like healthcare, industry, and space highlights their versatility and far-reaching implications.

Future research directions emphasise the importance of innovation in materials science, the development of adaptive designs, and the integration of AI for real-time optimisation. A multidisciplinary approach that combines insights from various fields will be vital in overcoming current limitations and unlocking the full potential of RIS.

Ultimately, as RIS technology continues to evolve, it will play a crucial role in meeting the demands of next-generation communication systems, contributing to the realization of smarter, more connected environments that enhance our daily lives.

REFERENCES

Basar, E. (2020). *Reconfigurable Intelligent Surface-Based Wireless Communications: Antenna Design, Prototyping, and Experimental Results.* Available at arXiv: [2002.04068](https://arxiv.org/abs/2002.04068)

Björnson, E., Sanguinetti, L., & Debbah, M. (2021). *Reconfigurable Intelligent Surfaces: A Signal Processing Perspective.* Foundations and Trends® in Signal Processing, 14(3), 180–328. DOI: 10.1109/JSAC.2020.3007211

Di Renzo, M., Zappone, A., Debbah, M., Alouini, M.-S., Yuen, C., de Rosny, J., & Tretyakov, S. (2020). Smart Radio Environments Empowered by Reconfigurable Intelligent Surfaces: How It Works, State of Research, and Road Ahead. *IEEE Journal on Selected Areas in Communications*, 38(11), 2450–2525. DOI: 10.1109/JSAC.2020.3007211

Huang, C., (2019). Reconfigurable Intelligent Surfaces for Energy Efficiency in Wireless Communication. IEEE International Conference on Communications (ICC), 1–6. DOI: DOI: 10.1109/ICC.2019.8762025

Liang, Y.-C., (2019). Reconfigurable Intelligent Surfaces for Smart Wireless Communications: Opportunities, Challenges, and Solutions. *IEEE Wireless Communications*, 27(2), 118–125. DOI: 10.1109/MWC.001.1900383

Renzo, M. D., (2020). Smart Wireless Communication Environments via RIS: Opportunities, Challenges, and Future Directions. *IEEE Communications Magazine*, 58(6), 40–47. DOI: 10.1109/MCOM.001.2000074

Wu, Q., (2020). Reconfigurable Intelligent Surfaces: Pathways to 6G. *IEEE Transactions on Wireless Communications*, 20(1), 496–515. DOI: 10.1109/TWC.2020.3032112

Wu, Q., & Zhang, R. (2019). Towards Smart and Reconfigurable Environment: Intelligent Reflecting Surface Aided Wireless Network. *IEEE Communications Magazine*, 58(1), 106–112. DOI: 10.1109/MCOM.001.1900107

Chapter 3
Reconfigurable Intelligence Surface-Aided Secured Wireless Communication Networks

N. Ashokkumar
https://orcid.org/0000-0002-4034-484X
Mohan Babu University, India

R. Pavithra
V.S.B. Engineering College, India

Ashok Vajravelu
Tun Hussein Onn University of Malaysia, Malaysia

Sheshang Degadwala
Sigma University, India

ABSTRACT

The customizable intelligent surface, also known as RIS, is a relatively new technology that could change how radio connections work if it works well. The RIS technology can be used both inside and outside, so the setting in which the sender sends signals can be changed on the fly. There is a different method used here than the active stepped antenna array that is usually used. The function of RIS has been tested in a large number of different real-life situations so that its main idea can be used. This will not only talk about the possible uses, like the system and channel model, the information theory analysis, and the physical implementation and design, but it will also go over the most important signal processing methods, like precoding and

DOI: 10.4018/979-8-3693-8099-4.ch003

channel estimation, as well as prototyping. Finally, we talk about some interesting study ideas that might one day be useful in the area of RIS-assisted messaging.

1. INTRODUCTION

Wireless communication systems have changed a lot because smart changes to the environment where the transfer takes place have made them stronger and better at what they do (Jian et al., 2022; Wang et al., 2024). Metamaterial technology was used to make the RIS, an electric surface structure that was made by people. An electromagnetic field can be designed in real time, and it is made up of many passively reflecting parts. Besides that, it can change electromagnetic waves, which improves the range and delivery of signals (Jian et al., 2022; Yuan et al., 2021). RISs can improve communication lines a lot because they can carefully change the phase shifts of the signals that bounce off of them (Cao et al., 2023; Chen et al., 2024). They are small, cheap, flexible, and easy to set up, which makes them a great choice for current radio frequency communication systems (Khoshafa et al., 2024). Crosstalk can be cut down, signal strength can be increased, and the whole network can work better thanks to RISs (Basharat et al., 2024; ElMossallamy et al., 2020; Pan et al., 2022). Researchers have looked into how smart surfaces that can change on their own could be used to make more wireless power. Communication networks have been set up so that Internet of Things devices can connect to each other more and more (Hameed & Koo, 2024). Smart surfaces (RISs) that can be changed have come a long way in the last few years, making this possible. (Sikander et al., 2023) Someone has come up with a unique way to make RIS beamwidths better and interference inside the beams less of a problem. It was said that the communication methods linked to Reconfigurable Intelligent Surfaces were looked into (Renzo et al., 2022). A lot of different things were looked into in this study, from how to make radio networks work better to the electric qualities of the materials. It's a big step forward for wireless technology that radio frequency identification systems (RISs) can change and adapt to different interactions. A lot of people believe that RISs will play a big role in building sixth-generation (6G) wireless networks (Pan et al., 2022).

A lot of research has been done on how to connect radio transmission devices to RIS. Now it's safe to use channel models and descriptions, and for fully passive RIS-aided systems, it's now possible to make sparsity-aware channel predictions. (Hameed & Koo, 2024) It was also looked into how to model channels for 6G communications with RIS help. Right now, most of the study being done on interference in RIS-assisted wireless communication systems is focused on how to handle interference between users who are allowed to use the system. Sikri A. and his friends used the °–¼ fade model in a research project that looked at how mobile noise affects

RIS-aided systems in general. You learned how to use numbers to talk about routes that are linked to each other. As a rough guide, the failure chance (OP) and the average bit error rate (ABER) were given. For the OP, this paper gives closed-form results for an active STAR-RIS-aided two-user downlink communication system when the users are linked and when they are not. (Sikander et al., 2023) How can the transfer be more stable when there are problems with the receiver hardware in a RIS-assisted secure communication system? That's what this study looked into. This study looked at how well multi-hop systems work with both Free Space Optics (FSO) and Radio Frequency (RF) lines and RISs to make them work better. To make wireless communications work better without having to worry about crosstalk, the main goal of the study we just talked about is to find ways to use RISs to do that.

Right now, the study is mostly about things that get stuck when real people link to RIS-based systems. People said that it's not possible to use a set RIS as an opening in dangerous situations like war, even though people have thought about how to do this on purpose. Unmanned aerial vehicles (UAVs) are speeding up the study of radio communication. Drones make this possible because they can do many things to help and change based on the situation. Drones that fly through the air (UAVs) improve network performance by improving link speeds and range when things aren't going well. (Renzo et al., 2022) An awful lot of new research has been done on how to link RISs to UAV-based radio communication systems. After that, UAVs with RISs can send and receive waves to strengthen the connection. A full study was done on robotic aerial vehicle (UAV) systems that use RIS. It looked at a lot of different things, like streamlining, ways to talk to each other, deep reinforcement learning, performance in secret, making things run more easily, and the Internet of Things. With the tests we talked about above, we didn't look into how unmanned aerial vehicles (UAVs) could mess up the RIS-assisted dual-hop system. Talked about relay radio links that need help from RIS.

In this paper, we look at a two-hop wireless system that combines RIS with an unmanned aerial vehicle (UAV). The system's sensor is being blocked by a bad jammer that is a UAV with RIS. The talk before gave us the idea for this study. These methods are very helpful in many areas. As an example, they keep people safe in the military and security, and they fix links in places where natural disasters have caused damage. (Dao et al., 2021) The writers don't know of any studies that look at RIS-Equipped UAV dual-hop wireless communication systems when RIS-Equipped UAVs are blocking signals. Here are some of the most important things to me about this work:

In the RIS-Equipped UAV dual-hop wireless communication systems, we made and shared a closed-form way to get the probability density function (PDF) of both the valid channel and the hacking channel. Another thing we did was make a new closed-form equation for the PDF of the received signal-to-jamming ratio (SJR).

We were able to figure out how to get the ABER for both BPSK and QAM modulation by using the Gauss–Laguerre Approximation of the signal.

We used data from both research and modeling to find out how the ABER worked in different channel conditions, how likely it was to get jammed, and where the right RIS could be found.

The following parts of this work are given in this order: There are models of jammers and channels in the second part, which gives a review of radio communication systems (Nguyen et al., 2021). Part 3 is all about getting the closed-form formulas for the ABER in both the BPSK and QAM ways. In Section 4, you can see the number of results and what they mean. The last part of the study is Part 5.

2. FOUNDATIONS ON RISS

This section provides an introduction to the core ideas of RISs, including their hardware architecture and the rules that govern their operation. Additionally, it offers a fundamental explanation of a communication system that is helped by RIS, which includes the end-to-end setup as well as the channel model structure.

2.1. RIS Hardware Architecture

A flat array that uses electromagnetic waves is one type of RIS. Its job is to control the flow of information. A metasurface is another name for a RIS. This is because it is made of a metal that can be changed with computer code. The main parts of a RIS are put together in Figure 1. The metasurface is on the left side of the picture. It's a flat group of things called meta-atoms or reflected elements. (Rodriguez et al., 2022) The RIS parts, which include its form, size, length, direction, and order, change the electromagnetic (EM) wave effect. If the meta-atom reflection coefficient is changed, it is possible to make wireless channels that change over time and still work. With this tech, you can talk to each other without wires.

Figure 1. RIS hardware architecture, detailing reflecting elements and equivalent circuits.

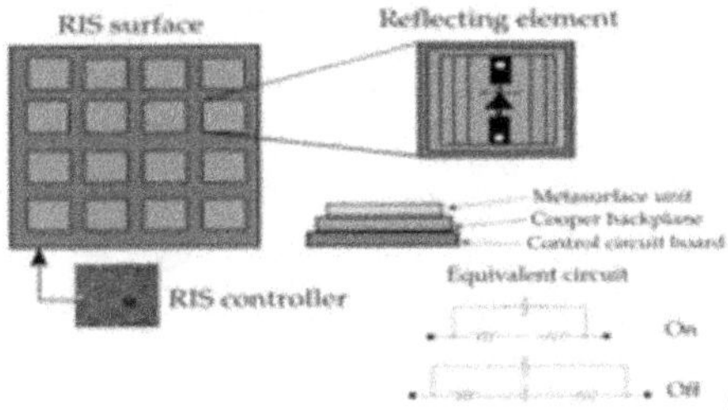

Figure 1 shows how to put together a part that shows two sides. Each part has a PIN diode built in that changes the state from one to the other (on and off). With the PIN diode, you can change the DC power that tells each part what to do. The phase shift difference between the two states is **g** radians, which is the same as in a circuit with two states. If you change the volts that go through the RIS parts, you can get different phase changes. The amount of resistors built into each part can also be changed to change the strength of each signal that is mirrored.

The RIS hardware is split into three parts, which can be seen in Figure 1. On the top layer, there is a base that keeps heat in and has metal spots on it. People sometimes call these spots "shinier features." After being written to the base, these changes go straight to an input signal. (Castillo-Soria et al., 2024) The next row is made up of a copper plate, which comes after the metal pieces. It also keeps the signal alive while it's being sent. On the inside of the last part is a control circuit board whose main job is to change each signal's strength or phase shift. The micro-controller is a popular type of computer device that can be used as a RIS control. The field-programmable gate array (FPGA) is another type.

2.2. RIS-Assisted Wireless Communication Systems

The electromagnetic (EM) wave is changed in several ways in places where radio messaging is useful. Waves move through the radio channel and hit different things along the way. This is called multipath propagation, and it lowers the power. These deals all take place at the same time. Both the sender and the listener can join to the channel through the RIS in this case. (Castillo-Soria et al., 2023) RIS surfaces help the base station (BS) and the target talk to each other better because they bounce waves back and forth. This is possible because different parts of the surface send data in a certain way. The end user's device might be able to get a better connection since this is the case. At the end of the day, this means that the unsure wireless channel changes into a model that is sure and is run by software. On the other hand, RIS devices work well and provide extra ways for the author and users to talk when direct lines aren't available.

Figure 2. RIS-aided wireless communication system scheme.

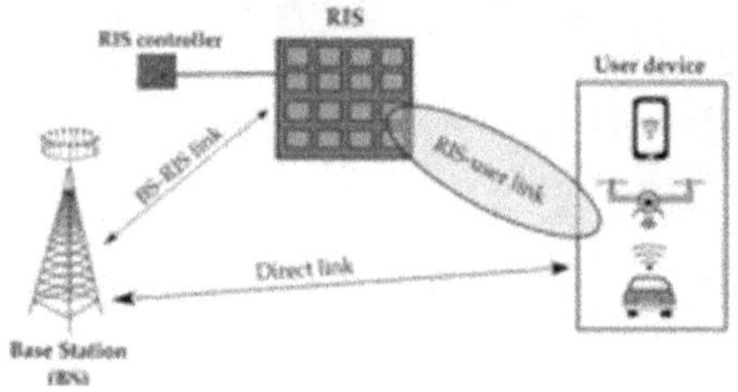

Another idea for a messaging system with RIS is shown in Figure 2. Everything that will be a target for K users has a broadcast antenna (Tx) built into the base station. (Di Renzo et al., 2020) Since there are N RISs with shiny parts, the system makes use of them. That is, $H_{AE} \in \mathbb{C}Nr \times Nt$, where k is any number from 1 to n. This grid shows the straight line that connects the BS to the kth person. The nth RIS is linked to the kth person by a matrix $H_{E}, k \in \mathbb{C}Nr \times Ne$. The link between the BS and the nth RIS is shown by $H_{EE} n \in \mathbb{C}Ns \times Nt$, where $\boldsymbol{S}$ is a number from 1 to n. To help you understand, let's say that each link in the plan is pretty much fixed and has a Rayleigh fading distribution. This spread is made up of many complicated Gaussian random variables, and each one has a mean of 0 and a standard deviation of $\boldsymbol{y}\mathcal{N}(0,1)$. But to get a better idea of how well the system will work, routes that are more like real life could be used.

2.3. RIS-Assisted Communication System Model

The emitter at the base station (BS) has m inputs, as shown in Figure 2. It's written as $a_E = \{dn\}mn=0$, where $d\boldsymbol{S}$ is a number between 0 and 1. The messages xi are what come out. They come from machine parts that have been changed. One kind of modulation that the machine can do is M-ary quadrature amplitude modulation (QAM). It says that the signal that the kth user gets through the nth RIS and the straight path is (Tang et al., 2020).

Figure 25. Math graphic 1

$$y_k = \sqrt{\gamma_k}\mathbf{H}_{1,k}\boldsymbol{\Theta}_1\mathbf{G}_1\mathbf{H}_1^{SR}\mathbf{x}_k + \sqrt{\gamma_k}\mathbf{H}_{2,k}\boldsymbol{\Theta}_2\mathbf{H}_2^{SR}\mathbf{G}_2\mathbf{x}_k + \cdots$$
$$+\sqrt{\gamma_k}\mathbf{H}_{l,k}\boldsymbol{\Theta}_l\mathbf{H}_l^{SR}\mathbf{G}_l\mathbf{x}_k + \cdots + \sqrt{\gamma_k}\mathbf{H}_{L,k}\boldsymbol{\Theta}_L\mathbf{G}_L\mathbf{x}_k + \sqrt{\gamma_k}\mathbf{H}_k^{DP}\mathbf{x} + \mathbf{n}_k,$$

You can write ɟ*l* as diag(*eεl*,1,*eεl*,2,⋯,*eεl*,*Ne*) to get the lth matrix of phase shifts. The phase shift of the ith reflected element of the lth RIS is given by *eεl*, and the noise is given by n*k*. The noise samples are thought to be separate and evenly spread out with respect to *y*𝒩(0,2). A different name for *r*−−√ is signal-to-noise ratio (SNR). You can also write (1) as

Figure 26. Math graphic 2

$$\mathbf{y}_k = \sqrt{\gamma_k}\left(\sum_{l=1}^{N} \mathbf{H}_{l,k}\boldsymbol{\Theta}_l\mathbf{G}_l + \mathbf{H}_k^{DP}\right)\mathbf{x}_k + \mathbf{n}_k.$$

I want to look at Equation (1). The grid channels H_E change the signal y_E in a direct way. It is tough to guess how strong these links are if you only think about a RIS that isn't doing anything and doesn't handle signals. There are a few different ideas about what the channel is. Two of them are the Rayleigh fading channel and the Rician fading channel. (Wang et al., 2020) But channel guess is one of the main problems that needs to be fixed before relations with RIS can get better. There is an ON/OFF method suggested by the authors of to find the cascaded channel H_E. The process is split into steps, and each step only works out one column vector that is connected to one RIS element. To be more specific, the cascaded channel is shown by 1 column

$$H_{l.k} = \left[h_{l,k}^{(1)}, h_{l,k}^{(2)}, h_{l,k}^{(3)}, \ldots, h_{l,k}^{(Ns)}\right] \tag{3}$$

In this case, h(i)i,i refers to a column vector that is connected to the channel that goes with the ith RIS element. At the ith stage, only the ith RIS part is on; all the other parts are off. The steps below should be taken to get an idea of the channel: For a rough idea of the direct channel Hm, users send wireless signals to each other in a normal way, like captain messages. After that, the related channels of the RIS element are found in a number of different ways, as shown by $h(i)_E$. (Hellström et al., 2022) So, the channel forecast might not be as good since the pilot signal is only sent to the baseline station by one RIS element. There's another way to guess that takes each RIS element's power into account. But the channel matrix H_E,*m* is still broken up into columns to help with the forecast.

You can change the matrices G_E and ɟ*l* in equation (1) to change the timing and strength of the signal as it hits the RIS elements. (Zappone et al., 2019) People believe that beamforming is one of the best things about RIS communications. Beamforming is one method that makes RIS encounters much better by letting you direct the signal bounce in the direction you want. This makes receiving better. These are the main ways that RIS beamforming is carried out: You need extra active parts or circuits that can change the phase and strength of the RIS-impinging signals in order to do active beamforming. Changes to the messages happen right away, but the system gets harder to understand and needs more power. Because it makes the most of what RISs can do, the device can do passive beamforming without adding any extra power lines. (Wang et al., 2020) The amount of power they use and how hard they are to use are both less than active beamforming.

Hybrid beamforming technology lets you combine active and passive systems in a number of unique ways. A lot of the time, when the RIS is used, it is meant to appear very reflecting. (Sejan et al., 2022) Since each part only needs a certain number of control bits, it would be better to make different levels for amplitude and phase shift. This will make the process go better. It's no surprise that phase-shift control costs more than amplitude control. And we can say this because the best phase-shift techniques should line up all of the RIS's signals, no matter how strong they are, with the BS's pulse (Faisal & Choi, 2022). These signs make a clear mixed signal when they are lined up right. This makes the sound stronger when it gets to the person listening.

3. IRS TECHNOLOGY FUNDAMENTALS

Within this area, we will familiarize you with the general architecture of the IRS hardware as well as the essential principles of operation.

3.1. IRS Hardware Architecture and Its Working Principle

Meta-material technology that is two-dimensional and can be written on digitally is used by the IRS. These changes are based on the meta-surface. The meta-surface must be at the right subwavelength frequency in order to work. The meta-surface is a flat grid made up of many meta-atoms that are longer than they are wide. By setting the elements up correctly, you can change the way each IRS meta surface atom element sends signals. Something like a mirror, a change in strength, or a phase shift can happen. Parts' sizes, directions, patterns, shapes (like square or split-ring), and more are all part of this. The wireless channels change as the user moves (UE), and each part of the system needs to be able to change its reflection coefficient to work with those channels. There are three main ways to handle IRS reflection mechanical reduction, as shown in the books. Some of these ways are making things like PIN diodes, FETs, MEMS switches, useful tools like liquid crystal and graphene, and moving and spinning things mechanically. A lot of people like electronic tools these days because they work quickly, don't lose much power when they reflect light, and don't cost much to power. The IRS needs to connect to the internet so they can learn how to talk to people outside of work and make sure their real-time flexible mirror works. Lastly, we need reflection factors that can change at any time in order to make the IRS parts.

An IRS tool that is often chosen is shown in Figure 3. This shows that the IRS is made up of three levels and a smart driver. The first layer has a lot of shiny spots that can be moved around. It is on a base that keeps the heat in. By making these changes, you can change the texts that go straight to the device. Under this layer is often a copper plate that stops the signal energy from leaving while the IRS reflects it. The job of a control circuit board is to change when and how strong each image is. It's also what makes the part reacting happy. Another smart thing the IRS has is something that turns the mirror on and off. The link between user equipment (UE) and base stations (BSs) is managed by a field-programmable gate array (FPGA). Different wireless lines let other parts of the network, like UE terminals, access points (APs), and base stations (BSs), talk to and work with it as well. This lets them share info at a slow speed.

Figure 3. The structure of the PIN diode-based changeable reflective element with its different levels.

There is a PIN diode in the middle of each part of the circuit in Figure 1. This shows how a structure can be made up of different parts. The biasing voltages sent to the PIN diode through a direct-current line can be changed to make it work in "ON" or "OFF" mode. That's how the element can add a phase-shift change to the information it gets. It is the job of the smart processor to set the correct biasing voltages. In this way, different parts of the IRS can go through phase changes in their own step. According to [30], the PIN diode element can change frequencies of up to 5 megahertz (MHz), which is the same as 0.2 microseconds (μ second). It only takes a small amount of time for the channels to become coherent, which is good for a phone where the channels change a lot. You can change both the phase shift and the amount that each IRS part stands out. Being able to change the shape of the copy signal gives you more ways to talk to each other. This method is also helpful because it makes it easy to balance the mirror's speed with the gear's price. For some reason, amplitude control is often less expensive than phase control. There are several ways to change the volume of the IRS network. A common action is to change the load resistance in every part of the IRS.

In this case, let's look at this: the signal's energy hits a point where it turns into heat because the resistance of each IRS part changes. This means that the reflection amplitude's size can be in the [0, 1] band. You can change the load resistance to change the power of the mirrored signal as long as the RFID tag is not functioning. It's a lot like tags. To make the mirrored design work best, you should also be able to change the amplitude and phase of each IRS part separately. Only having power over yourself is important. To do this, we need to use more complex technology than what we've already talked about, like making a surface with more than one layer. It would be great if the amplitude and phase shift of each part could be changed separately and easily during the IRS reflection process.

3.2. IRS Reflection

Figure 4 displays a possible setup for the channel model for radio communication based on the IRS. In this model, the lines that connect the base station to the user equipment (UE) are made up of three different parts. These parts are put together in each IRS piece. The channel link between the UE and the IRS, the copy of IRS elements, and the channel link between the IRS and the UE make up the three parts. The way this channel works is very different from a normal point-to-point straight channel. It is called a "multiplicative" channel model because each part of the IRS gets more than one signal from the base station (BS). Then, the IRS stretches out the linked signal, which can change its intensity and/or phase, like a single point source.

Figure 4. A wireless transmission system with a BS and UE channel link that is helped by IRS.

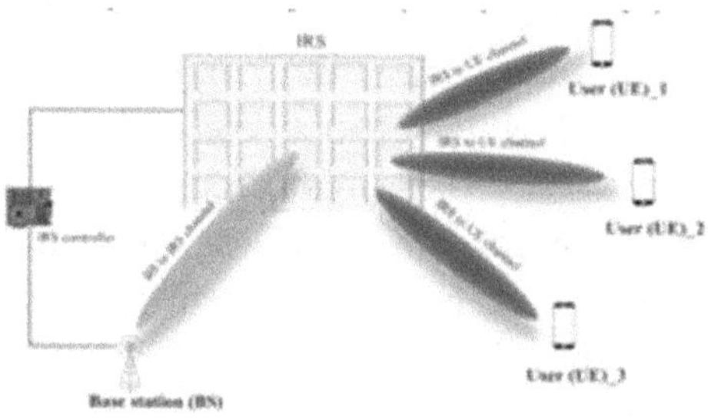

Here is a math method that can be used to find the complex reflection coefficient of the signal that the nth element of the IRS shot back:

$$h_n = (\beta_n e^{j\theta_n})x_m, \quad n = 1, ..., N \tag{4}$$

The signal that enters the IRS is named xn, and the signal that the nth element reflects is named hn. The IRS has N reflecting elements, and βn and θn are the reflection factors that decide how strong and in what direction the signal is mirrored. You can change the channel setting to help you meet a measure goal, like the system's doable rate or spread. To do this, you need to use the right IRS phase shift designs, amplitude designs, or both. You can do this if you use the right art. It has been shown that a set of transceivers is the only thing that can be used to build and improve radio communication systems. Things are not at all like this in what is written here.

The phase change and the size of the picture are linked in a simple way. We can use math to understand the phase shift. The IRS-backed tools for cell phones are set up so that this link can be made. The model works with most of the processors that are used to run the IRS. The phase change and its amplitude are what we're interested in. This is written as Ln=²n(n)ej n. βn can be -1 or π, and βn(θn) can be 0 or 1. The reflection amplitude, which is shown by ²n(²n), can be written in this way:

$$\beta_n(\theta_n) = (1 - \beta_{\min})\left(\frac{\sin(\theta_n - \phi) + 1}{2}\right)^{\gamma} + \beta_{\min} \tag{5}$$

where βmin≥0, ≥0, and γ≥0 are factors that are specific to the circuit design being examined. Figure 5 clearly shows that βmin is the lowest amplitude, is the distance between −π/2 and βmin, and γ is what makes the function curve steep. In the case where γ = 0, equation (5) is the same as the perfect phase-shift model. This is clear from the equation ²n(n)=1. In this case, n stands for "n times." Once an IRS link has been made, it can't be changed. Any tool that fits curves should be able to find these numbers without much trouble. The IRS lines are also set.

Figure 5. The phase shift model and the different settings that go with it.

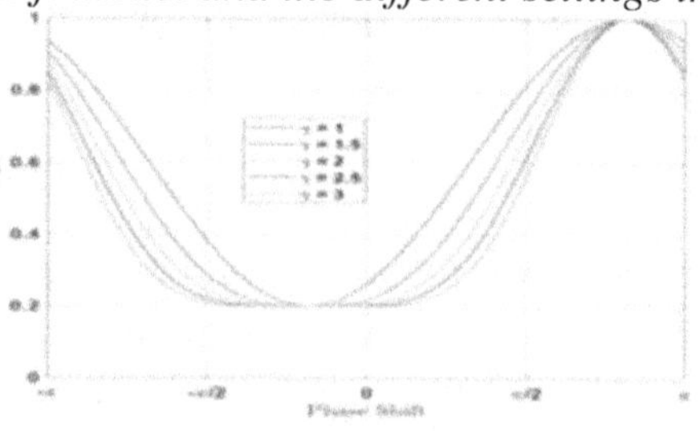

3.3. Channel Model of IRS

Figure 4 shows how to make a plan for a system. Users' equipment (UE) can join to a base station (BS) through an IRS network. It is thought that the base station has M antennas that are all lined up in a straight line, and the IRS has N parts that return signals. It is thought that each person in the service area has one receiver for each dish. Some people might say that the IRS's message was

$$y_{irs} = h_u \Psi H_b x + n \tag{6}$$

That which is sent is shown by x, that which is received at the user equipment (UE) by yirs, and that which is sent from the base station (BS) to the IRS by Hb. The channel vector from the IRS to the UE is shown by hu. At the user, the additive white Gaussian noise (AWGN) is shown by n~CN(0,π2). There is a number grid ¨ that displays the phase shift values. Any of the following can be z's value: [< 1ej 1, 2ej 2,..., Nej N]T isn't in CN1. This is one way to talk about the whole road that includes touch:

$$y_t = h_u \Psi H_b x + H_d x + n \tag{7}$$

It is part of the C1M set and is called “Hd.” It is the straight path between the user device and the base station. This is one way to write the BS-IRS channel Hb:

Figure 27. Math graphic 3

$$\boldsymbol{H}_b = \sqrt{\frac{MN}{P_1}} \sum_{p_1=1}^{P1} \alpha_p^1 \mathbf{t}\left(\beta_p^1, \theta_p^1\right) \mathbf{r}\left(\beta_p^1\right)$$

The straight link between the base station and the user equipment is shown by the letter "hd," which is part of the C1M set. One way to write the BS-IRS channel Hb is like this:

Figure 28. Math graphic 4

$$\mathbf{t}\left(\beta, \theta\right) = \frac{1}{\sqrt{N}} \Big[1, \ldots, e^{j\frac{2\pi}{\lambda} d_r (n_1 \cos(\theta) \sin(\beta) + n_2 \sin(\theta))}, \ldots,$$

Figure 29. Math graphic 5

$$\mathbf{r}\left(\beta\right) = \frac{1}{\sqrt{M}} \left[1, \ldots, e^{j\frac{2\pi}{\lambda} d_a (m \sin(\beta))}, \ldots, e^{j\frac{2\pi}{\lambda} d_a ((M-1) \sin(\phi))} \right]^T,$$

The number λ stands for the signal frequency, and the numbers dr and da show how far apart the IRS parts are and how far apart the antennas are. One through nine are also shown.

Figure 30. Math graphic 6

$$\mathbf{h}_u = \sqrt{\frac{N}{P_2}} \sum_{p_2=1}^{P2} \alpha_p^2 \mathbf{r_2}\left(\beta_p^2, \theta_p^2\right),$$

The symbol λ stands for the signal's wavelength, and the symbols dr and da stand for the elements' spacing in the IRS and the antennas' spacing, respectively. Also, it's important to note that the numbers 0 through 1 are

Figure 6. An IRS-based model for transmission between a base station and a user.

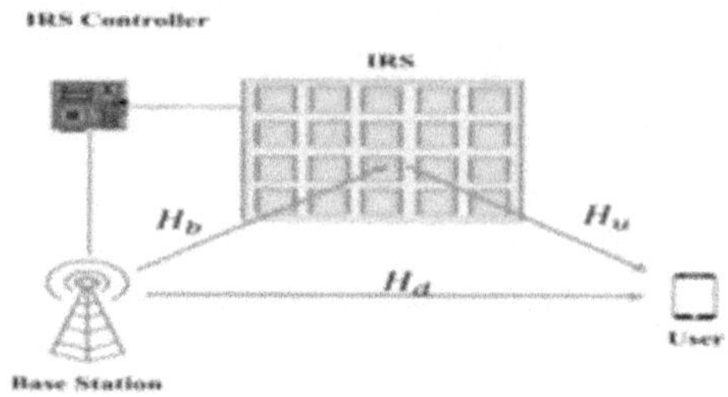

This is how you can describe the flow channel from BS to UE:

3.4. IRS Based Communication Advantages

Radio contact is hard because it's hard to find a way to talk that gives you a clear line of sight. In real life, there may be a number of things that get in the way and allow a weak signal to be sent. The service might end because of this. The IRS and the decode-and-forward (DF) transmission method are shown next to each other in Figure 7. The main things that this comparison is about are how fast and how much power each system needs. The IRS can clearly offer a faster data rate while also using less power when compared to the DF relay. Here are a few of the best things about it:

If you can't get in touch with the Internal Revenue Service (IRS) directly, they may offer another way to do so. The source and destination talk to each other while the virtual route is being set up.

This is because IRS devices are quiet, so they need less power to work than relay transfer devices. Without power boosts, a signal that has been received can be made stronger and sent to a different place. The reflecting parts, on the other hand, change the phase shift of the signal so that it can reach the user devices.

If you don't have enough speed for 6G transmission, millimeter frequency communication might be able to help. Tax authorities (IRS) might be able to do a better job if they could talk to each other better.

IRS can be used to make up for problems with channel rank in places where there aren't enough ranks.

Software can be used to change the way electromagnetic waves move. There is a chance model of a wireless channel that can be changed to a predicted model in this way. The Ralyleigh fast fading method can be changed into the Rician slow fading method because the IRS works by returning messages through guided beamforming.

IRS might be able to help find a good and quick way to fix the issue of co-channel and inter-channel confusion in radio communication.

Figure 7. The comparison between energy efficiency and data rate for DF relay and IRS.

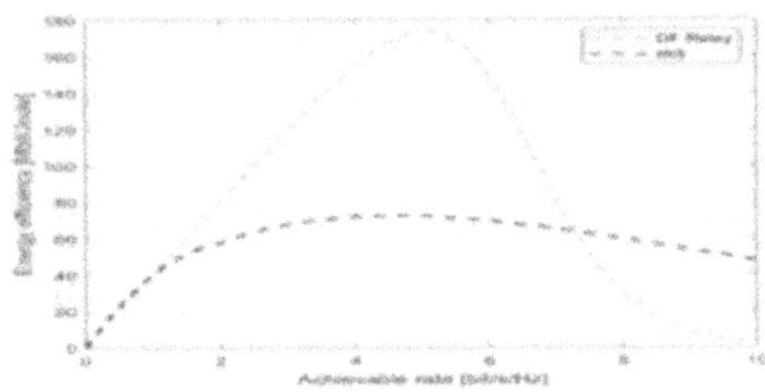

4. SYSTEM MODEL OF RIS-AIDED COMMUNICATION SYSTEM

4.1 Signal Model

RIS is used in this radio transmission system, which you can see in Figure 8. An N-element radio interface system (RIS) lets a big multiple-input multiple-output (MIMO) base station (BS) with M transmitters serve a user with just one listener. When the downstream spread is taken into account, the system model looks like this:

Figure 8. A system model of downstream transfer in a single-user communication system that uses RIS

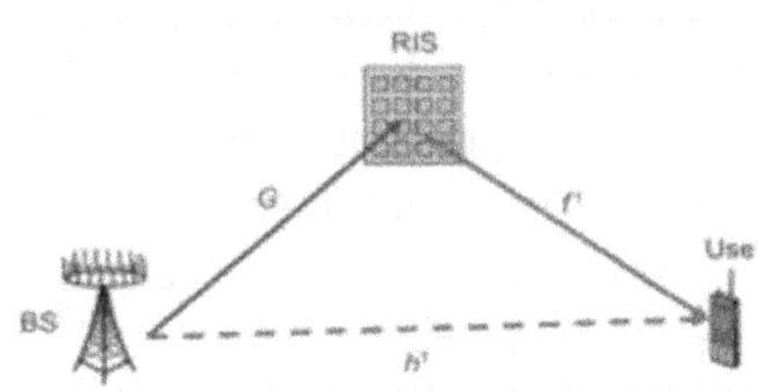

$$y = (h^{T} + f^{T}\Theta G)wx + n \tag{12}$$

There are terms like "y" for the signal that the user receives, "h" for the direct channel between the BS and the user, "f" for the channel between the RIS and the user, "Θ" for the tuning coefficient matrix at the RIS, "G" for the channel between the BS and the RIS, "w" for the precoding vector at the BS, "x" for the signal that is sent at the BS, and "n" for the additive noise. In RIS hardware, the setting value can only be changed by a certain amount. These limits can be seen in the tuning coefficient matrix, which is written as Θ=diag(θ), where θ is the RIS tuning coefficient vector. Yes, a good RIS should have a unit number for its factors that only changes the phases of the signals it gets, not their speeds. You can write ¸=[ej 1,ej 2,...,ej N], where ¸ is a positive number and ¸1, 2,..., N are the phase changes.

For users with K antennas, it's easy to change Equation (13) and write it as

$$y = (H^{\mathrm{T}} + F^{\mathrm{T}} \Theta G) wx + n \tag{13}$$

where y∈CK×1y∈CK×1, H∈CM×KH∈CM×K, F∈CN×KF∈CN×K, and n∈CK×1

4.2 Channel Model

It is thought that the Rayleigh-fading channel is the easiest to understand, but there are others. There is a complicated Gaussian distribution for h, f, and G that is circularly symmetric and different for each one. This is important to remember when you look at the Rayleigh-fading channel model. In this case, h is a CN(0,π2hIM×M), f is a CN(0,π2fIN×N), and vec(G) is a CN(0,π2GIMN×MN). The changes π2h, π2f, and π2G are connected to the distances between the base station (BS) and the user end (UE), the RIS and the UE, and the BS and the RIS, in that order. Because the lengths don't always match up, all three lines can't be adjusted at the same time. Before a Rayleigh-fading channel for models can be made, π2h, π2f, and π2G must be set to the right values. Second, the model for pathways with more than one path: There are two main types of multipath channels: Rayleigh-fading channels and Saleh-Valenzuela channels. Users and the base station can talk on channel h. Channel f is for the person and the information system that is far away. Channel G is for the router and the base station.

Figure 31. Math graphic 7

$$\boldsymbol{h} = \sqrt{\rho_h} \sum_{l=1}^{L_h} \alpha_l^h \boldsymbol{b}\left(\vartheta_l^{rx}, \psi_l^{rx}\right)$$

The factors that change with distance are ρh, ρf, and ρG. There are a certain number of lines in each channel that lead to each number, Lh, Lf, and LG. There have been changes to the complicated gains for each channel. ±hl}Lhl=1, \fl}Lfl=1, and \±Gl}LGl=1 make them clear. When you plug in h, the numbers {ϑrxl}Lhl=1 and {ψrxl}Lhl=1 show the position and tilt angles at the base station (BS). Setting {txl}LGl=1 and {ψtxl}LGl=1 will give you the elevation and azimuth angles at the BS for G. It's written as {ϑtl}Lfl=1 and {ψtl}Lfl=1, which show the RIS's direction of elevation and azimuth for f. These numbers, {ϑrl}LGl=1 and {ψrl}LGl=1, show the top and bottom angles of G at the RIS. For the RIS, the standardized array steering is shown by a(ϑ,ψ)∈$CN^{2}1$, and for the BS, it is shown by b(ϑ,ψ)∈$CM^{2}1$. The direction vector a(ϑ,ψ) of the N1 array can be written in this way [17]. The N1×N2 (N=N1×N2) uniform planar array (UPA) is what this is for.

Figure 32. Math graphic 8

$$a\left(\vartheta,\psi\right)=\frac{1}{\sqrt{N}}\left[e^{-j2\pi d\sin(\vartheta)\sin(\psi)\boldsymbol{n}_1/\lambda}\right]\otimes\left[e^{-j2\pi d\cos(\psi)\boldsymbol{n}_2/\lambda}\right]$$

n1 is equal to [0, 1,..., N1−1].The range of values for Tn1 is [0,1,...,N1−1].T, n2 = [0, 1,..., N2−]The range of values for Tn2 is [0,1,...,N2−1].dd is the distance between two parts that come right after each other. It usually follows the equation d=»2d=»2[18]. The signal wavelength is T, and the signal wavelength is λλ. If you change NN, n1n1, and n2n2 to MM, where m1 is equal to m1 and m1 is equal to [0,1,...,M1−1], you can write b(ϑ,ψ) in the same way as Equation (6).T[0,1,...,M1−1] T, and m2 = [0,1,...,M2−1].The range of values for Tm2 is [0,1,...,M2−1].M = M1 M2M = M1 M2 is the same as T.

The channel models in this part only work for the far-field release of electromagnetic waves. This is important for you to know. When two parts are close to each other, the path that links them has more complex models. The results of this study are being gathered right now and will be shown in the next section.

5. SYSTEM MODEL

We look at an outdoor situation with n+1 valid nodes to find the best model for testing how well the proposed method works. In this case, a source n0 sends data to a target n+1, which could be the base station. The data travels along a path ©=(n0,n1,...,n+1) made up of K RISs and n+1 listeners, as shown in Figure 9.

Figure 9. The plan for a system that can work with software-defined networking (SDN). The control plane is made up of useful blocks that can talk to the nearby RIS controller.

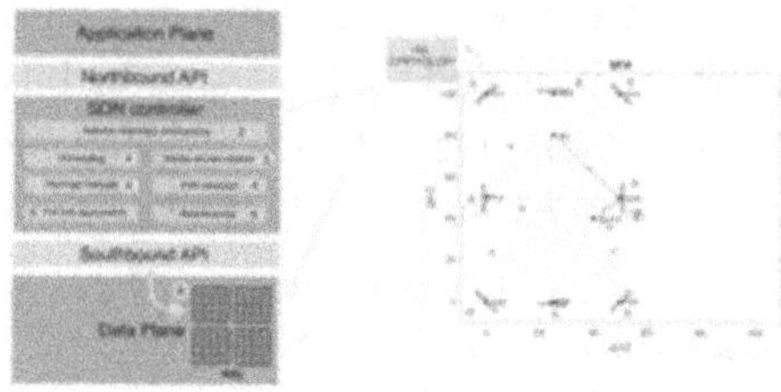

We really think that different RISs should be added to a big system for SDN (software-defined networking). We use the usual layout of the Application, Control, and Data Planes (AP, CP, and DP) to do this. This layout has been talked about before, but it is more focused on security than on normal communication. It turns out that RISs are part of the DP and are linked to the SDN Controller in the right way. It is possible for control messages to go both ways inside the gateway because it has a downward link. This is done to keep an eye on and manage the sharing of e2e info. The OpenFlow protocol is expected to be used since this link is already connected.

As part of the management code for the controller shown in Figure 1, the following steps were also thought about:

In Network Discovery and Tracking, wavefronts of RISs that are packed together are used to find out where the device is. Device Access Control is meant to put people into groups when they use it. The Topology Manager is in charge of keeping an accurate picture of the network as a whole. Path Selection, RIS Link Deployment, and End-Device Beamforming all do the same thing, just like the Decision Engine.

It's important to remember that these tasks depend on the SDN Controller getting the right messages, which are then looked at by a program to see how the global network is doing. Figure 1 also shows the steps that SDN Control uses to set up the whole process of control.

It is important to keep track of devices, decide who can access them, manage the network's structure and pick the best route, set up RIS links and beamforming at the end device, send e2e data flows, and set up the device's unit. The network is generally shown as a named grid.

The software-defined networking controller (SDN Controller) hides the structure of the network in a weighted graph so that no one can listen in. These are the valid nodes. They are in the vertex set, and the links between them are in the edge set. Most of the time, RISs can offer different ways in case something blocks the straight path or the channel quality is bad. In the end, the e2e ability of a source to a goal move along a certain path To find ©=(**B**0,D1,...,DK+1), use the following equation:

The signal-to-noise ratio (SNR) at the target point is given by d. Along the same lines, this is how you could describe the wiretap channel's e2e mode:

The e2e SNR for the listener is shown by the symbol γ**A**³E. If we think about the fact that there are people who listen in on conversations without permission, it is important to pick the way that provides the best level of privacy. This can be described as follows:

$$C_{\Omega}^{S} = \max\left(C_{\Omega}^{d} - C_{\Omega}^{E}, 0\right) \ [\text{bps/Hz}] \tag{16}$$

Someone could say that a safe link is only possible if the real user has a bigger channel bandwidth than the person who is listening in. In this case, the only way to communicate safely is to follow these steps.

When radio waves travel through empty space, a perfect isotropic transmitter antenna sends a signal to a lossless receiving antenna that is d miles away. We know that the power that is received is:

$$P_{rx} = \frac{A}{4\pi d^2} P_{tx} \tag{17}$$

Mtx stands for the power that is sent out, and *βd*=**N**4 *d*2 for the free-space channel gain, which is also called pathloss. **D**= 2(4) shows the area of an isotropic receive antenna.

First, we'll look at an uplink connection that is handled by a single RIS. In the RIS, there are M reflection devices that are not in use. You can also call these tiles. The signal that comes in can be changed by these parts, but they don't make the signal stronger. What this means is that each tile independently controls the time and strength of each reflected signal. Our first guess is that the signal moves along a line of sight (LoS). To make a signal that looks like the one that was sent to the target, follow these steps:

$$\log_2(1 + SNR_{RIS}) \tag{18}$$

Very much like the idea in, which also includes the far-field guess. With the assumption that the source is d degrees away and the target is degrees away, you can get a good idea of the SNR. It is also thought that the source and the target are both in the far field of the RIS. This means that *d*cos( )>>*M*B––– and cos( )>>AB–––.

$$SNR_{RIS} = M^2 \varsigma_{d,\eta} \varsigma_{\delta,\omega} \frac{P_{tx}}{\sigma^2} \tag{19}$$

$$\varsigma_{d,\eta} = \beta_{d\cos(\eta)} \cos^3(\eta) = \frac{A\cos(\eta)}{4\pi d^2} \tag{20}$$

$$\varsigma_{\delta,\omega} = \beta_{\delta\cos(\omega)} \cos^3(\omega) = \frac{A\cos(\omega)}{4\pi \delta^2} \tag{21}$$

5.1. Multi-RIS Setup

We think that many RISs are set up to make it easy for clients to connect by giving both the source and the target multiple e2e LoS tracks. This fits with what was said before. Each RIS in the path adds a passive beamforming gain of *M*2M2, but it also makes the multi-reflection pathloss term bigger. =(D0,D1,...,D_A+1)·=(A0,A1,...,AK+1) shows how much power is going through the path being looked at. This can also be written as:

Figure 33. Math graphic 9

$$|h_{A_0,A_{K+1}}(\Omega)|^2 = \frac{M^{2K} N \beta^{K+1}}{d^{\alpha}_{A_0,A_2} d^{\alpha}_{A_K,A_{K+1}} \prod_{k=1}^{K} d^{\alpha}_{a_k,a_{k+1}}}$$

Figure 34. Math graphic 10

$$\kappa(\Omega) = \frac{\sqrt[\alpha]{\beta}^{K+1}}{d_{A_0,A_2} d_{A_K,A_{K+1}} \prod_{k=1}^{K-1} d_{a_k,a_{k+1}}}$$

The number of antennas that can be used at the base station is N. The symbol δ stands for cos( )4 [2], which is the same as Acos(ω)4π. The symbol *B* stands for (/4)2A, which is the same as (λ/4)2.

5.2. Problem Statement

This is the goal. To make the physical security policy in an SDN system work better, we use Graph Theory and an SSPP problem that is similar to the one that was talked about in [10]. We can improve the app with this. We want to find the best e2e route that keeps things as secret as possible. This is one way to describe the general optimization problem (OP): worth.The goal must be bigger than or equal to the C-S amount for this to work.If γd is less than γth, then • Given ($_E$<1,y1), (xK+1,yK+1), (x***RUSi***, y***RUSi***), (xe,ee),∀e=1,...,E,∀i=2,...,K−2Given (x1,y1), (xK+1,yK+1), (xRISi,yRISi), and (xe,ye), where e = 1,...,E and i = 2,...,K−2. If you plug in (B1,D2,...,D_A) = (A1,A2,...,AK), you get the path with the best e2e secret rate.

The most was set in the OP. The goal at this stage is to make sure the target has enough channel space. We will not look at paths that are lower than this level.

Let's say the SDN Controller knows where people who aren't supposed to be on the network are and where people who are are. The IRS saves any unwanted uplink signals it gets with a certain southbound interface control sub-protocol and a general location method. This helps us reach our goals. Compressed sensing is a method that works in a similar way. It is often used in systems with passive sensors that are spread out.

With a directed weighted graph $\mathscr{G}$D=($\mathscr{V}$,$\mathscr{E}$)GE=(V,E), we can hide how our network is set up from people who are seeing it. SPA is the fastest way for us to deal with the OP. Next, we need to make a good model for the weights of the graph. To be exact, the vertex set (V) shows the real network nodes, and the edge set (E) shows how these nodes are connected to each other. You need to find the weight of the edge that links points i and j if you have this equation:

$$\mathrm{W}_{l,f}^{E} = \xi * \mathrm{W}_{l,f} + \hat{e} = \xi * \ln\frac{d_{l,f}}{M\sqrt{\beta}} + \hat{e} \tag{24}$$

There is a lot of extra work to do for control because the problem is hard. Here are five methods that make it easy to use the weights Wi and jE in the end graph. In more detail, the methods are arranged by how hard they are to figure out. The main goal is to find a balance between the first term's contributions from distance and route loss and the second term's effects from eavesdroppers on each link (e^). Here are some suggestions to help you do that:

HEU1

ξ=$cost$ξ=cost, $\hat{e}$ =$cost$=1∀e∈Ee^=cost=1∀e∈E detected in the coverage area.

HEU2

ξ=$cost$ξ=cost, $\hat{e}$ =$\rho\cdot X$ ∀e∈Ee^=ρ·X^∀e∈E detected in the coverage area. ρ=$cost$ρ=cost and where X =maxi,j=1,...,|V|Wi,jX^=maxi,j=1,...,|V|Wi,j, thus relating WEi,jWi,jE to the higher weight in Wi,j∀i,j=1,...,|V|Wi,j∀i,j=1,...,|V|.

HEU3

ξ=costξ=cost, ê =2·X log10(maxePi,e)e^=2·X^log10(maxePi,e), where maxePi,e,e∈EmaxePi,e,e∈E is the maximum value of the pathloss term in the channel between the transmitting node and each eavesdropper.

HEU4

ξ=costξ=cost, ê =2·X |log10(∑Ee=1Pi,e)|e^=2·X^|log10∑e=1EPi,e|, where ∑Ee=1Pi,e∑e=1EPi,e is the sum of the pathloss terms between the transmitting node and each eavesdropper.

This is what

HEU5

ξ=cost, ê =log10(∑Ee=1Pi,e−−−−√)e^=log10∑e=1EPi,e.

Finally, we talk about two different types of working cases.

Eavesdroppers can talk to each other to put together the signals they've picked up and figure out which ones belong to real users.

- People who listen in on talks without being told what's going on don't work together; they do it alone.

This means that the total signal-to-noise ratio (SNR) across the road is (A0, A1,...,AK+1) = ($_E$0, N1,...,DW+1).

Figure 35. Math graphic 11

$$(c)\quad \gamma_{\Omega}^{E} = \sum_{r_{i,j}\in\Omega} \sum_{e=1}^{E} \gamma_{i,j,e}^{E}$$

Figure 36. Math graphic 12

$$(nc)\quad \gamma_{\Omega}^{E} = \max_{1\le e\le E} \sum_{r_{i,j}\in\Omega} \gamma_{i,j,e}^{E}$$

It is important to remember that in the joint situation, people who are listening in work together by sending each other notes. In theory, this means that an SDN Controller could keep an eye on them, find them, and tell where they are.

6. NUMERICAL RESULTS AND DISCUSSIONS

There was a lot of noise in the study's wireless network, so these numbers show how different system and channel features affect RIS's ability to work. What the Monte Carlo method found is the same as what the mathematics study found. This shows that the closed-form methods were correct. This is how the system and channel are set up if nothing else is made clear to be different: If we look at these numbers, we can say that: R = 2, P I = 15 dB, N p = N I = 2, m p = m I = 2, ^{3} p = ^{3} I = ^{3} R = 5 dB, — = 2.5, ρ = 0.6, β = 0.8, A = 1, d p = 2 m, d I = 10 m, and d AR = d RD = 2 m.The number of receivers (N p) and the number of interferences (N I) show how likely it is that a system will fail (Figure 11). We learned from our research that the system doesn't work as well when N I goes up. It works better as the number of antennas N p at the PB goes up. This is because as N p goes up, so does the power that gets to area p. Figure 12 also shows how the system works when the PB link and the interferer link's fading factors, m p and m I, are changed. In this case, it shows how the system works when the interference's broadcast power P I changes. These things can happen because of the th.ings listed above. Now that we know this, we can say that the system doesn't work as well when the output power P I of the interferers goes up. The system works better when the fade settings on the PB and interferer links are raised, on the other hand. This picture (Figure 13) shows how well the method works. This picture shows how N I has changed. Through the tests, we can see that the system does not work as well when there are a lot of interferences in the area.

Figure 10. Shows how the system in question handles outages for different N numbers.

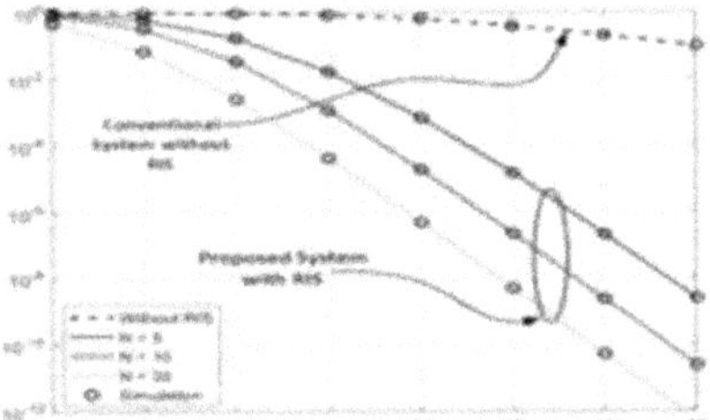

Figure 10 displays how the chance of a system loss changes based on the number of reflected parts in the RIS. The RIS works a lot better when the power goes out after more reflecting parts were added. It is also known that adding N makes things work better. It was found that the new system with RIS works better than the old system without RIS.

Figure 11. The success of the outage chance at different N I and N p values when N = 10.

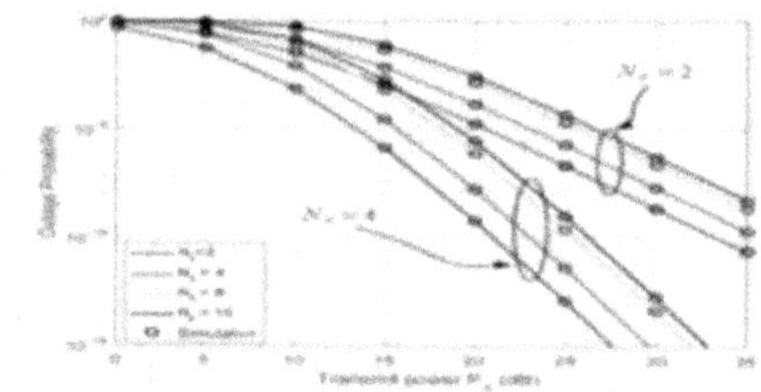

Figure 12. shows how P I changes the performance of the system's failure chance at different m I and m p values for N = 10.

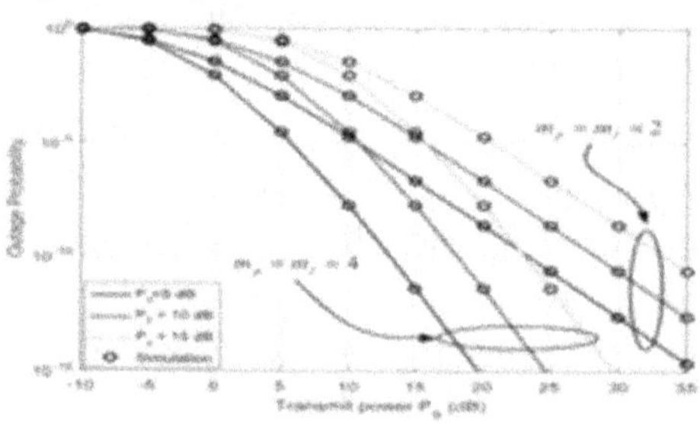

Figure 13. 13, 960 11 of 16 2 p.m., figure of the Symmetry 2021 Collection = = = 4p I'm m Figure 4 displays how the value of P I alters the likelihood that the system will fail for a variety of m I and m p numbers when N is 10. Figure 5 shows what happens to the average flow when N = 10 things are in the way.

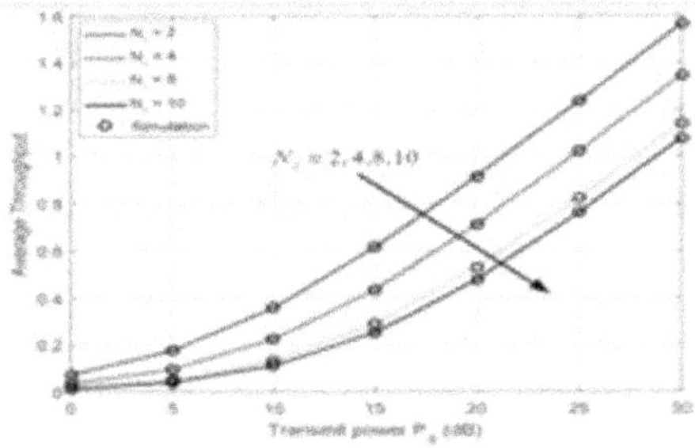

In Figure 6, you can see how the system's output changes based on how many receivers are at the PB (N p). Now it's clear that as N p rises, so does the system's average return. The results also show that the system's output rises when it gathers energy for a longer time, as shown by ρ. When ρ is just right, on the other hand, the most output is seen. One reason for this is that the system gets more power from the PB, which makes it work better over time.

So, Figure 7 shows how the number of reflected parts can change how fast RIS works. It turns out that the suggested system with a RIS is faster than a normal system that doesn't have a RIS as N more network nodes are added. When there are a lot of reflecting parts in a RIS and the PB and interferer links aren't strong, ABER changes. This is correct when the BPSK binary modulation method is used. Figure 8 shows these outcomes. It's clear that increasing N had a big impact on the ABER. The machine worked better when the value was bigger. Also, we can learn that the system might work better if the fade settings are turned up on both the PB link and the interferer link. Also, it's clear that the system that was given, which has RIS, works better than the normal system that doesn't have it. The number of interferers changes the system's average bit error rate (ABER) when the M-QAM coherent modulation method is used. Figure 9 shows this. As there were more interferences at the target, the system's average bit error rate (ABER) went down. The ABER of the system worked better when the number modulation order M was lowered.

Figure 14. Shows the average flow vs. the time it takes to collect energy (ρ) for different Np values at N = 10.

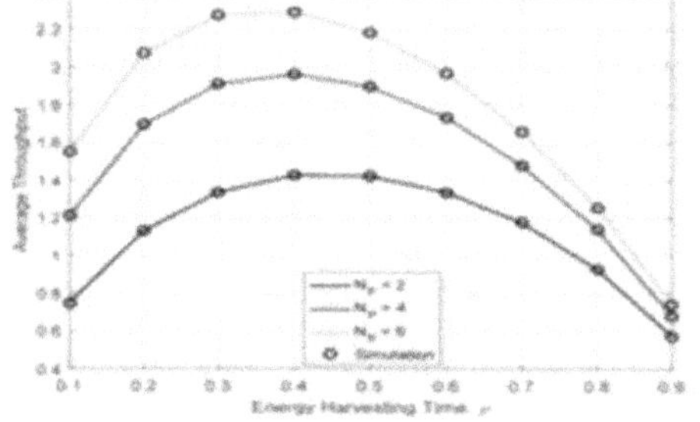

Figure 15. Shows how the average flow changes depending on the number of reflective elements.

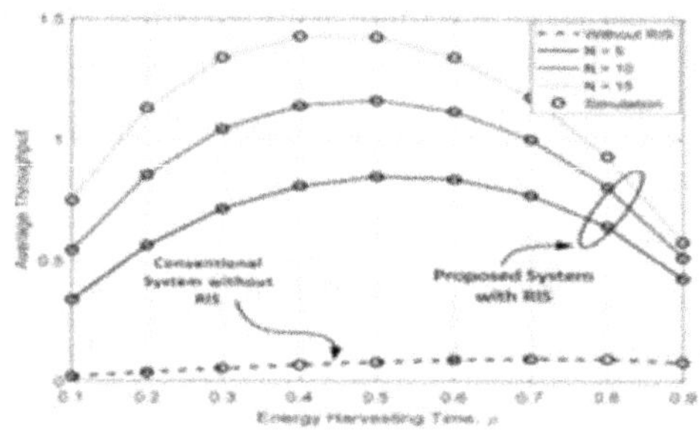

Figure 16. Shows how the average bit error rate (BER) changes depending on the number of reflecting elements in the RIS and the fading factors. This is in a binary modulation scheme.

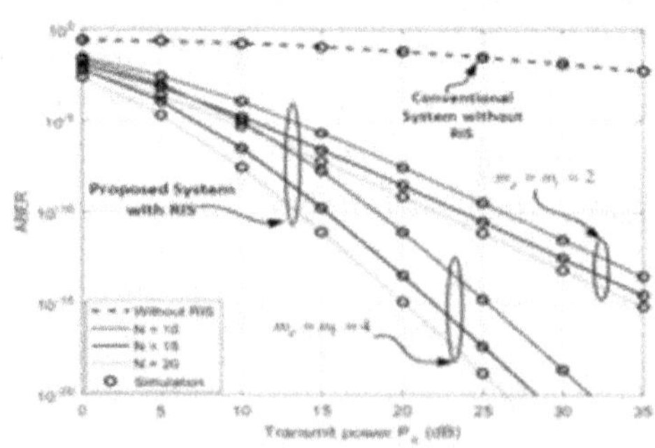

Figure 17. Shows how the average bit error rate (BER) changes depending on the number of reflection elements in the RIS and the order of modulation. This is shown using the M-QAM coherent modulation method.

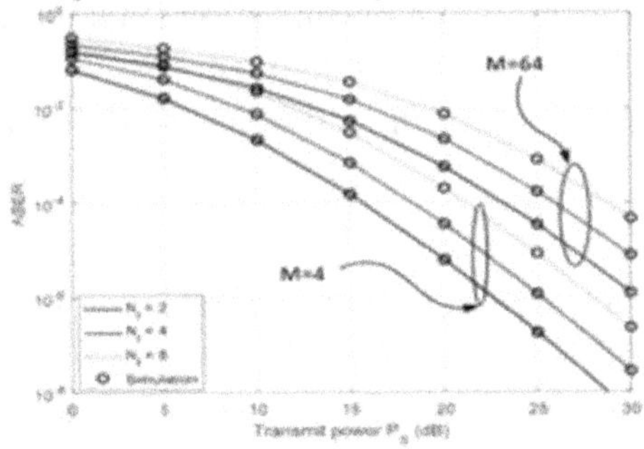

7. EXPERIMENTAL RESULTS

MATLAB was used to sketch out the network. Next, we used OAA to figure out how to solve the issue shown in Equation (8a). The number 20dBW stands for the same thing. There is 10 dBm between dBm and dBm. It needs to send and receive at least 0.1 bits per second per hertz. The fewest users (n=2) and the most users (n=300) are both possible. Table 1 shows a list of the modeling factors that were used to plan the network.

Table 1. Simulation parameters.

Parameter	Value	Unit
*PBS*PBS	20	dBW
*PUE*PUE	10	dBm
*PRIS*PRIS	10	dBm
*Rmin*Rmin	0.1	bits/s/Hz
*minUE*minUE	2	–
*maxUE*maxUE	300	–
ℕN	32	–
𝕂K	16	–

Here are some other methods that we compare the one you suggested to make sure it works well. There are also the genetic algorithm (GA), the firefly algorithm (FA), the artificial bee colony (ABC), the cuckoo search (CS), the teaching swarm optimization (PSO), the social group optimization (SGO), the teaching learning-based optimization (TLBO), and the basic open-source nonlinear mixed integer programming (BOMIN). Another full study of how well that way works can be seen in Figure 18.

Figure 18. A test of how well the suggested method works is shown. 1. The bit rate (Mbps) as a percentage of the base stations; 2. The percentage of base stations as a percentage of the bit rate (Mbps) and the delay (ms); 3. The channel efficiency (ms) as a percentage of the bit rate; and 4. How resources are shared among base stations.

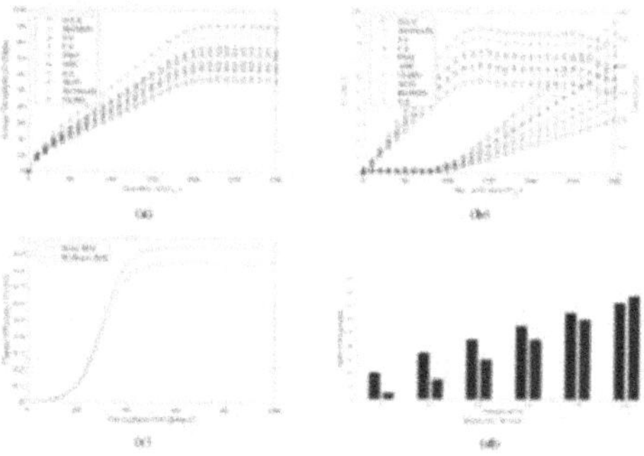

Figure 18b shows that the EE and delay of the network can be looked at in a number of different ways. 1. The OAA works better than other ways, as shown in Figure 18b. It is possible for the other methods, especially NOMAD, to reach a maximum $_{AA}$ of 8.5 Mb/J, but OAA can always reach its perfect **AA** of 9 Mb/J. It's clear that OAA has put a lot of work into b. NOMAD can only reach a low of 1 millisecond in terms of delay, but the OAA can reach a high of 0.5 milliseconds for 90 milliseconds. NOMAD is faster than other methods. But OAA works better than NOMAD. Our group used OAA to figure out the RIS and NLOS channels' channel efficiency (ℂ𝔼). sees the CE as a share of OAA, and the RIS-assisted network has an overall ℂA value of 95%. On the other hand, the network without RIS has only 85% degree of freedom. This shows how important the RIS surface is to the network. Based on the RIS, Figure 5d shows how the RIS is spread out. To keep the NLOS from getting blocked, more T are sent to the RIS as the amount of T does. IRIS can only connect with each other if there are at least ten of them. But if there are more than 10 users, the RIS route changes because the mobile communication network can't figure out who all the users are. This lets the RIS channel handle the extra users. When Algorithm 2 is used, the prediction should come true. By adding to the service, the network stays stable and all users can use the best tools. Next, we'll talk about the end.

Table 2. Approaches and objectives

Approach	Objective					Contributions
	EE	Throughput	SE	Latency	Channel Allocation	
Search with gradient fall	✓	×	×	×	×	300% higher EE
The alternating descent method	✓	×	×	×	×	A rise in the EE of cell-free networks that matters
DRL	✓	×	×	×	×	By adding more RIS elements, EE got up to 77% better.
Dinkelbach method	✓	×	×	×	×	The method made a big difference in the EE of the D2D network.
Increasing the projected incline	✓	×	✓	×	×	Higher EE was reached in the MIMO network.
How much power is used	✓	×	×	×	×	In terms of both rate and EE, the RIS network is faster than relay one.
Approximation Method	✓	✓	×	×	×	Getting the most speed and EE out of the RIS-assisted wireless network
OAA	✓	✓	✓	✓	✓	Increased the network's speed, EE, and SE and lowered its latency.

8. FUTURE RESEARCH TRENDS FOR ML-BASED IRS-ASSISTED WIRELESS COMMUNICATION

Digital transfer methods that are based on machine learning work really well to improve the service. What are the different ways that an IRS-based radio communication system could be used? Find them in Figure 19. There are times when someone can't see the system because something is in their way. The IRS might be able to help you get in touch with them. Figure 10b shows that IRS can be used to build very big machines. If you use IRS and managed beamforming together, you can make the physical layer safer. It is thought that setting up mmWave wide area transmission will be easy if IRS is used as shown. If you need to send power electronically inside your home, Figure 19e shows that IRS might be useful. Next, we'll talk about some possible fields of study that use both taxonomy and machine learning.

Figure 19. IRS-aided wireless communication applications.

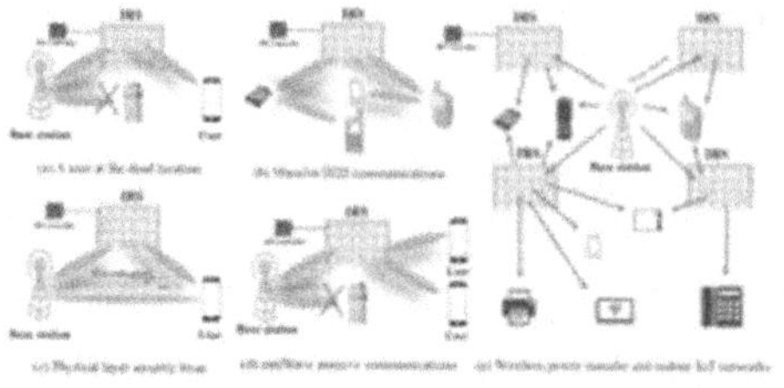

8.1. Optimal Placement of IRS

Setting up or fixing the mirrored surface means that it can't be moved or moved around in any way. Because of this, it is important to get enough information about how IRS works in the context before putting it to use in the real world. Looking for the best possible level of performance, you should think about where the IRS device should be placed based on how well it works. This is something that is thought about when the IRS is placed inside and outside the country. It is possible to get the best results when guessing certain things by using machine learning models. Because each installation situation is likely to be different, new models might help with guessing how well something will work.

8.2. Dynamic Hybrid Beamforming

The BS signal can be sent back by IRS devices to control beamforming. Because beamforming works with the IRS device's idle parts, it can make a shape that is unique to the user's tools. You can use machine learning to guess how well beamforming for the user equipment (UE) service will work. It's possible for mobile UEs that the dynamic beamforming method will also take into mind comments from the UE. To get the best results, you should try out new training data and models.

8.3. Data Collection and Model Training

When you train a machine learning model, you need to collect data. The data tells you how well the model works. It's hard to use machine learning-based methods in IRS-based communication because it's hard to get accurate data, which can lead to models that don't work. When predictions are made, signal recognition, CE, and the receiver's beamforming design are the key things that are thought about. In the future, people might study the best way to gather data so that machine learning-based methods can be used effectively.

8.4. Constrained System Modeling

When used in this way, IRS-based communication can improve a device or system that doesn't have lot of features without adding more cost. It is possible to use machine learning to look into business communication systems that are limited by time and energy. You can also get a good signal-to-noise ratio for communication even if you don't have a clear line of sight. These kinds of tools can benefit from being able to talk to each other more easily thanks to the IRS. This can be looked into again.

8.5. Channel State Characterization

It has taken a lot of different models to help people figure out how to talk to groups. The IRS has been helpful with this. One of these models is based on physics, another on an impedance network, a fifth on independent diffusive scatter, and the last one is based on a tile and code-book structure. CE is now an area of study built on machine learning, and people have tried many ways to guess what will happen with channel modeling. But most of the methods for improving are hard to use on computers. There may be times when people want to find new ways to do things better and faster.

8.6. IRS for IoT Network

The number of Internet of Things devices is growing at an extremely fast rate. This is because they can be used for so many different things and are in such high demand. This fourth industrial shift is moving faster thanks to 6G and the Internet of Things (IoT). Using wireless connectivity based on IRS makes it possible to connect more Internet of Things devices both inside and outside. Because Internet of Things devices are spread out, it's hard to tell how well the wireless system will work. So, methods built on machine learning can be changed to more accurately mimic how things work in the real world. Sensor nodes are what make up most Internet of Things gadgets. They collect data about different real values. Machine learning methods can be used to find out more about how IRS affects the link between things that are connected to the internet.

8.7. Protection Against Eavesdropping

You can take the right steps. As more study is done, this method may be looked into further. A lot of attention has been paid to physical layer security (PLS) over the last few years as security has become a very important issue in communication systems. There are a number of reasons why PLS is better than the coding method. It takes less computing power to run PLS, and a 6G network is not controlled. This means that managing keys for all the devices in the network will be hard. Because the IRS came up with beamforming, security can be maintained while data is being sent. Machine learning techniques make it possible to model many types of environmental risks, and

8.8. MmWave Communication

People believe that millimeter wave (mmWave) technology might be a good way to meet the needs of 6G transmission, which needs a lot more data. Just a few studies have used IRS to help with mmWave connection. We need more studies to fully understand how IRS works, though. Some things that could be looked into further are radio array design, plan algorithms, signal patterns, and the use of machine learning methods for the IRS that has been put in place.

8.9. EDGE Intelligence

Pushing numbers to the edge of a network instead of to the server is what "edge computing" means. This means that processing is sent to the source that is closest to the network. Edge computing has many benefits when it comes to connecting user equipment (UE) to the server. These include low delay, low cost, privacy, limited energy use, and high data utilization. A network function system that works well might be possible with edge computing that is built on machine learning. It is thought that IRS and ML will be able to make edge computing work better, but this is something that will need more research in the far future.

8.10. Hybrid Communication Implementation

Another way to help increase the RF bandwidth is through VLC, which stands for visible light streaming. Together, VLC and radio frequency (RF) technology might help people study new areas with the help of communication based on IRS. One idea for a mixed communication method is to use sound waves to talk to each other. In the ground, optical communication can be used to send data. Near the surface, radio frequency communication can be used to send data to a computer. This is possible because IRS works by returning signals, so it can use both light and radio waves.

9. SIMULATION RESULTS

This part's simulation of the process shows that the idea is good. Researchers have also looked into how likely it is for networks that use distributed RIS to join the air and ground to fail, how often they do so, and how much energy they use. In this study, we look at four different ideas. Scheme 1 shows how to set up a distributed air–ground fusion network with parts that aren't perfect. Rules can be seen in Schemes 2–4. They are Scheme 1, Scheme 2, Scheme 3, and Scheme 4.

This study says that the way a distributed RIS-assisted air–ground fusion network talks to each other makes it work even when things aren't just right. It's called k2_RIS, which means "direct link transmission method" between the user and the base point of the UAV. The name of this plan is k2_noRIS, and it only looks at good things. It doesn't look at things like odd people, broken gear, or bad channels. Best RIS is a "communication scheme using AF relay but with hardware impairments for the user and the UAV base station." This method is called Ideal RIS. This plan's full name is AF_k2.

Most people believe that all lines follow the Nakagami-m distribution. The base for models was made with MATLAB 2021a. You have set the broadcast bandwidth to 100 MHz and the strength of the white Gaussian noise to -174 + 10log10D decibels per hour. The power loss is set to -30 dBm, and the path loss exponent is set to 3. The random radius R of 10 meters and the random radius r0 of 1 meter will be picked at random. This area has three points of entry from outside. These are where they are: (21 m, 12 m), (51 m, 36 m), and (80 m, 40 m). The unmanned aerial vehicle (UAV) and the middle of the users who are spread out randomly are nowhere to be found. They are (90 m, 10 m) apart. Each RIS part only has one amount of power, but the person has five. The computer looks at a lot of important things, which are shown in Table 3.

Table 3. You can see the simulation's parameters.

Simulation Parameters	Value
Path loss exponent	3
The number of RISs	N = 3
The power consumption of the user	P1 = 10 dBm
The power consumption of RIS element	PRIS = 1 dBm
The radius of disc	R = 10 m r0 = 1 m
Bandwidth	100 MHz

In Figure 19, you can see that the UAV's send power is linked to three different types of service loss. These are pure _RIS, k2_noRIS, and k2_RIS. When we think about these situations, we know that the hardware limit factor is k2 = 0.01, and the range of values for each RIS unit Ln is from 10 to 20. Figure 19 shows that the modeling curve and the theory approximation curve are now in sync with each other. This shows that our research work is correct. People who use the RIS are k2_noRIS less likely to lose their link than people who don't use the RIS. This was shown by the study. Cascaded lines make it less likely that transmission will go wrong, which is the main reason for this. It's also possible for more than one RIS unit to fail. As the number of RIS units rises, the chance of a failure falls.

Figure 20. The relationship between outage probability and the UAV transmit power variation, where N = 3, Ln = {10, 20}, and k2 = 0.01.

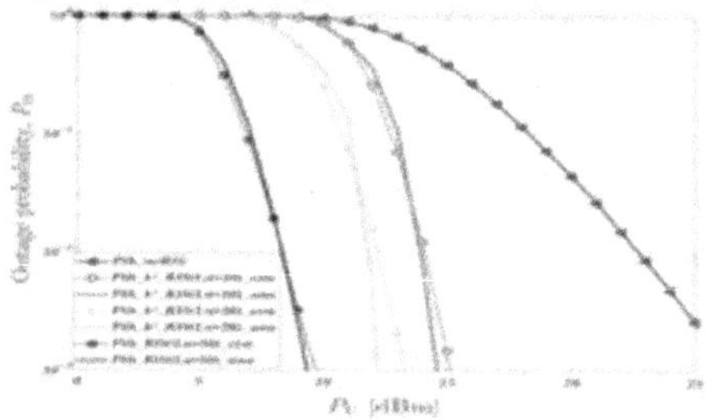

Check out how likely it is that k2_RIS (Ln = {10, 20}) and ideal_RIS (Ln = 10) will go down. Broken hardware makes it more likely that there will be an outage, and adding more RIS units doesn't fully make up for that. In our study, this shows how important it is to think about things that don't always go as planned. Gear that is getting older is a big reason why transmissions break down more often.

Figure 20 shows the link between the ergodic rate and the change in the send power of the UAV. This link can be seen in three different ways: k2_noRIS, k2_RIS, and perfect _RIS. When we look at these values, we see that the hardware disability factor is k2 = 0.04 and that each RIS unit Ln is equal to 10 to 20. It is clear that when the UAV's communication power goes up, the ergodic rate of k2_RIS is higher than that of k2_noRIS (Figure 21). RIS can help the system work faster, as shown here. Also, k2_RIS (Ln = 20) and k2_RIS (Ln = 10) are not the same. It shows that the ergodic rate rises as the number of RIS units increases. When the UAV's send power goes up, those system measures that change because of hardware problems tend to come back together. The ergodic rate can only go as high as it can because of broken technology. Before they join, the k2_RIS and perfect _RIS (Ln = 10) forms look a lot alike. This means that hardware wear and tear doesn't have a big affect when the UAV sends less power. This is what would happen if the same things keep happening. In other words, the UAV base station can still do its job even if the UAV's send power is low, which can happen when hardware fails in real life.

Figure 21. The relationship between ergodic rate and UAV send power change. N = 3, Ln = [10, 20], and k2 = 0.04.

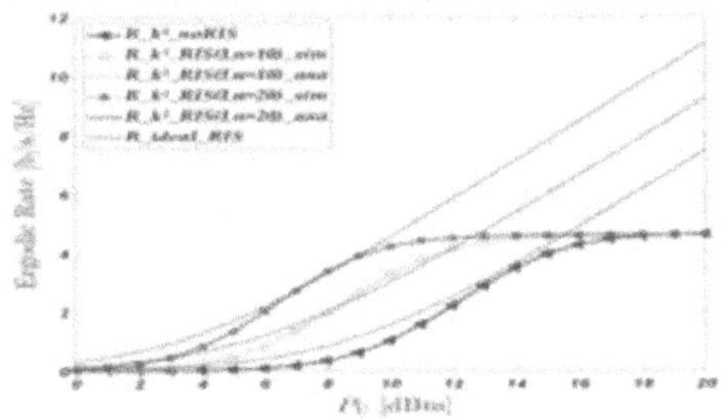

People think that there are always 10 RIS units. The hardware impairment k2 and the parameter m, as well as the change in the UAV's send power, are shown in Figure 22. It was also shown that the upper limit of the ergodic rate and the hardware drop are related in a bad way. This means that the upper limit of the ergodic rate goes up as the hardware impairment goes down. This is because the rate changes because of the drop, and the rate slowly goes up. The measure m goes up and the average rate goes down when the unmanned aerial vehicle (UAV)'s send power is low. The main reason for this is that the ergodic rate drops a lot when things go wrong with the channel. Still, one might think that the damage to the gear has something to do with the fastest moving speed.

Figure 22. The connection between hardware problems and changes in channel parameters, with N = 3 and Ln = 10.

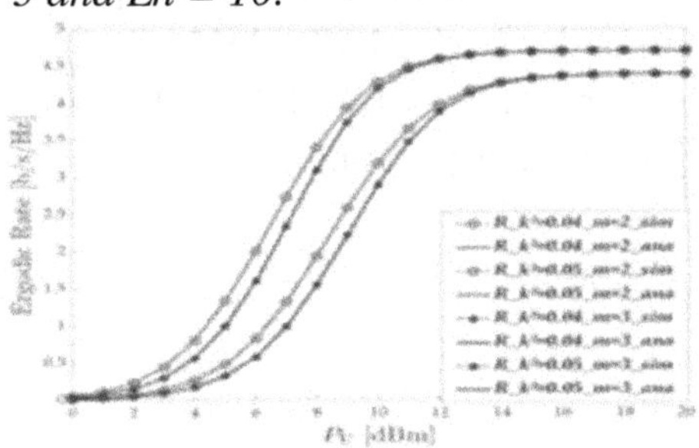

The ergodic rate changes with the amount of RIS units, as shown in Figure 23. This is because technology can break. The study looks at the difference between the ergodic rates of a RIS that is centralized (N = 1) and one that is spread out (N = 3). It takes all the parts of a centralized RIS and puts them all together into one RIS. As the number of units goes up, there is no question that the spread RIS has a better ergodic rate than the centralized RIS. This shows how important it is to think about how spread-out RIS systems would really work.

Figure 23. The graph shows how the ergodic rate changes with the number of RIS reflection elements, where k2 = 0.01 and PU = 9 dBm.

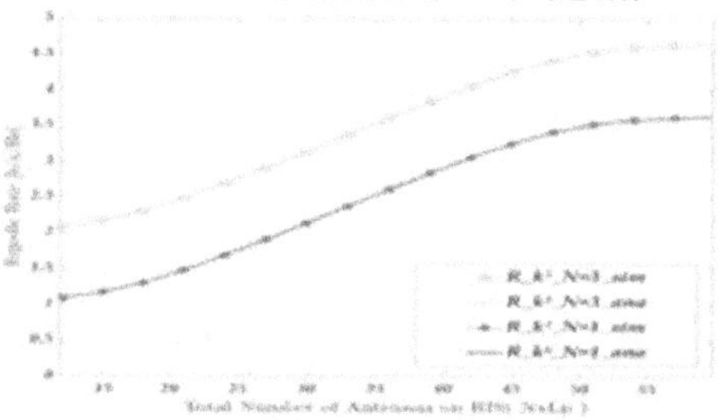

The relationship between the UAV's send power changing and how well it uses energy is shown in Figure 24. It shows three different situations: k2_noRIS, k2_RIS, and AF_k2. The study is based on the following assumptions: k2 = 0.05 and 10 RIS units are fixed. The modeling results make it very clear that k2_RIS uses less energy than k2_noRIS and AF_k2. This might be because RIS makes it possible for transmission speeds to be faster. One more thing that can be seen is that energy efficiency levels level off as k2_noRIS and k2_RIS get closer to their greatest values. This merger was made possible by improving the UAV's send power, which kept energy use from going down. The findings in Figure 6 show how spread RIS-assisted transmission between robotic aerial vehicles (UAVs) could help save energy. Also, these results show how important it is to find the UAV's ideal send power level so that the system doesn't lose its energy economy.

Figure 24. The relationship between how much energy a UAV uses and how much power it sends, where k2 = 0.05, P1 = 10 dBm, and PRIS = 1 dBm.

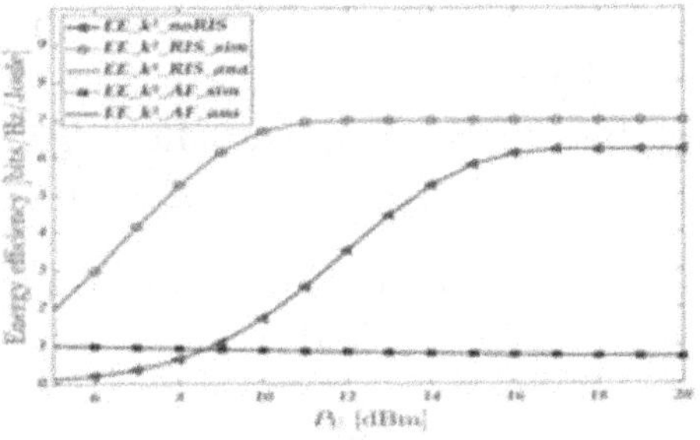

10. CONCLUSION

This article talks about how RIS technology could be used in 6G connections in the future. It gives a short summary of these choices. Let's start by quickly going over the main idea behind RIS. After that, we looked at the different ways RIS could be used, including the asymptotic analysis, precoding, CSI capture, examples, and how it could work with other technologies. Below are some cases that show how important RIS technology is for the growth of future wireless systems. This is so you can fully understand the issue at hand. Finally, we talked about the biggest issues and some possible answers so that readers could think about where RIS-assisted communication research could go in the future. A lot of people are interested in studying RIS in the coming years because it might lead to new wireless ways of talking.

REFERENCES

Basharat, S., Hassan, S. A., Pervaiz, H., Mahmood, A., Ding, Z., & Gidlund, M. (2021). Reconfigurable intelligent surfaces: Potentials, applications, and challenges for 6G wireless networks. [Google Scholar] [CrossRef]. *IEEE Wireless Communications*, 28(6), 184–191. DOI: 10.1109/MWC.011.2100016

Cao, K., Ding, H., Lv, L., Su, Z., Tao, J., Gong, F., & Wang, B. (2023). Physical Layer Security for Intelligent Reflecting Surface aided Wireless Powered Communication. [Google Scholar] [CrossRef]. *IEEE Internet of Things Journal*, 10(20), 18097–18110. DOI: 10.1109/JIOT.2023.3278238

Castillo-Soria, F., Gutierrez, C., Kumaravelu, V., & Garcıa-Barrientos, A. (2024). RIS-Assisted Non-orthogonal Multiple Access System Based on SSK. [Google Scholar] [CrossRef]. *Wireless Personal Communications*, 134(4), 2391–2412. DOI: 10.1007/s11277-024-11055-w

Castillo-Soria, F. R., Macias-Velasquez, S., Kumaravelu, V. B., Ramos, V., & Azurdia-Meza, C. A. Multiple Parallel RIS-Assisted MU-MIMO-DQSM System; Blind and Intelligent Approaches. Available online: http://www.cic-chinacommunications.cn/EN/10.23919/JCC.ja.2023-0695 (accessed on 13 May 2024).

Chen, Q., Guo, Z., Meng, W., Han, S., Li, C., & Quek, T. Q. A Survey on Resource Management in Joint Communication and Computing-Embedded SAGIN. arXiv 2024, arXiv:2403.17400. [Google Scholar] [CrossRef]

Dao, N.-N., Pham, Q.-V., Tu, N. H., Thanh, T. T., Bao, V. N. Q., Lakew, D. S., & Cho, S. (2021). Survey on aerial radio access networks: Toward a comprehensive 6G access infrastructure. [Google Scholar] [CrossRef]. *IEEE Communications Surveys and Tutorials*, 23(2), 1193–1225. DOI: 10.1109/COMST.2021.3059644

Di Renzo, M., Zappone, A., Debbah, M., Alouini, M. S., Yuen, C., De Rosny, J., & Tretyakov, S. (2020). Smart radio environments empowered by reconfigurable intelligent surfaces: How it works, state of research, and the road ahead. [Google Scholar] [CrossRef]. *IEEE Journal on Selected Areas in Communications*, 38(11), 2450–2525. DOI: 10.1109/JSAC.2020.3007211

ElMossallamy, M. A., Zhang, H., Song, L., Seddik, K. G., Han, Z., & Li, G. Y. (2020). Reconfigurable intelligent surfaces for wireless communications: Principles, challenges, and opportunities. [Google Scholar] [CrossRef]. *IEEE Transactions on Cognitive Communications and Networking*, 6(3), 990–1002. DOI: 10.1109/TCCN.2020.2992604

Faisal, K., & Choi, W. Machine learning approaches for reconfigurable intelligent surfaces: A survey. IEEE Access 2022, 10, 27343–27367. [Google Scholar] [CrossRef] DOI: 10.1109/TCCN.2020.2992604

Hameed, I., & Koo, I. (2024). Enhancing Throughput in IoT Networks: The Impact of Active RIS on Wireless Powered Communication Systems. [Google Scholar] [CrossRef]. *Electronics (Basel)*, 13(7), 1402. DOI: 10.3390/electronics13071402

Hellström, H., da Silva, J.M.B.Jr, Amiri, M.M., Chen, M., Fodor, V., Poor, H.V., & Fischione, C. Wireless for machine learning: A survey. Found. Trends® Signal Process. 2022, 15, 290–399. [Google Scholar] [CrossRef] DOI: 10.1109/TCCN.2020.2992604

Jian, M., Alexandropoulos, G. C., Basar, E., Huang, C., Liu, R., Liu, Y., & Yuen, C. (2022). Reconfigurable intelligent surfaces for wireless communications: Overview of hardware designs, channel models, and estimation techniques. [Google Scholar] [CrossRef]. *Intell. Converg. Netw.*, 3(1), 1–32. DOI: 10.23919/ICN.2022.0005

Khoshafa, M. H., Maraqa, O., Moualeu, J. M., Aboagye, S., Ngatched, T., Ahmed, M. H., Gadallah, Y., & Di Renzo, M. RIS-Assisted Physical Layer Security in Emerging RF and Optical Wireless Communication Systems: A Comprehensive Survey. arXiv 2024, arXiv:2403.10412. [Google Scholar]

Nguyen, V.-L., Lin, P.-C., Cheng, B.-C., Hwang, R.-H., & Lin, Y.-D. (2021). Security and privacy for 6G: A survey on prospective technologies and challenges. [Google Scholar] [CrossRef]. *IEEE Communications Surveys and Tutorials*, 23(4), 2384–2428. DOI: 10.1109/COMST.2021.3108618

Pan, C., Zhou, G., Zhi, K., Hong, S., Wu, T., Pan, Y., Ren, H., Di Renzo, M., Swindlehurst, A. L., & Zhang, R. (2022). An overview of signal processing techniques for RIS/IRS-aided wireless systems. [Google Scholar] [CrossRef]. *IEEE Journal of Selected Topics in Signal Processing*, 16(5), 883–917. DOI: 10.1109/JSTSP.2022.3195671

Renzo, M. D., Danufane, F. H., & Tretyakov, S. (2022). Communication Models for Reconfigurable Intelligent Surfaces: From Surface Electromagnetics to Wireless Networks Optimization. [Google Scholar] [CrossRef]. *Proceedings of the IEEE*, 110(9), 1164–1209. DOI: 10.1109/JPROC.2022.3195536

Rodriguez, J., Verikoukis, C., Vardakas, J. S., & Passas, N. (2022). *Enabling 6G Mobile Networks*. Springer. [Google Scholar] DOI: 10.1007/978-3-030-74648-3

Sejan, M. A. S., Rahman, M. H., Shin, B. S., Oh, J. H., You, Y. H., & Song, H. K. (2022). Machine Learning for Intelligent-Reflecting-Surface-Based Wireless Communication towards 6G: A Review. [Google Scholar] [CrossRef] [PubMed]. *Sensors (Basel)*, 22(14), 5405. DOI: 10.3390/s22145405 PMID: 35891085

Sikander, M., Butt, F. A., Chattha, J. N., & Naqvi, I. H. Improving Wireless Communication through Beamwidth Optimization in Reconfigurable Intelligent Surfaces. In *Proceedings of the 2023 10th International Conference on Wireless Networks and Mobile Communications (WINCOM)*, Istanbul, Turkey, 26–28 October 2023; pp. 1–5. [Google Scholar] DOI: 10.1109/WINCOM59760.2023.10322929

Tang, W., Chen, M. Z., Chen, X., Dai, J. Y., Han, Y., Di Renzo, M., Zeng, Y., Jin, S., Cheng, Q., & Cui, T. J. (2020). Wireless communications with reconfigurable intelligent surface: Path loss modeling and experimental measurement. [Google Scholar] [CrossRef]. *IEEE Transactions on Wireless Communications*, 20(1), 421–439. DOI: 10.1109/TWC.2020.3024887

Wang, C. X., Di Renzo, M., Stanczak, S., Wang, S., & Larsson, E. G. (2020). Artificial intelligence enabled wireless networking for 5G and beyond: Recent advances and future challenges. [Google Scholar] [CrossRef]. *IEEE Wireless Communications*, 27(1), 16–23. DOI: 10.1109/MWC.001.1900292

Wang, J., Gao, D., Li, J., Huang, L., Ding, H., & Zhou, S. (2024). Analysis and Mitigating Methods for Jamming in the Optical Reconfigurable Intelligent Surfaces-Assisted Dual-Hop FSO Communication Systems. [Google Scholar] [CrossRef]. *Electronics (Basel)*, 13(9), 1730. DOI: 10.3390/electronics13091730

Wang, J., Jiang, C., Zhang, H., Ren, Y., Chen, K. C., & Hanzo, L. (2020). Thirty years of machine learning: The road to Pareto-optimal wireless networks. [Google Scholar] [CrossRef]. *IEEE Communications Surveys and Tutorials*, 22(3), 1472–1514. DOI: 10.1109/COMST.2020.2965856

Yuan, X., Zhang, Y.-J. A., Shi, Y., Yan, W., & Liu, H. (2021). Reconfigurable-intelligent-surface empowered wireless communications: Challenges and opportunities. [Google Scholar]. *IEEE Wireless Communications*, 28(2), 136–143. DOI: 10.1109/MWC.001.2000256

Zappone, A., Di Renzo, M., & Debbah, M. (2019). Wireless networks design in the era of deep learning: Model-based, AI-based, or both? [Google Scholar] [CrossRef]. *IEEE Transactions on Communications*, 67(10), 7331–7376. DOI: 10.1109/TCOMM.2019.2924010

Chapter 4
Reconfigurable Intelligent Surface-Empowered Smart Radio Environments Design Perspective

Geetha M. Sri
https://orcid.org/0000-0002-5248-9368
Amrita Vishwa Vidyapeetham, India

Ashok Vajravelu
Universiti Tun Hussein Onn Malaysia, Malaysia

A. Rajasekar
Sri Sai Ram Institute of Technology, India

Kiran Kakade
Symbiosis Institute of Management Studies, Symbiosis International University, India

ABSTRACT

A lot of people are interested in intelligent surfaces that can be modified. These surfaces are also known as intelligent reflection surfaces (IRSs) or big intelligent surfaces (LISs). This is because they can change the surroundings in which wireless messages move in smart ways to make networks faster and more reliable. RISs are thought to be a tool that could be useful for the sixth generation (6G) of computer networks because of this. How the RISs work, how to measure their performance,

DOI: 10.4018/979-8-3693-8099-4.ch004

how to design beamforming patterns and manage resources, how machine learning can be used to make RIS-based wireless networks better, and how RISs can be mixed with other new technologies are analyzed. In this study, how well multiantenna assisted RIS systems work are tested. Then, the main ideas behind RISs from the point of view of both physics and messaging are discussed. The current plans for RIS-enhanced wireless networks from the points of view of information theory, performance improvement, and performance analysis are carried out.

1. INTRODUCTION

There are a lot of problems with the current cellphone networks because so many people want high-quality digital services that can be reached by a lot of people. Better mobile broadband (eMBB), ultra-reliable, low latency communications (URLLC), and massive machine-type communications (mMTC) are some of the things that this new generation of communication systems (5G) is designed to do. At the same time, the sixth generation (6G) of wireless communication devices is likely to bring about big and important changes. As part of these changes, there will be data-driven, instant, ultra-massive, and broad WiFi links, as well as linked intelligence (Wu & Zhang, 2020). Now that there are these brand-new services and uses, we need new gearbox technologies to meet their needs. Smart surfaces that can be customised (RISS) have a group of reflecting parts built into their structure. These surfaces are also known as intelligent reflecting surfaces (IRSs) (Liaskos et al., 2018) or big intelligent surfaces (LISS] (Di Renzo et al., 2020). This lets them change the way signals hit them.

RISs have been studied a lot in the field of wireless communications because they can change the surroundings of wireless transmission before it happens. In order to fix a lot of problems that different kinds of wireless networks have (Farrell, 2018), this is done. The following is a list of all the good things that RISs do.

The wireless industry is excited about Beyond-5G (B5G), also called 6G, technological advances because there will be a lot of new uses and applications that need wireless communication and sensing networks with a lot more flexibility, coverage, and throughput (Jiang et al., 2022). This is happening at the same time that the Fifth Generation (5G) network is being set up all over the world more and more. Fifth-generation wireless networks are being used more and more for things like self-driving cars, haptic remote touch, and virtual reality. Massive Multi-Input-Multi-Output (MIMO) began as an interesting research topic ten years ago and went on to be a key part of the switch from 4G to 5G in commercial use. For these kinds of uses, we will need new technologies that can perform ten times better than 5G counterparts. This is how MIMO turned out to be.

Both academics and business people are very interested in customisable intelligent surfaces (RISs), which are possible technologies that are still being studied in their early stages (Mwase et al., 2022). A wireless mobile communication network would often change the broadcast methods of the receiver ends to get the best performance. This is done to either make the dynamic multipath transmission setting better or better use it. These kinds of networks, on the other hand, can't control the surroundings. RISs can help create a smart radio system that can be set up using software. This is a good option. An idea behind a RIS is that it is a big, thin metasurface made of metal or insulating material. This is made up of a group of passive sub-wavelength scattering pieces with a specially designed physical structure. The electricity (EM) properties, like phase shift, of the return of radio frequency (RF) signals that hit it can be changed by controlling its parts in a way that is described by software. If you use a combined phase control of all the scattering components, you can change the mirrored radiation pattern of the radio frequency signals that come in at any time. This creates more degrees of freedom that can be used to make the wifi network work better generally. A lot of the time, the phase-only version of the Reflective Image Sensor (RIS) is called the Intelligent Reflecting Surface (IRS) (Gams et al., 2019).

Because a RIS is passive and flexible, it can be simply attached to surfaces that are already there, like walls and furniture. This improves the surroundings of current wireless networks without getting in the way, which leads to better energy and frequency efficiency. It's possible to do this without changing the wireless standards or designs of transceivers that are already in use. RISs can help with tracking, figuring out where something is, and sending power wirelessly (Zhang et al., 2020). Not only that, but they also make computer contact networks that use wireless technology work better. This could be especially helpful for networks that work in high-frequency ranges, like millimeter-wave and terahertz, where performance problems like dead zones and channel rank deficiency are common. To get around these problems, networks would have to set up ultra-dense Base Stations (BS), which would use more power and cost more for backhaul (Pavia et al., 2021).

More than one review study has been done on RIS. Newer ones go into more detail and include the most recent information on surface and network models, the best way to design at the EM, physical, and network layers, comparisons with older technologies (like massive MIMO and active relay), as well as the problems that the field is facing now and where it wants to go in the future. (Wu et al., 2021) talked a lot about how important it is to use surface touch models that match the way things work in the real world. The methods for putting these models together and figuring out what they mean were based on the idea of surface electromagnetic theory (SEM). It is still very new to know what the highest performance limit of a RIS network with active ends and spread-out passive surfaces is and how that performance is determined by the design of the network (Di Renzo et al., 2019). It's clear that study and modelling have been key to most of the progress made so far. Most of the time, this has been done on systems that are working perfectly, like when the state of the world is known for sure, there is infinite-resolution phase control, and the electromagnetic model of the surface is made simpler. Also, the added work that comes with protocols and calculations when a network is used in the real world has not been taken into account.

Another reason for this is that the spreading world is a lot bigger than the contact ends. Power concentration on receivers used to be done by colocated coherent endpoint beamforming (for example, massive MIMO). Now, it needs to be done through many spread-out passive surfaces over a much larger area in order to pick up enough energy to change the surfaces. It's important that the cost per unit area and power use are low enough that the total cost of ownership is about the same as a normal MIMO network that is spread out or co-located. That is the only way a rollout can make financial sense. For this tech to work in the real world, we need a surface that is both cheap and doesn't use a lot of power. It's also important to improve the physical and network systems in certain ways to get the fastest speeds while also cutting costs and power use. So, to reach this goal, it's important to know more about the most recent experiment (Huang et al., 2020). Aside from that, the review papers above did talk about Machine Learning (ML) and Artificial Intelligence (AI) methods, but only very briefly and in different ways. Adding these methods to model-based techniques in RIS system-level planning has been talked about. This is especially true for the large-scale use of cheap surfaces that don't work with parametric models.

There are three things that must be true for a real-world network operation to last: performance, cost, and power use. The goal of this review piece is to show a number of RIS research areas that are important to a choice between these three needs at the system level. So, along with explaining the basic ideas of RIS, the channel model, and how to make the Physical layer (PHY) design work better, we also go into great detail about how machine learning and artificial intelligence are used in a RIS system (Basar et al., 2019). Of course, we also give a full review of the hardware design and execution.

This study adds to what is already known because it shows the latest changes in RIS-assisted technologies and metasurfaces based on big surfaces, along with the pros and cons of each (Liaskos et al., 2018). Also, it talks about some different ways that RIS can be used, such as active RISs and the signal models that work with them. It then compares how well these work to an inactive RIS. The changes between them are looked at when the link budget for communication is looked at.

We first talk about the problems that come with RIS and then stress the possibilities that could come up if these problems can be fixed. This makes the system work better as a whole. We also look at a number of RIS designs that use field-programmable gate arrays (FPGAs) and are controlled dynamically (Wang et al., 2021). In the end, we take a close look at RIS-assisted communication systems and compare how well they work, what they can be used for, and the research that is being done now and will be done in the future.

2. TECHNICAL CONSIDERATIONS FOR RIS

2.1 RIS Standardization

High power consumption, hard-to-get spectrum, and costly deployment and maintenance costs—especially for network operators—are some of the current problems with 5G networks. Setting guidelines is the main job of 3GPP and other SDOs. You can also get new tools and ways of working that might work with 5G networks and later ones, and they use less technology, power, and frequency (Danufane et al., 2021). IRIS has recently become better known as a possible fix for B5G and 6G wireless networks and other next-generation networks. Channel-adapting, which changes the channel for better performance using modulation and coding (Subrt & Pechac, 2012), is something that many people in science and business agree can totally change the way people talk to each other. After that, it goes to global SDOs like 3GPP, where detailed technical needs are written down in Work Items (WIs). In the end, the international norms are made public and legally recognised.

SDOs mostly focus on initiatives connected to RIS in China, Europe, and the United States. The IEEE Communication Society in the United States has created one Emerging Technology Initiative (ETI) and two Special Interest Groups (SIGs) (Zheng et al., 2021; Karami et al., 2016). Scientists who work on things like signal processing, metamaterials, and radio transmission are being asked to work together by these groups. Businesses like network infrastructure owners and providers are also being asked to help customers see things from their point of view and talk about how RISs could be used and how much they cost.

The FuTURE Forum was created in 2005 by well-known mobile phone companies, device makers, research centres, and universities from around the world to work together on non-profit projects. Chinese companies and institutions set up the CCSA as a formal standards group to work on standardising ICT. The 55th TC5-WG6 meeting in 2020 agreed to make a plan to make a SI for RISs (Lam et al., 1997). The FuTURE Forum is looking into ways to add RIS to next-generation wireless communication, and the CCSA is looking into how to make wireless communication systems that use RIS work (Yu et al., 2011). This talks about beamforming with RIS, RIS and AI, channel models, channel predictions and feedback, and RIS-compatible networking protocols.

2.2 RIS Operation Modes and Types

The new type of network unit is called a RIS. It is made up of a surface that can send and receive data. Figure 1 shows the options the RIS device has for moving the surface passively or actively. As a result, the wireless world stops being inactive and becomes smart with the ability to design channels. As a result, the RIS can work in the network in six different ways: it can broadcast, receive, backscatter, refract, absorb, and reflect (Table 1). The traits of each one are described in.

Figure 1. RIS aided wireless communication network example.

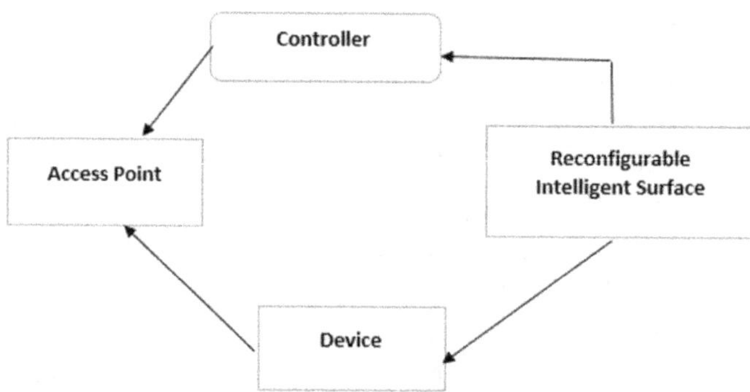

Table 1. RIS operation modes.

Mode of Operation	Qualities and Roles
Thoughts:	To improve range, block out noise, and free up more room, it could be used as a screen in a wifi setting.
Refraction	Changing the phase of incoming electromagnetic waves as they travel through the RIS allows them to be refracted in various directions. Since the RIS panel lacks a protective layer, electromagnetic waves may go through it. Electromagnetic waves are sent into certain parts of a house through the RIS, which acts as a window from the outside to the inside.
Absorption	It is perfect for radio waves with a certain middle frequency and bandwidth to hit and be fully absorbed, getting rid of any waves that bounce back. Information security, privacy, and interference mitigation—using RIS on the outside of a structure to block electromagnetic radiation, separating them from one another inside and out or from various internal areas.

2.3 Channel Modeling

A communication model that may be used to assess the performance of wireless networks is created by combining channel modelling, interference analysis, and RIS modelling. This crucial aspect of RIS research aids in our comprehension of the technology's potential applications in wireless networks of the future. You need to make sure that your channel models are reliable and consistent so that you can try them in real life when you use next-generation wireless communication technologies. The physical effects of propagation change in different places, like how path loss and scattering change in a big office versus on the street. It is also important to think about how RIS affects the end-to-end channel model, like how an adjacent RIS may change the near-field effects and spatial correlation, as well as how RIS elements relate to each other.

Aside from that, the channel model is greatly affected by the various frequency ranges, bandwidths, and RIS designs, like those that are built on metasurfaces or mirrored arrays. You may also get updated LOS probability and route loss indices by taking into account the possible locations of RISs in an actual wireless environment. You often have to decide between accuracy and complexity when creating channels. The majority of ray tracing-based channel modelling techniques, for example, only include one reflection, which simplifies the model but compromises its accuracy. Conversely, higher-order reflections may improve the accuracy of channel models, but they are computationally and experimentally expensive. In Table 2, you can see the 3GPP's channel model, which can be used to look at the RIS in channel models. In towns, the Uma/Umi model is often used to switch from a base station (BS) to a RIS channel model. It is based on the BS–RIS's height. The RIS–UE link might show up as InH/InF when it is inside.

Table 2. 3GPP channel models that are applicable to RIS use cases.

Situations	Model of Channel BS-UE	BS-RIS Channel Model	RIS-UE Channel Model	The RIS Channel Model
RIS is under UE control for urban rollout	Uma or Umi	Uma or Umi	LoS	Not at all
This is about the same as BS and RIS in towns.	Uma or Umi	LoS	Uma or Umi	Umi
Launch coverage will be widened in towns.	Uma or Umi	Uma or Umi	Uma or Umi	Uma or Umi
An inside place	InH/InF	InH/InF/LoS	InH/InF/LoS	InH/InF

Propagation channel models have recently been studied in order to include a RIS and alter the radio propagation environment. As of late, end-to-end channel modelling has been getting a lot of attention. A lot of study has been done on modelling the route loss of RIS-enabled communication systems. Another important thing to think about is the end-to-end channel planning method. There are different ways to model channels. Some use physics to explain them, while others use statistics to look at the physical effects of things like scattering and reflection. There are also statistical methods that look at outdoor scattering and cluster statistics. A channel model based on electromagnetic (EM) waves has also been put forward. Channel models have also been suggested that take into account EM features like how subwavelength unit cells can touch each other. The current study on RIS-enabled channel modelling (Figure 3a) shows that there are many main types of channel and spread modelling methods. Since it's not easy to tell one approach from another in this group, channel modelling studies that use a lot of different methods might give a more complete picture. One way to describe a RIS channel is with the 3GPP defined channel model.

Table 3. RIS channel modeling research trends.

Abstract
• Huygens-Fresnel principle and physical optics (PO) are used to study path loss in both close and far fields. • It is possible to model RIS as a thin EM sheet.
• Determine route loss using a physics-based method. • Determine the near and distant fields' electric fields for a finite-size RIS.
• Determine the passive reflectarray RIS's route loss. • Determine route loss in relation to RIS unit cell real gain, link geometry, and RIS size.

Figure 2. RIS channel modeling. Classification of approaches

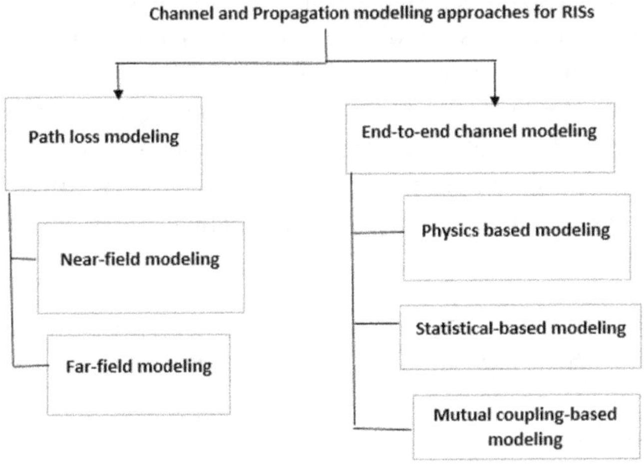

2.4 Hardware Implementation of RIS

The idea behind RIS is different from the receiver technology used in wireless networks because it is based on the idea of a changing metasurface. So, you can change the wireless world by making the electromagnetic response to radio waves and controlling it.

Reconfigurable reflectarray antennas often resemble RISs in form. A number of papers on customisable metasurfaces have shown how to change the unit cell reflection properties on the fly. Two common ways to change how customisable reflectarray antennas respond to reflection are to use delay lines that can be tuned and resonant components that can be changed. In the first method, the wave-sending device is connected to a communication line whose electrical length can be changed. This lets the phase-shift properties change before the wave is connected and sent out again. There is a second way that changes the resonance frequency of the radiating part by using switching or reactive loading. This is now the method that RISs use most often.

In a RIS, there are two main types of metasurface layers. One way to do this is to use a structure that is kind of like an array, with array members placed in an area of λ/2. Another method is to use a dynamic metasurface structure with parts that last much less than λ/2 of a second. There are three different ways to change the shape of changeable metasurfaces: circuit reconfiguration, physical reconfiguration, and material reconfiguration. To rearrange the circuit, the resistance of unit cell parts like switches and capacitances that can be changed can be changed. The best things about RISs are that they are cheap, have low delay, and use little power. In a RIS unit cell, a PIN diode may work as a switch when it is on or off. This could cause one, two, or more phase states and different resonant modes. In RIS systems, PIN diodes can only provide a single phase. However, varactors can be used to change phases all the time. We can change the electrical shape of a gadget by turning and deforming parts of it. This is called geometric transformation. Material rearrangement is the process of changing the substrate's conductivity and other material qualities. Table 4 shows ways that different RIS change tools and methods are better or worse.

Table 4. Comparison of reconfigurable techniques (+: good, o: normal, –: poor, M: microwave, T: THz). Adapted with permission from. Copyright 2014 IEEE.

Type	Technology	Stability	Integration	D/A Control	Cost	Loss (M/T)	Bias Power Consumption	Linearity	Switching Time
Circuit	Pin diode	+	–	D	+	–/–	–	o	+
	A variac	+	–	D	+	–/–	+	–	+
	MEMS	o	+	D	+	+/o	+	+	o
	LC	o	o	A	o	–/+	o	o	–
	Using light to	o	–	A	o	–/–	–	–	+
Mechanical	Fluid	o	–	A	o	o/+	+	o	–
	Small motors	–	–	A	–	+	o	+	–

Based on the phase control technique, RISs may be classified as either digital or analogue (continuous). Digital RISs, which are being thoroughly studied because to their quick reaction times and simple control circuits, are made using MEMS and PIN diodes. It is not possible to ignore the phase quantisation mistake that digital phase control devices cause, though. For 1-bit phase control, the phase quantisation error leads to a huge loss of gain, and the phase error itself leads to a loss of gain of more than 3 dB. When you build analogue RISs with analogue changeable devices like varactor diodes you can get rid of the gain loss that comes from quantisation error in digital RISs. An analogue RIS may reduce the quantisation loss caused by phase shift because it changes the capacitance of the varactor diode to use continuous phase shift.

You need a bias network to change how the unit cells of the RIS reflect light. This network will be built in a different way depending on how it is changed and how much control you want over the data being sent. For every PIN diode in a column, the bias can be set to either forward or backward by 0° to 180°. An FPGA is generally what runs this kind of network. How tough the RIS bias circuit is is based on the change scheme, the amount of unit cells, and the beat. To use a varactor for continuous setting, you need a DC voltage that can be anywhere from 0 V to 30 V. This way, each part of the unit cell can be run on its own using an audio input. These devices need digital control signals, like those found in SPI and I2C, along with the hardware and time signals that go with them in order to work. Digital tools, such as computer IO ports, are often used to cut down on size and cost. For the unit cell to work as it should, there needs to be an electrical gap between the bias circuit and the RF sources that make it up. To keep it electrically separate from the transmitters of the unit cell, either an RF choke or an optical bias circuit must be used.

The change control circuit for beam steering is another part of large-area RISs that could be hard to use. One way to make it easier is to load the cells one at a time and collect them all at once. he use of programmable reflectarray antennas with row/column control is a new way to make bias networks more efficient. There are even more phase changes where rows and columns meet. It doesn't need N × M manual control lines like a metasurface that is N × M does. Instead, it only needs N + M. A 1-bit row/column control method is another idea in this work. As the working frequency goes up, the bias circuit gets not as easy to use. A hypersurface is an idea that has been put forward to solve this problem. Each unit cell's size and phase may be changed by adjusting its resistance using a small ASIC. In this case, a frequency of 60 GHz may be attained with a unit cell size of around 1 mm × 1 mm.

As previously stated, the RIS consists of one RIS processor and three layers. The topmost layer controls the incoming electromagnetic radiation by acting as an antenna. This layer is made up of several conducting patches that may be moved around on an insulating base. This layer could additionally include a dedicated radio frequency sensor to pick up radio waves in the real world. In order to facilitate intelligent RIS setup operations, this sensor may instantly identify adjacent radio waves. The intermediate layer is then equipped with an RF choke structure to stop excessive RF signal leakage. Finally, the bottom layer has control components that change the reflecting parts' configuration in real time. FPGAs, industrial microcontrollers, or SDR devices like USRPs are often used to create these control components. To receive and carry out RIS control directives, they connect to the whole network, as seen in Figure 4. The RIS and network devices, like base stations, can share information through a bridge link. The RIS receives information about its unit cells via a wireless transceiver, usually connected to a microcontroller, via the gateway interface. In response to orders from the network equipment, the CPU sends setup signals to the bias network. It might also examine data from onboard sensors or problem detection devices before sending the data to the network controller.

Figure 3. Example of control structure in RIS. Adapted with permission from. Copyright 2023 John Wiley and Sons.

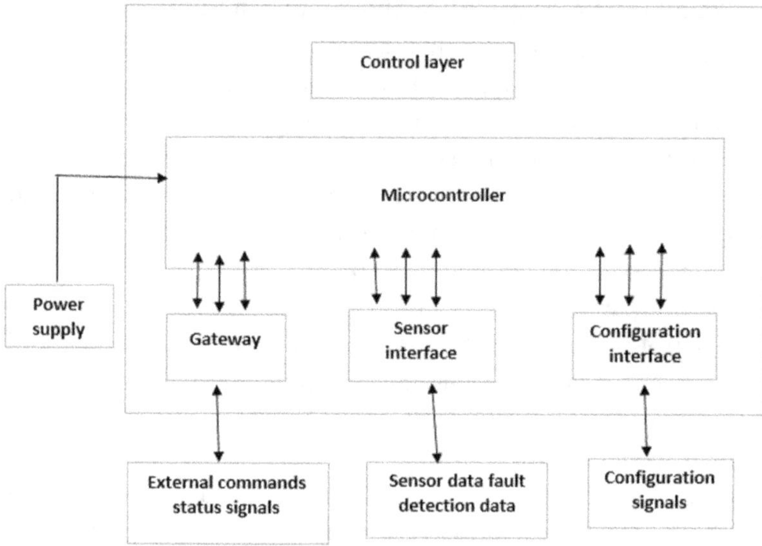

There isn't a lot of research on designing and characterising RISs in millimeter-wave bands. Most of the writing about RISs is just academic and talks about idealised RISs. Also, most of the tests have been done in frequency ranges below 6 GHz. There were steps taken to improve the signal in both the close and far fields, and a RIS plan that would work was put forward. The device could only handle a bandwidth of 27.5 to 29.5 GHz, which was not enough for the 5G millimeter-wave range, which needs a wide bandwidth. They made a 25X32 RIS prototype and a small unit cell that can work at 28.5 GHz, but they didn't say how wide the range was. They showed how to use an Arduino CPU and 144 shift registers set up in a RIS phase system to handle 1152 PIN diodes. The shift registers were set up in 6 series. This research suggested a 20x20 RIS that will work in all 5G n257 (26.5 to 29.5 GHz) and n258 (24.25 to 27.5 GHz) bands in 2023. The developed RIS can guide beams at an angle of 50° and has a wide frequency range of 22.5–29.5 GHz (26.9%).

The total efficiency of most PIN diodes and varactors is further diminished by their significant insertion loss. This may be decreased by using diodes with low insertion loss, but their manufacturing will be more expensive. In situations when the EM function doesn't need to be changed often, electrically controlled RISs use a lot of energy since they can only perform a set EM function. In order to improve signals in static areas, including dark areas for indoor communications, electrical RISs have been suggested. Recently, many ideas have been put forward for metasurfaces that can change electrically and physically using micromotors, piezoelectric actuators, and spring devices. It is easy to change and works really well. Metasurfaces and mechanical devices that aren't being used, on the other hand, can only work in different phase states and are hard to set up. We need to find a cheap, simple, and flexible electrical RIS system that can be wavefront tuned in a number of ways. A RIS that runs at 2.55 GHz was made in 2023.

Figure 4. Unit cell of an electromechanical RIS. (a) Schematic diagram. (b) Driver module of the RIS supercell. (c) Photograph of the driver module.

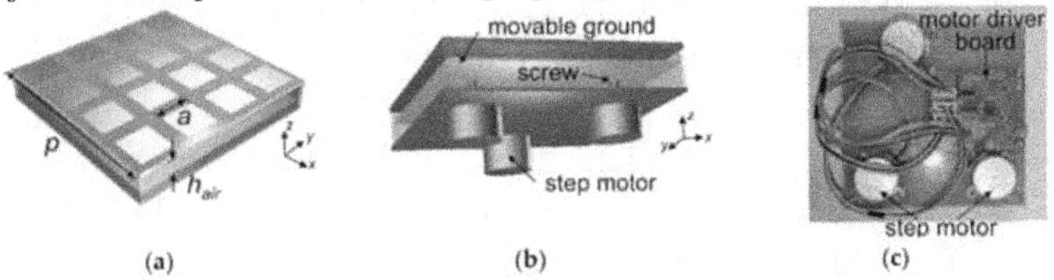

To make the smart wireless environment idea work, a lot of RISs made of flexible metasurfaces need to be placed in different areas to control the signal routes. For this to work, there needs to be good communication that doesn't ruin the look of the RISs, if that's even possible. RISs that are optically clear could be useful when a lot of mobile end devices need exceptional optical clarity and top-notch connection services. RISs that are perfectly clear can have solar panels put on their backs so that they can work on their own. It was suggested that a 12 GHz ITO film metasurface that can be changed and is electromagnetically clear work. With the addition of a solar screen on the back, this building that is 80% see-through became a power source. The issue with ITO is that a lot of electricity is lost when the material is first made. In 2022, they made a 2-bit photonic RIS out of a metal mesh that was very sensitive and worked at 3.5 GHz. This gadget did a great job of moving beams from 0° to 60°, but it only let 49.3% of light through. For this reason, vanadium dioxide (VO2) is sometimes used instead of this material because its performance can be changed by temperature and current.

A metal mesh lattice pattern was added to make it 79% better. On the bottom, a metal mesh stripe pattern was added to keep the microwave properties of the RIS device in the 3.53–3.7 GHz frequency range. A lot of the technology we need to make designs that are easy on the eyes is still not here. We need to work on more metal vias and stacked structures. We also need to put parts on the surface and set up direct current (DC) supply networks. We also need to do other design and production work. In other words, it's still hard to make RISs that are clear, flexible, and have the same electromagnetic qualities as regular metasurfaces.

There are two clear surfaces that make up the transparent dynamic metasurface. You can change between full transmission, total reflection, and split mode by moving them farther apart. A metamaterial pattern that has already been made is cut into one mode. There is only about 1 dB of transmission loss when the metasurface is in full transmission mode. About 10 dB are lost when it is fully reflective. In 2021, tests were done on an improved, clear, dynamic metasurface that directs millimetre waves. By about 24 dB, the metasurface may make the SNR better at the focus point than standard glass. Millimetre waves may be able to reach more places inside by putting a repeater near the metasurface's focus point, as shown in Figure 20c. This makes the connection between an outside millimetre wave base station and a device inside the person more secure.

Additionally, the worldwide manufacturer TowerJazz and the American company Lumotive announced beam steering integrated circuits (ICs) for LiDAR systems in self-driving cars in 2019. These integrated circuits use liquid crystal metasurface technology as their base technology. Pivotal Commware came up with a new technique in 2019 called Holographic Beamforming (HBF). Software-defined antennas (SDAs) are used. The suggested holographic beamformer is better than other broadcast technologies like large MIMO and phased array since it is less expensive and uses less power. Greenerwave said in 2023 that they will be providing an active RIS in the 27.5–29.5 GHz frequency range with a bandwidth of 500 MHz. The RIS is marketed as an extension for AP.

The following factors need to be taken into account before RIS devices can be put together and used for interactions in the real world. In order to investigate the efficacy of a RIS, we will first go over how to set up an experiment. We know that the near-field and far-field areas are equal to 2D2/λ when the array antenna size is D. On the other hand, smart houses and other near-field indoor applications are growing in popularity. It may be possible to improve the internal operation of RIS systems by comprehending the physics of electromagnetic field dispersion in nearby and far-off fields as well as the factors affecting their design.

Another concern is if the phase control has to be more exact or whether it should be 1 bit or 2 bits. It also often has isolated phase changes because the RIS prototype only has one working cell built into its hardware. The strength drops by more than 3 dB because the antenna isn't very good at what it does. There are two phase states that a 1-bit RIS can be in at once. To make up for the speed loss caused by 1-bit phase quantisation, a number of experts have created multibit RIS boards. The cost of hardware goes up with multibit designs, though, and systems become more complicated. Two-bit phase quantisation strikes a good mix between the performance of the parts and the complexity of the system. It has great sidelobes and less than 1 dB of antenna gain loss. So, when experts are making RIS, they should think about both how hard it is to build the system and how much more accurate it is when they use more than one bit. RIS prototypes may be able to make "supercells" with continuous phase change, though, by connecting blocks of different cells and making the connections customisable.

The third question is how to make things that can connect to the internet. According to Shannon's theory, bandwidth determines how fast a message can be sent. In order to work at high speeds, wireless systems need a RIS that has a large operating bandwidth. To make the frequency wider, unit cell designs with two resonant parts and an air gap are being created.

2.5 Differences Between Relay and RIS

In cell messaging, switches are now used to find other ways to send texts that can get the same information right away. This is done to make the lines less likely to break. A single link that doesn't have a clear line of sight (NLOS) can be quickly changed into several lines that do. There is a problem with switches; they cost more and use more power, which is not good. How well a relay-enabled system can use the network's spectrum rests on how the outputs are duplexed. The sender and the rebroadcaster can't both stream from the same real resource at the same time when H.264 relaying is used. An alternative solution is to use FD relay protocols, but they have some issues. They also cause high loopback self-interference at the relay because it sends and receives signals at the same time, which makes signal processing more complicated and uses more power at the relay.

A lot of the time, relays are busy parts that need a lot of power. Power amplifiers, DACs, ADCs, and low-noise receiving amplifiers are some of the active electronic parts that make them up. This is especially true in frequency bands with millimetre waves and submillimeter waves. Also, it was already said that using FD relays makes things a lot more complicated because loopback self-interference needs to be removed using special receivers and methods for handling analogue and digital signals. For another thing, the extra noise that the active electronics used in relays make makes current relay methods less effective. In this case, noise gets louder at the relay when AF relays are used. DF relays may lessen the affects of additive noise because they decode the signal and then re-encode it at the relay. But they need more power and data processing that is more complicated. But the FD relay method doesn't work as well because of loopback self-interference that is still there. The duplexing method a relay-assisted gadget uses affects how well it works with frequencies. Most of the time, an HD relay can send data at twice the speed because the sender and rebroadcaster need different physical resources to do so. It is possible for loopback self-interference to hurt an FD relay, but it is the transmitter and the relay sending at the same time that cause interference to the receiver. An FD relay can't go twice as fast as it is now.

Simple electrical circuits like switches or varactors are used in reconfigurable RISs to save power. If you want to make a lot of low-cost large-area devices, it's easier to set up RISs than relays because they don't need special power amps, mixers, or DACs/ADCs. RISs that are passive or almost passive don't boost or restart signals, so they don't pick up noise either. But phase noise might make them less useful. If you set up a RIS as a mirror, it won't be harmed by loopback self-interference or HD limits like relays are. Reconfigurability can be reached with only active parts that use little power, like switches or varactors. And this means that passive parts can be used to create even metasurfaces that can't be changed. RISs are great for systems that don't need to do much. Switches and varactors work well in devices that collect almost no energy because they don't use a lot of power. People often think that in relay-assisted systems, the sender and relay share all the RF power to make sure there is no extra power. The transmitter in a RIS uses all the RF power. A better metasurface can be built to make this happen. The gearbox also controls how much power the RIS spreads and reflects. The radio waves that hit the RIS should have the same total power.

There's a way to think about the RIS as a full-duplex MIMO relay that doesn't talk to itself. Different choices have different ideas about speed, as well as different types of systems. And finally, from the operator's point of view, the RIS and relay are set up and used in different ways. RISs can be used in a wider range of situations than relays because they can work in an idle or almost passive mode. It's possible that adding more reflective features could help cut down on power and efficiency loses. RISs' ability to work without a power source is their most important trait. This lets 6G networks be set up whenever needed and in different ways. In addition to better coverage, it gives customers and traffic on-demand services. RISs must be controlled from afar over the network in order to reach the goals mentioned above. The way control is done and how information works need to be looked at again. Low complexity and low power usage are two more conditions for the machine that was made. Relays may be able to process and handle things more complexly because they are not limited by electricity.

The RIS makes sure that wireless contacts can be made over non-line-of-sight (NLOS) links and increases the communication range through controlled scattering and multipath components when transmission conditions are bad, like when LOS connections are blocked or stopped. On the other hand, reflective RISs only mirror signals in the front half of the area. Users behind them can't get signals. Because of this, banned people don't do well. Because the nodes are only in certain areas, the service spread is limited to 180°. This makes RISs less useful and might stop the rollout of 6G tiny cells.

According to a recent use case, RIS placement may be helpful for customers who are on the sides of a surface that doesn't have receivers on it. STAR-RIS stands for "simultaneous transmit and reflect RIS," which is a name for a group of different RIS systems. Unlike traditional RISs, hybrid RISs can both broadcast and reflect. This lets users on both sides of the surface communicate wirelessly in any direction because the surface covers the whole space. For reflective RISs and STAR-RISs, different surfaces and metal parts are used. In the case of a reflected RIS, the base uses conductors to keep radio waves at the working frequency from getting into the RIS. The base of a STAR-RIS is difficult to build, though, because it has to send and receive wireless messages at the working frequency at the same time.

2.6 Passive, Active, and Hybrid RIS

There is an array structure in the RIS made up of several devices that are not in use but still bounce electromagnetic signals in a way that is similar to how signals move in a wireless world.

Even though the RIS uses passive signal processing, it is not totally idle; it needs moving parts to drive and change. For example, a RIS is not working if it has a PIN diode inside that controls the phase and needs electricity to stay on when it is connected to parts that are working.

One great thing about a passive RIS is that it gives you a "square-law" array boost with almost no noise. This means that when there are a lot of RIS unit cells (N), the receiver SNR that can be reached by passively beamforming the RIS is almost directly related to N2. The coordinated mix of the received signals makes it clear that N2 is linked to the signal strength that the listener wants. What the minimum noise assumption does, on the other hand, is make sure that the receiver's bad noise power stays the same. It's true that these benefits only really happen when the person sending the message and the person getting it don't know each other well or at all. In other words, the square law says that the RIS will make wireless systems much more powerful. the RIS needs more than one part to make up for this loss. When there are strong straight links, the RIS may not seem to add much.

In other words, the path loss between the transmitter and the receiver is not the sum of the path losses between the transmitter and the RIS and the receiver. It is the product of those path losses. This route loss can be a thousand times worse than the straight link. This is due to "multiplicative fading," which means that the RIS makes path loss twice as strong. To give a real capacity boost with a passive RIS in many wireless settings is very hard because of the "multiplicative fading" effect. The RIS study doesn't always see this effect because it only looks at situations where the straight link is seriously at risk. The "multiplicative fading" problem makes the RIS less useful in future 6G wireless networks. To make it more useful in those networks, the problem needs to be fixed. An active RIS was suggested as a way to fix these issues. One of the things that makes an active RIS unique is that it can fix phase shift and bring the weak signal back to its normal level of strength at the first hop. In this way, an active RIS may be able to avoid the twice route loss reduction. The tech inside an active RIS and an inactive RIS is very different. To make the signal stronger, active RISs use phase shift circuits and reflection amplifiers, which are also known as current inverting converters. It's no longer possible to ignore the power use, since the amps in active RISs and BSs may need about the same amount of power as those in low-power inactive RISs.

2.7 Noise and Power in RIS

RIS have thought that its noise power is very low since it doesn't make a lot of noise and works quietly. RIS, on the other hand, makes noise since most of the patch antennas inside the RIS unit cell are what send out signals. This makes sounds even when the RIS is not working. It is said that the Johnson–Nyquist noise does not put more noise power into any one system than what is theoretically possible (pn = kTpB). To find this, multiply the working bandwidth B by the Boltzmann constant k, which is equal to 1.3807 x 1023 JK−1, and the standard value of the thermal noise temperature *Tp*. As shown in equation (1), the antenna sends the SNR0 signal to the receiver. The input signal is *Si*, the mismatch loss is τ, and the noise temperatures of the antenna and receiver are *Ti* and *Ta*, respectively.

$$SNR_0 = \frac{Tn, St}{kB(TnrTa + (1 - nr)Tp + Tr} \tag{1}$$

Equation (1) shows that the signal power is greater than the internal noise plus kTpB if the RIS has enough gain and directivity. But the RIS's output signal-to-noise ratio, or SNR0, would go down if there was more noise power than that. This could be caused by mismatch losses, poor radiation efficiency, or other things. In fact, system noise makes each unit cell's noise power bigger. The noise power should be at least -54 dBm for a RIS with 10,000 unit cells and a frequency of 100 MHz. It gets hard to deal with the noise that the RIS makes once there are a certain number of unit cells. It starts to lower the channel capacity.

The RIS silently echoes the signal that comes in, so it doesn't need high-power parts like power amps and ADC/DACs. This means that it uses almost as little power to phase shift as more standard methods like huge MIMO and relays. So, it is expected that the RIS will greatly reduce the amount of power that will be needed for future 6G connections. As an example, it is thought that a RIS with 1600 1-bit unit cells would need 11.2 W of power to work. Also, more power is needed for beamforming design, channel estimation, channel feedback, and RIS rearrangement in order to support the RIS processor and send control messages for RIS beamforming. To make sure that future 6G networks use less energy, researchers need to look into RIS designs with low power and related communication methods. This is because the installation itself could not use very little power.

The 2019 study finds two easy ways to use computers to find the best RIS reflector values and BS send power distribution for both active and passive beamforming systems. These ways help solve the EE maximisation problem. The whole picture of how much power RIS-based MISO systems use was shown by these ways. Normal phase changers use a certain amount of power. The author said that the amount of power used by RIS depends on how many unit cells it has and how accurate each one is. The RIS's power use can be found by PRIS = NPN(b). They made a general model of how power was used so they could see how much power was used and how the different CSI reports and calculations were done. It wasn't very detailed about how the RIS's power usage model worked, though. Exact numbers are given for how much power RIS needs. The writers measured how much power different types of RIS use and found that they use different amounts. This will change the phase shift of each unit cell from 0 to π. The DC power source doesn't send out much energy when all the diodes are "OFF." This is 0.8 A for a 50×34 RIS. In the "ON" state, this shows that every unit cell uses 0.33 mW of power. The device that tells the RIS what to do actually uses energy as well. The number of control messages the controller sends and the way its circuits are set up affect how much power it needs. A controller made for an 8x32 small RIS that ran at 4.2 GHz used about 0.72 W of power. A controller made for a 50x34 large RIS that ran at 10.5 GHz used about 10 W. It is looked at how much power a new varactor diode-based RIS uses. The writers say that when the built-in RIS board is not being used, the Xilinx ZYNQ7100-based FPGA control board uses 1.5 W. The fact that this kind of RIS doesn't use a lot of power shows how useful it could be in gearbox systems in the future.

In the past, people thought that RIS would improve connection even more if the area or number of RIS unit cells went on forever. For some reason, though, as the area got bigger, the RIS used more power generally, which led to a higher RIS hardware power usage. So, the link between the number of RIS units, speed, and power use needs to be improved. The table below shows how much power the control circuit uses when the change technique is used.

2.8 RIS Systems and Channel Models

2.9 Reconfigurable Surface and Basic Functionalities

When a wireless signal, or electromagnetic wave in general, crosses the boundary between two isotropic mediums, in essence, Snell's law governs how the angles of incidence, reflection, and refraction relate to one another. Recent developments in this technology have made it possible to alter the impedance of the surface and induce a specific phase shift between incident and dispersed waves in reflectarrays with metasurfaces. When the surface is divided into many closely spaced components and each metasurface element is set up to have the appropriate phase shifts, the EM wave is tuned to a different angle instead of the symmetric reflecting wave based on Snell's law.

2.10 RIS Systems

The RIS panel reflects the transmitter's incident signal. With a programmable phase shift on each component, the RIS alters the channel environment from the transmitter to the receiver from the perspective of a wireless communication system. The resulting channel between the transmitter and receiver now consists of two parts: the direct channel from the emitter to the receiver without using the RIS, and the channel with the RIS interaction. Using appropriate designs for the RIS phase shifts, amplitude, or both, a specific metric target, such as the system achievable rate or the coverage, may be improved by changing the channel environment. This is fundamentally different from typical wireless communication research, where the design and optimisation choices are restricted to the pair of transceivers.

3. PRINCIPLE OF OPERATION OF RIS

Instead of being governed by the standard principles of reflection and diffraction, they are governed by the wavefront and phase of the radio waves that impact them. They could make it possible to modify the electromagnetic response of the surrounding environment in real time if they are applied to architectural features, walls, or facades. To increase the effectiveness of the received signal, the reconfigurable electromagnetic material sheets, or RISs, also closely monitor air propagation.

A PIN diode is one of the many cheap parts that make up a RIS. These parts may be able to control electromagnetic waves and change them in ways that aren't possible with naturally occurring materials. Furthermore, with little use, these "intelligent" surfaces may be intricately and sensitively phase-adjusted. Enhancing communication performance is the goal. RISs have an advantage over conventional reflect array/ transmit array antennas since each utility's characteristics, including the focal point's position and the direction of reflection or refraction, may be dynamically changed.

These surfaces produce reflection anomalies when included into the physical environment (such as walls), allowing for more cost-effective and energy-efficient regulation of the radio propagation channel. In addition to limiting radio interference and gradually lowering the necessary transmission power, this will enable the RIS to serve a user in an unprotected region and even enhance radio connection security against passive eavesdropping. The RIS was also suggested as a way to increase coverage and connection quality effectiveness.

As seen in Figure 6, when the surface is separated into several closely spaced components (supercells of scattering particles), Snell's law allows the EM wave to be modified to any angle other than the symmetric reflected wave. The phase changes of each metasurface element are shown as they should be. To produce a reflected beam at any angle, the phase shift of each metasurface element should ideally be adjusted to any value. Because of its many benefits, the RIS is made up of meta-surfaces that act as programmable reflectors. To create the reflected beam with incoming electromagnetic waves, the surface element's phase shift must be suitably or cleverly managed. The so-called reconfigurable intelligent surface may be built using machine learning-based methods or signal processing design methodologies. Implementing continuous or analogue phase transitions might be difficult. Quantised phase shift adjustable RISs have recently been developed, according to. If the signal coming in or going out is a plane, the following steps can be used to explain what RIS is all about:

(1) The process that causes a planar wave to deviate from its initial transmission route is called reflection or refraction;
(2) absorption, when a planar wave's amplitude is significantly diminished;
(3) focusing/collimation, which converts a spherical wave from a point source into a single plane wave or compresses a plane wave to a single point;
(4) shift in polarisation, whereby steps (1) or (3) often include converting the linear to circular polarisation of the incoming wave. The comparison of RIS performance attributes is shown in Table 5.

Table 5. RIS performance parameters comparison.

Performance Criteria	Values
Chance of an outage	3×10^{-3}, M = 100 3×10^{-1}, M = 1600
A steady rate	3.8, M = 100, K = 5 4.5, M = 1600, K = 20
Spectral performance of the network (bps/hz)	(T = 50 and t = K) 85, M = 100
Ergodic System Spectral efficiency (bps/hz)	(L = 0.5, K = 20) 65, M = 100
Variance of data rate	L = 0.5, K = 20; 0.2, M = 1600

Figure 5. Schematic of RIS system illustrating basic principle (Zhang et al., 2020).

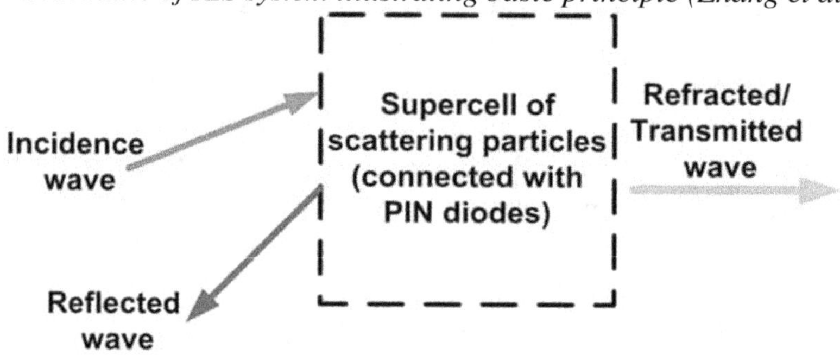

3.1. Limited Phase Shifts Problem and RIS Assisted Communication Model

The reason for this is that the wifi conditions change all the time. Things could go wrong and you could lose things you didn't expect. A lot of people around the world want RIS to help them figure this out. RIS is used to mirror the signal and make QoS better so that delivery can happen in the best way possible. Figure 6 shows a cell phone transfer network with M–N RIS parts that are handled electrically. The user's straight link to the base station (BS) has been broken.

Figure 6. RIS assisted communication model for uplink cellular network.

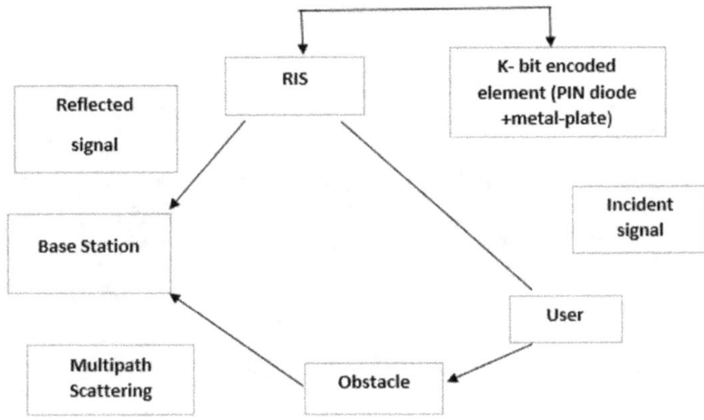

4. MEASUREMENT TECHNIQUE OF OUR PROPOSED RIS

This way of measuring is shown in Figure 7, which also shows how the different factors were seen. Figure 5 shows that the 10x10 combination array, the 10x10 circular path array with a ring, or the 10x10 circle type patch array can be put in the middle of the main material that absorbs light. Between the main absorber and the ones on the left and right, there were a lot more. Another thing was that there were pillows on the ground. The goal of adding more screens is to cut down on both the radiation that escapes from outside and the radiation that bounces off the walls of the room. Two standard 3.5 GHz horn antennas are shown in Figure 5; these were used to figure out the different parts of the 10x10 grids. To find out how reflective the planned arrays were, links were made between the antennas and the vector network analyser (VNA). To find the reflection coefficients of three types of 10x10 arrays, the first thing that had to be done was to tilt the transmitting horn antenna 10 degrees. After that, the received strengths were checked with a similar horn antenna set at a -10-degree angle.

Figure 7. Measurement setup to measure different parameters.

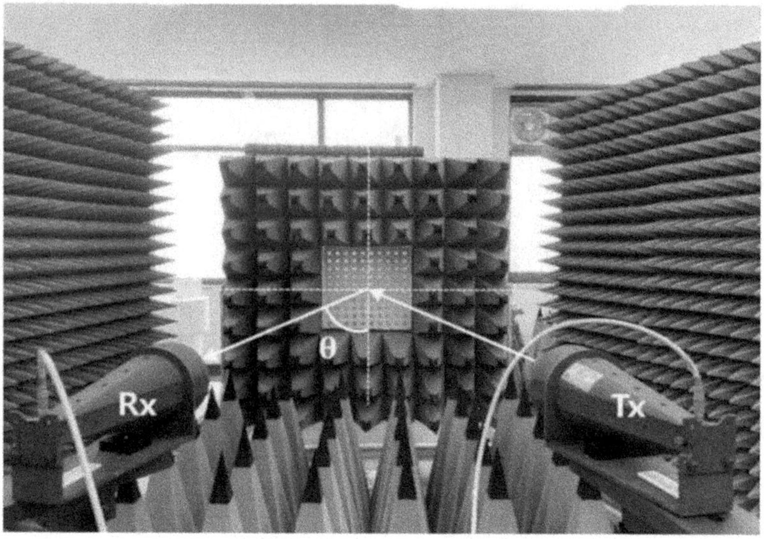

The layout we came up with is shown next to the RIS that is already in the books in Table 6. The table shows that there are some good things about the method we showed. There are no PIN or varactor diodes in the setup we showed you. The structure can be turned into a different shape by cutting turning it. Our plan is made up of just one layer, which is held together by the copper plate. Because of this, it's very simple to make and is also very cheap. Besides that, our building already has a complicated outside monitor, so we don't need one.

Table 6. A literature-based comparison of RIS kinds.

Nature of Unit Cell	Year	Tunning Method	Freq.	Comments
D-type patch	2022	Diode shifter	2.4 GHz	It costs a lot to make and needs an expensive external control board.
Normal patch	2021	Diode shifter	4.7 GHz	It costs a lot to make and needs an expensive external control board.
Patch	2020	Pin diode	1.2 GHz, 17..4 GHz	Structure is very complicated, and it needs an outside manager.
Patch	2021	Pin diode	17.4 GHz	Structure is very complicated, and it needs an outside manager.
Joined E-shaped	2022	Pin diode	4.3 GHz	Need a complicated outside tool

4.1 Channel Estimation in RIS-Powered Systems

To deal with electric fields in a smart way, you need the right channel state information (CSI). In order for RIS to work, CE has to guess a lot of channels at the same time. One user's equipment (UE) gets the straight channel from the base station (BS). The RIS is also linked to the BS and every UE. And there's the fact that the RIS channel prediction isn't always right. An awful lot of guessing channels have been added to the scale, and readings have been shared. In the near field, this makes it hard to guess the routes. Even so, it's tougher to stick to this plan when many of the spread RISs have hardware issues that don't happen in a straight line. Sometimes it's hard to figure out the channel because of bugs in the hardware, and beam squint can happen in broad channels. In many books and papers, people have written and thought about these things. Also, different people have come up with different ways to estimate RIS paths.

The BS-RIS and RIS-UE lines may have different specs when they are used for different tasks. The RIS being on the wall makes this clear. Half of the time, the BS-RIS line stays the same. It takes longer to join properly, and the LOS path uses up most of the energy. In other words, planning channels can last longer. Users' equipment (UE) can move around with RIS-UE, which makes it go much faster. The channel estimate doesn't need to be done as often as the BS-RIS channel estimate, though. RIS could also be used on the fast train windows. The RIS-UE channel doesn't change much, but the BS-RIS channel does. In this case, it's hard to figure out how to do high-dimensional BS-RIS fast-changing channel guess. UAVs and buses are two examples of things that can move and have RIS put on them. BS-RIS and RIS-UE don't have much time to get in sync when these things are moving. We need to know right away how much both lines cost.

A lot of study papers talk about how to set up CE for wireless systems that are built on RIS. The use of paired bilinear sparse matrix factorisation and matrix completion was found to be a general way to find channels in MIMO systems that use RIS. In a MISO setting, a way to manage that turns on some RIS parts was talked about. To find a straight square, you can use the channel vectors that this method changes. The first time the plan was used was in a two-timescale channel guess plan. This was possible because there is some noise on the BS-RIS line. It takes a long time to guess the high-dimensional BS-RIS channel. Several low-dimensional RIS-UE CEs can be done at the same time with a single BS-RIS channel estimate. There are two base nodes close to the RIS to help people figure out how to get from UE to BS to RIS. As part of this plan, two prediction models for channels that work together have been given. Many people think it's good that everyone shares the BS-RIS channel. One way to get a feel for a part of the BS-RIS channel during the coherence time is to train and get comments from speakers. Users' test signals are used more often after that to learn more about how the cascaded channel works. You need to find the important routes of anchor points Al and A2 before you can use the second way.

RIS-based OFDM devices can now talk to each other for RIS CE, and they can set up quickly. The RIS mirror picture that was talked about, on the other hand, needs to be turned on and off, which can really cost a lot. Plus, the guess isn't quite right because only some parts of the RIS are working during the CE time. We now know of an interesting new way to send data that makes use of signs that show up almost every so often. The tools quietly figure out the beam's direction and shape it at the same time. There are several ways to use this process. You can change the time between training symbols that are next to each other and put RIS parts together. This way, you can send data as quickly as possible while also taking as little time as possible. This will work if your message doesn't like to wait. When RIS is used, the channel size grows a great deal. This is why it's important to quickly try a channel predictions and feedback system that doesn't cost a lot of money.

When making the real RIS channel estimate method, this is one of the most important things to keep in mind. One way to cut down on the pilot is to put RIS parts together. Each group of RIS elements should be made up of close elements that all have the same return function. This is a new way to group elements. The idea was put forward this way since every group is different. It takes a lot less work to teach because only the main route of each group needs to be found. The same thing happened with a new hierarchical teaching mirror system. Putting RIS pieces in groups and then separating them is one way to get a better picture of the RIS-enabled lines that go through several blocks. The test could also go faster if the guess was based only on the channel from the base station to all users, rather than each channel coefficient separately. It is the instant combined CSI that gives the two-timescale system its BS beamforming. Test sequences only need to last as long as the number of users, not the number of RIS components. Test runs should last longer than the number of people who are there.

5. SYSTEM-LEVEL CONFIGURATION

This part will talk about the setup that was picked for the C-RAN system-level test. It will also go into detail about the test cases that will be looked at. The C-RAN splits the network into three TRP sets that are close to each other using the static clustering method. Because of this, each user has at least one TRP, and the other TRPs stop users from talking to each other. When it comes to setting up, the C-RAN is the same as any other smart phone. There are three TRP sets next to each other. When the cluster size is three, three TRPs next to each other take care of each person. That's because of the other groups. People don't understand each other.

5.1. Link-Level Simulations

One of the things we did was compare how well the chosen scheme, which includes the shared design of the precoder and phase-shifts of the RIS parts, worked with a MIMO–OFDM link to other MIMO–OFDM systems. Based on the model of the system we just talked about, computer games called Monte Carlo were played. There is a space between the driver and the receiver. This tells us how well the bit error rate (BER) works. In Figure 5, the x-axis shows how far away the generator (Tx) and the receiver (Rx) are from each other along the y-axis. The amount of power that Puser sends is 10 dBm, which is what we choose to compare it to. For each of the five possible sets ups, we can see a different shape. What numbers did you use to do this maths? Ntx is 64, Ns is 3, Nrx is 16, Nris is 144, which is 576 parts, and Nc is 1. Between the speaker and the user, there should be both direct and indirect links. The shapes you see are made with the AM-SVD-APG method that we talked about. When there is less space between the sender and receiver, the BER numbers need to go down at the same speed. Putting these two things this far apart makes the opposite happen. It gets to a certain point where the received signal strength is so weak that it leads to more than 0.001 mistakes.

5.2. Simulations at the System Level

We used $SNV=(\boldsymbol{Q}s/N0)+10\boldsymbol{TU}b(Vs/N)$ dB to get the signal-to-noise ratio (SNR) at the system level. Rs tells you how many symbols were sent by each receiver and user as a whole. Broadband as a whole is given by B. No. 0 smart city got 50 MHz at 3.6 GHz and 28 GHz, and No. 3 smart cities got the other frequencies. $Ms/N0$ is the ratio of the energy of the symbols to the noise in the spectrum. This is what the link-level BER study taught us about $\boldsymbol{Q}s/N0$. All of the QPSK changed symbols were talked about in this paper. FDD stands for frequency division duplex, which is one type of frame layout used in 5G New Radio (NR). TDD stands for time division duplex (Liaskos et al., 2018). The FDD is used when there are two messages. The TDD is used when there are no signs. For this job, we decided to use TDD all the time.

The most symbols that can fit in each spot, the transmission time, and the cycle prefix (CP) are all set by this new 5G NR scalable OFDM counts. In some of our tests, we used the numbers 0 and 3. For the smart city, the number 0 (4G) was chosen because the most important KPI is the growth of Internet of Things devices, not their speed. Number 3 was chosen for the stadium with the field and the plant inside in the mmWave and THz bands.

In the smart city case, various UPA antenna arrays on the receiver side (BS) were studied, including ones with Nrx = 144, 256, and even 576 elements. Ntx is 1, Ns is 1, and Nc is 1 on the back of the Internet of Things device. What does $\Delta f = 15$

kilohertz and Bprb = 180 kilohertz mean in terms of numerology? It means that each PRB has 12 bands. That's 168 subframes, or 12 times 14 letters. Every one of them lasts one instant. It's not always the same in real cell phone networks that use big MIMO. The channel doesn't change much at this point, so pilot reference symbols need to be sent again. This is what a regularity gap looks like:

$$\tau_c = T_C \times B_C$$

For low-power Internet of Things sensors that move around and stay in one place, Tc = 50 ms is a good coherence time. This band width, 1.697 MHz (9.42857 × 180 kHz), works well for a big city like a smart city. Adding 50 to 168 gives you Nsc, which is 79,200. This is the largest number of subcarriers that can be in τc. That's 9.42857. One half of the coherence gap can be used to send data upstream. You can use the other half going downstream. In the uplink, we need orthogonal reference pilots. These are in the bit patterns that are sent from the cable. Nsc/4 must be equal to 79,200/4. In other words, we need 19,800 subcarriers for the data and reference planes that are coming in. When we divide 19,800 into twelve groups, we get 1,650 subcarriers. This is because there are three areas and four base spots. Each of these subcarriers is an IoT device that can send in all four areas at the same time without messing up the cell. The extra choices for the smart city case can be seen in Table 7. In real life, each place may have a lot more Internet of Things gadgets than what is shown here. This is because people don't use these things very often. Things that know what they're doing don't have to send packets all the time.

Table 7. Extra smart city simulation settings.

Parameter	**Value**
Coherence time Tc	40 ms
Coherence band Bc	0.586MHz
Maximum time Tp	01.4 ms
Maximum N_{sc}	68,100
Max. active IoT devices $N_{sc}/4$	08,700
Number of sectors	01
Number of IoT devices/sector	0540

In the case of a smart city, we are only interested in the upload. This is what connects the objects on the Internet of Things to both the BS and the RIS. A lot of Internet of Things devices receive the same amount of power when link power management is done right. You might not get the data sent by Internet of Things devices that are too far away from the sensors. These devices are called BSs and/or RISs. Some of the power coming in must be stronger than the receiver's sensitivity for files to be received correctly. If the BER is less than or equal to 5 × 10–4, it means that there are no bit mistakes in the packets that were sent. There are 25.000 hertz in Δf, and in a 5G network, 168 subcarriers are sent in every PRB for one millisecond. Every Internet of Things device must use a single carrier and QPSK codes to talk to each other. Each device must be able to send and receive at least 2 kilobits per second (Kbps) and have a range of at least 2/15 bits per second per hertz (bps/Hz). The most output can be found with the help of a scientific equation. If power control is perfect and there is only one cell that doesn't get any help from other cells, then this equation makes sense.

$$Th_{max} = 0.5 \times B \times (1 - \tau_p) \times \log_2(1 + SINR) \quad (2)$$

In this case, ϱp is the time it takes to send a common way to measure things.

After that, we put the stadium through its paces with a pitch scenario test that included a number of different factors. We decided to add more subcarriers (shown by Nc), more spatial streams, and higher broadcast powers because of this. Every ms, it sends 112 characters, and the number for 5G NR is three. 400 MHz is the band width. F is the frequency of the frequency, which is 120 kilohertz. It is possible to use up to 3,300 subcarriers. Throughput, which is the number of contacts per unit of area, is the major key performance indicator (KPI) in this case. On the transmission side, Ntx = 64 is always set on the base station (BS), and on the listener side, Nrx = 16 is always set on the user. Two different numbers of symbols sent per subcarrier—Ns = 2 and Ns = 10—were used in the tests. There are several ranges for the amount of OFDM subcarriers (Nc) that can be used in the tests. Twelve, sixty, one hundred, eighty, one hundred and twenty-four are some of these numbers. When you add these up, you get 12, which is how many subcarriers a PRB has in 5G NR.

A trend has been found for putting people near TRPs that are close to the four corners of the football field. For more information, see Figure 4 and Table 2. About 46% of the people are sitting on the grass, and 54% are sitting in the stands of the field. There are twelve sections, four base stations (BSs), and no RIS in this set-up. This is a normal cell phone setup. In this case, the users are pretty spread out within a 57-meter circle (see Table 2 for more details). If both BSs and RISs are active at the same time, however, people connected to BSs will be fairly spread out in a circle 34 meters in diameter. There are 17-meter circles with RIS screens spread out evenly across the scene. Inside those circles, RIS users are also spread out evenly.

Figure 9, which shows the overall flow for each area, again takes into account what's going on in the stadium. Each user's number of PRBs and the highest broadcast power of 24 dBm are also taken into account. It doesn't matter how many PRBs different users are given; 3300 is always the most subcarriers that can be used in a sector. We only look at a certain number of PRBs to figure out the sector flow:

one PRB, five PRBs, ten PRBs, fifteen PRBs, and twenty PRBs. Every 12 users generates 12 PRBs. Every 5 users generates 60 PRBs. Every 10 users generates 120 PRBs. Every 15 users generates 180 PRBs. And every 20 users generates 240 subcarriers. All of these numbers come from the fact that every person has a certain number of subcarriers. Three thousand twelve hundred, or 275, is the number of people in each area if everyone has their own PRB. The pitch has 12 parts, which means that 3300 people use it all the time. For every area, there are 3,300 people, or 13.75 people. As a whole, 165 people are using the field. This is based on the idea that everyone has 20 PRBs right now. Take a look at Figure 8. The BS + RIS curve also has a top capacity. This is something that is very clear to us. The black curve (BS + RIS) shows that the rate goes down for both 5 and 10 PRBs for each person. The red curve shows that the rate stays the same. When there are 20 PRBs instead of 15, the flow slows down. But as the number of PRBs went up, the total flow per area went down by a very small amount. For each area, the total flow for the BS + RIS slope is about the same. Users send bigger packets when they send more PRBs. This makes it more likely that the packets will go through deep fading. Another thing is that the C-RAN's spread of RIS and BS links is enough to handle the predicted drop in flow in this case.

Figure 8. Stadium scenario maximum load sector throughput.

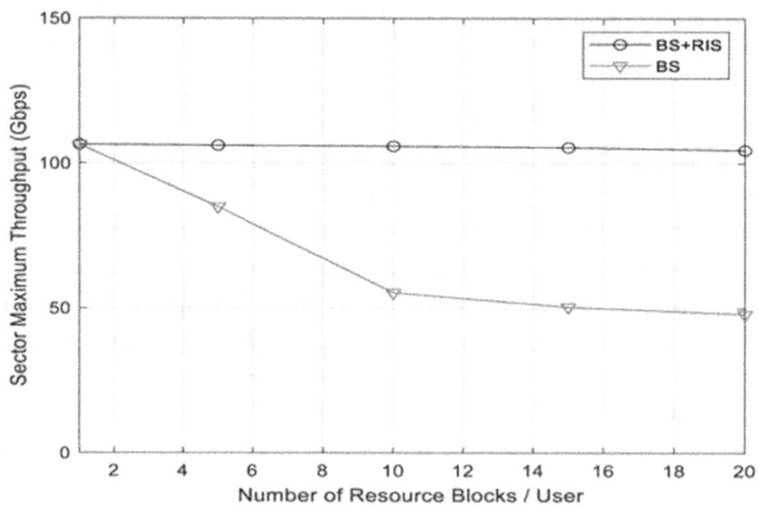

One of the three cases that were looked at has a higher frequency band in the range of less than THz. Only one of the other two does not. As the frequency drops below 100 GHz, we are able to use more bits at the RIS (NRIS = 576-2304). BS can only send 30 mW of power, which is 14.8 dB per metre. There are a good number of users spread out over two areas. They are on the TRPs, which have UPA antennas for each area. You can see how far these towers are from the house's walls by looking at Figure 5 and Table 3. In the Factory, there was a lot of noise because the BS antenna height was both low and high. It's set to 400 MHz, the 5G NR number stays at 3, and 112 characters are sent every millisecond to look at this case. It's like what took place at the stadium. The two cases can be directly compared, even though they take place in various locations and frequency ranges. First, there is a sending TRP and a UPA on the user's side with Ntx = 256 antennae and Nrx = 16. Nothing else is changed. Several different amounts of characters sent per subcarrier were talked about while the tests were being done. Finally, Ns = 3 was chosen. When the RIS screens are not being used, the plant only has two BSs. There are eight spots in all, since each of these BSs has three. In a 70-meter space, Table 3 shows that the users are pretty evenly spread out. We can then see how performance changes when RISs and BSs are both running at the same time. In a fifty-meter circle, users who are linked to BSs will be spread out evenly. The sixteen RIS screens are 25 meters apart and spread out evenly across the scene. There are a lot of people in a group who can use these screens.

To see the difference, look at Figure 13. It can be seen that the total output of each plant area is half of what the stadium can do. This might be because there are only half as many BSs in the plant as there are in the stadium. In the plant case, we can increase the total output in two ways: (1) we can add more BSs, or (2) we can combine the carriers. We have a choice between these two things. It is true that the 5G New Radio standard lets 16 bands, each with 400 MHz, work together. The whole 6.4 GHz band could be used if 100 GHz is used. Because of the first choice, the network needs more resources, but this method can't be used to save power. We need to spread out RIS surfaces around the scenario's BSs and group carriers together to get high sector flow and energy economy.

6. CONCLUSION

A lot of people believe that RISs are important for creating B5G, 6G, and other future wireless networks. As a way to make future networks use less energy, this article talks about OFDM and MIMO lines, which stand for multiple inputs and outputs. RIS screens will be put in place of base station panels as part of this plan. There are times when orthogonal multiple access lets more than one person talk to each other. Crosstalk can't happen in some rooms because of this.

We've talked about how to improve a smart radio and set it up. One thing we looked at was how many busy IoT devices there would be in a smart city, especially in the 3.6 GHz and 28 GHz bands. In this case, the coherence interval tells us how many busy people a smart city must have at the very least. This is the case because the gearbox needs reference guides that are not parallel to each other. This study looked at the 19,800 Internet of Things gadgets that were sending data. Every device had its own subcarrier for every 1-ms segment. If you use a C-RAN, you can cut down on passenger pollution between parts. It works this way. When IP devices were close to cell lines in this case, they worked better when they had RIS parts that worked at 28 GHz. This was still true even though they moved more slowly and used less power. It was possible to take care of about 7% more people.

We tried other smart cases and looked at how coverage and output got better in those. Two of these were the indoor plant at 100 GHz and the outdoor/indoor stadium with pitch scene at 28 GHz.

The busy time in the stadium was found to be when 3,300 people were there. While the game was going on, each of them sent one PRB with twelve subcarriers. We saw pretty much the same thing happen both times when we raised the power in the BS gearbox from 24 dBm to 35 dBm. Even though it takes more work to handle, C-RAN cut down on the amount of crosstalk between cells in different areas. When RIS was put in place at the cell ends of the BS on the football pitch and in the stands, coverage and flow went up by 97% to 110%. The speed and reach boosts, on the other hand, depend on how many PRBs each person gets. The gains go up as people get more PRBs.

It was thought that each user would get one PRB. This meant that there could be up to 1650 active users in the office. People looked at how well RIS parts with numbers between 576 and 2304 worked and found that the changes were not very noticeable. It became more clear that RIS did better with 2304 pieces as the number of PRBs per person went up. Based on how many PRBs are given to each person, coverage and yield go up by 29% to 50%. To get a better signal, the BS antennas should be put higher than the foliage height of the plant. It is possible for up to sixteen carriers to work together, which means that more data could be sent at 100 GHz. Each carrier would have a frequency of 100 MHz.

As part of our next project, we will use system-level models to check the speed and scope of our work. This is what we are going to do with brand-new semi-deterministic path loss models. These models can correctly guess path loss at 300 GHz (THz band) when there are more obstacles and plants that aren't set up in a straight line.

REFERENCES

Basar, E., Di Renzo, M., De Rosny, J., Debbah, M., Alouini, M.-S., & Zhang, R. (2019). Wireless communications through reconfigurable intelligent surfaces. *IEEE Access : Practical Innovations, Open Solutions*, 7, 116753–116773. DOI: 10.1109/ACCESS.2019.2935192

Danufane, F. H., Di Renzo, M., De Rosny, J., & Tretyakov, S. (2021). On the path-loss of reconfigurable intelligent surfaces: An approach based on green's theorem applied to vector fields. *IEEE Transactions on Communications*, 69(8), 1–1. DOI: 10.1109/TCOMM.2021.3081452

Di Renzo, M., Debbah, M., Phan-Huy, D.-T., Zappone, A., Alouini, M.-S., Yuen, C., Sciancalepore, V., Alexandropoulos, G. C., Hoydis, J., & Gacanin, H.. (2019). Smart radio environments empowered by reconfigurable ai meta-surfaces: An idea whose time has come. *EURASIP Journal on Wireless Communications and Networking*, 2019(1), 1–20. DOI: 10.1186/s13638-019-1438-9

Di Renzo, M., Zappone, A., Debbah, M., Alouini, M.-S., Yuen, C., de Rosny, J., & Tretyakov, S. (2020). Smart radio environments empowered by reconfigurable intelligent surfaces: How it works, state of research, and the road ahead. *IEEE Journal on Selected Areas in Communications*, 38(11), 2450–2525. DOI: 10.1109/JSAC.2020.3007211

Farrell, S., LPWAN Overview. Internet Draft (Work in Progress), Draft Name: Draft-ietf-lpwan-overview-10, 2018. DOI: 10.1109/JSAC.2020.3007211

Flamini R, De Donno D, Gambini J, et al. Toward a heterogeneous smart electromagnetic environment for millimeter-wave communications: an industrial viewpoint. IEEE Trans Antennas Propag. 2022 Oct;70:8898–8910. DOI: 10.1109/JSAC.2020.3007211

Gams, M., Yu-Hua Gu, I., Härmä, A., Muñoz, A., & Tam, V. (2019). Artificial intelligence and ambient intelligence. *Journal of Ambient Intelligence and Smart Environments*, 11(1), 71–86. DOI: 10.3233/AIS-180508

He, M., Xu, W., & Zhao, C. "RIS-assisted broad coverage for mmWave massive MIMO system", Proc. IEEE Int. Conf. Commun. Workshops (ICC Workshops), pp. 1-6, 2021. DOI: 10.1109/ICCWorkshops50388.2021.9473674

Huang, C., Hu, S., Alexandropoulos, G. C., Zappone, A., Yuen, C., Zhang, R., Renzo, M. D., & Debbah, M. (2020, October). Holographic MIMO surfaces for 6G wireless networks: Opportunities challenges and trends. *IEEE Wireless Communications*, 27(5), 118–125. DOI: 10.1109/MWC.001.1900534

Jiang, M., Cezanne, J., Sampath, A., Shental, O., Wu, Q., Koymen, O., Bedewy, A., & Li, J. (2022). Wireless fronthaul for 5G and future radio access networks: Challenges and enabling technologies. *IEEE Wireless Communications*, 29(2), 108–114. DOI: 10.1109/MWC.003.2100482

Karami, A. B., Fleury, A., Boonaert, J., & Lecoeuche, S. (2016). User in the loop: Adaptive smart homes exploiting user feedback – state of the art and future directions. *Information (Basel)*, 7(2), 35. DOI: 10.3390/info7020035

Lam, K.-W.. "Implementation of transmitarray antenna concept by using aperture-coupled microstrip patches." *Proceedings of 1997 Asia-Pacific Microwave Conference*. Vol. 1. IEEE, 1997.

Liaskos, C., Nie, S., Tsioliaridou, A., Pitsillides, A., Ioannidis, S., & Akyildiz, I. F. (2018). A New Wireless Communication Paradigm Through Software-Controlled Metasurfaces. *IEEE Communications Magazine*, 56(9), 162–169. DOI: 10.1109/MCOM.2018.1700659

Liaskos, C., Tsioliaridou, A., Pitsillides, A., Ioannidis, S., & Akyildiz, I. F. (2018). Using any surface to realize a new paradigm for wireless communications. *Communications of the ACM*, 61(11), 30–33. DOI: 10.1145/3192336

Mwase, C., Jin, Y., Westerlund, T., Tenhunen, H., & Zou, Z. (2022, June 1). Communication-efficient distributed AI strategies for the IoT edge. *Future Generation Computer Systems*, 131, 292–308. DOI: 10.1016/j.future.2022.01.013

Pavia, J. P., Velez, V., Ferreira, R., Souto, N., Ribeiro, M., Silva, J., & Dinis, R. (2021). Low Complexity Hybrid Precoding Designs for Multiuser mmWave/THz Ultra Massive MIMO Systems. *Sensors (Basel)*, 21(18), 6054. DOI: 10.3390/s21186054 PMID: 34577260

Renzo, Marco di et al. "Smart Radio Environments Empowered by Reconfigurable Intelligent Surfaces: How it Works, State of Research, and Road Ahead." *ArXiv* abs/2004.09352 (2020): n. pag. DOI: 10.3390/s21186054 PMID: 34577260

Subrt, L., & Pechac, P. (2012). Intelligent Walls as Autonomous Parts of Smart Indoor Environments. *IET Communications*, 6(8), 1004–1010. DOI: 10.1049/iet-com.2010.0544

Ullah Khan, W., Sheemar, C. K., Abdullah, Z., Lagunas, E., & Chatzinotas, S. (2024). Beyond Diagonal IRS Assisted Ultra Massive THz Systems: A Low Resolution Approach. arXiv e-prints, arXiv-2406. DOI: 10.3390/s21186054 PMID: 34577260

Wang, Y., Li, C., Liu, C., Liu, S., Lei, Y., Zhang, J., Zhang, Y., & Guo, Y. (2021). Advancing DSP into HPC, AI, and beyond: Challenges, mechanisms, and future directions. *CCF Transactions on High Performance Computing*, 3(1), 114–125. DOI: 10.1007/s42514-020-00057-2

Wu, Q., & Zhang, R. (2020, January). Towards smart and reconfigurable environment: Intelligent reflecting surface aided wireless network. *IEEE Communications Magazine*, 58(1), 106–112. DOI: 10.1109/MCOM.001.1900107

Wu, Q, Zhang, S, & Zheng, B Intelligent reflecting surface-aided wireless communications: a tutorial. IEEE Trans Commun. 2021 May;69(open in a new window):3313–3351DOI: 10.3390/s21186054 PMID: 34577260

Yu, N., Genevet, P., Kats, M. A., Aieta, F., Tetienne, J.-P., Capasso, F., & Gaburro, Z. (2011). Light propagation with phase discontinuities: Generalized laws of reflection and refraction. *Science*, 334(6054), 333–337. DOI: 10.1126/science.1210713 PMID: 21885733

Zhang, J, Björnson, E, & Matthaiou, M Prospective multiple antenna technologies for beyond 5G. IEEE J Sel Areas Commun. 2020 Aug;38(open in a new window):1637–1660. DOI: 10.3390/s21186054 PMID: 34577260

Zhang, L., Wang, Y., Tao, W., Jia, Z., Song, T., & Pan, C. (2020). Intelligent Reflecting Surface Aided MIMO Cognitive Radio Systems. *IEEE Transactions on Vehicular Technology*, 69(10), 11445–11457. DOI: 10.1109/TVT.2020.3011308

Zheng, C. Y., & Zhang, R. "Uplink channel estimation for double-IRS assisted multi-user MIMO", *Proc. IEEE Int. Conf. Commun.*, pp. 1-6, Jun. 2021. DOI: 10.1109/ICC42927.2021.9501057

Chapter 5
Reconfigurable Intelligent Surface-Assisted Backscatter Communication for Enabling 6G IoT Networks

Rajiv Iyer
https://orcid.org/0000-0002-1136-2061
Amity University, Mumbai, India

A. Saranya
Amity University, Maharashtra, India

Vedprakash Maralapalle
https://orcid.org/0000-0002-7770-1134
Amity University, Maharashtra, India

Dipak Raskar
Amity University, Maharashtra, India

ABSTRACT

As such, backscatter communication with RIS's support is the suitable technology for using in 6G IoT networks, according to the efficiency, reliability, security, and low energy consumption. The analysis indicates that the integration of the RIS with backscatter communication enables achieving additional enhancements of up to 2.5 dB in the channel gain and up to 17% in the data rate as well as up to 3 dBm/Hz in energy efficiency. In particular, this book chapter will be devoted to an introduction

DOI: 10.4018/979-8-3693-8099-4.ch005

to RIS-assisted backscatter communication, as well as to the elaboration of certain technical principles of this system, the definition of its possibilities in the context of improving large-scale IoT networks for the sixth generation. This chapter will also contain information concerning these problems, directions of further research, and potential applications of this technology.

1. INTRODUCTION TO RECONFIGURABLE INTELLIGENT SURFACES (RIS) AND BACKSCATTER COMMUNICATION

In Wireless Networking as the globe moves from 5G to 6G, some of the new trends that may be worth the effort include Reconfigurable Intelligent Surfaces (RIS) (Basharat, 2021). An RIS can be defined as a two-dimensional structure comprised of thousands of unit cells that can regulate EM waves by reflecting, refracting, or absorbing them. This capability enables the control of wireless signals to optimize signal quality, signal range, and wireless link utilization. Hence, the assumption on which the use of RIS is based on its capacity to change the propagation path of the wireless signals in a network. On the contrary, traditional recipes tend to consider the radio channel as an open environment, while RIS has the chance to turn into an interface to be interposed between the transmitters and the receivers. This helps in directing the signals to the desired directions hence improving its reception and reducing interferences where obstacles were observed to have negative effects on reception. RIS and backscatter communication are probably some of the more promising paradigms for the next generation of wireless communication. In Fig 1 the RIS is a technique to manage and ameliorate the wireless scenario, and the backscatter transmission is one of the efficient modes of transmission in terms of energy performance. Combined they can afford great possibilities to change the rhythm and organization of an increasingly wireless people connection and networking (Zhang, 2021). For more research work, the focal areas will also include areas of standardization, integration, and signal processing as the vital components that can determine the proper utilization of such technologies. The capability to control or reconfigure the environment through which the radio signal was being transmitted or could be transmitted to be more efficient, more reliable, and significantly more sustainable was the future wireless systems.

Figure 1. Reconfigurable intelligence surface structure

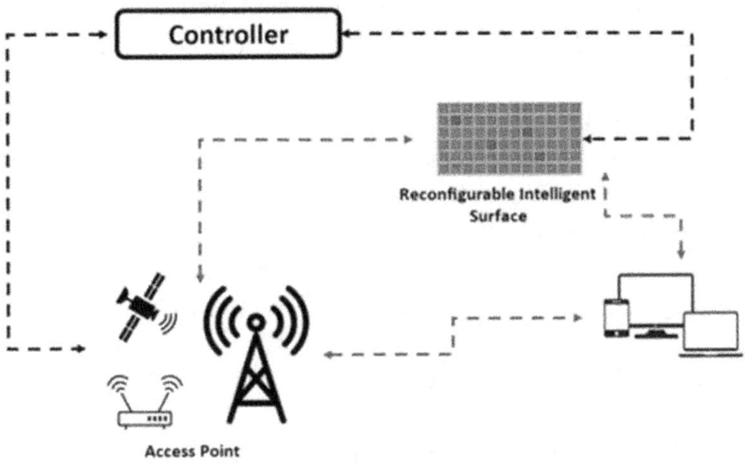

1.1 Structure and Functionality

A RIS consists of numerous passive subordinate elements known as meta-cells or unit-cells that can regulate the phase and amplitude of the reflected waves.

a. **Reflect Array Approach:** The method also uses phase shifters for each element as described for conventional phased arrays. In the region, the reflect array can also change the phase of the reflected signals with the help of a coherent signal-combining technique (Liaskos, 2021).
b. **Meta Surface Approach:** This was done by using thinner and denser structural members that offer additional freedom regarding geometries. Meta surfaces may change the direction of EM waves more systematically than direct reflection like data modulation and mathematical manipulation.

1.1.1 Applications of RIS

The uses of RIS are numerous and interesting and for this reason, RIS will play a fundamental role in the progression of future wireless communications. The applications of RIS are vast and varied, making it a key player in the future of wireless communications:

a. **Enhanced Coverage and Capacity:** That is the main reason why most of the time it is almost impossible to send a signal where it is supposed to be for other systems, RIS leads to improvement of coverage and capacity in urban environments.
b. **Energy Efficiency:** RIS can therefore operate in a passive mode where these nodes are energized by energy extracted from the received signals and this makes wireless networks power-conscious.
c. **Support for Advanced Technologies:** It is therefore expected that RIS will again be applied to the new area of WPT, ambient backscattering, and better positioning system and as such a useful tool in the wireless world.

1.2 Backscatter Communication

The novel technique is backscatter which allows devices to not emit their radio signals but rather reflect other signals (Liang, 2022). This is particularly helpful in those circumstances in which ascent power exists to perform regular communication means. Since the strength of the signals is reflected the backscatter devices allow transfer of information while utilizing lower power. Backscatter device on the other hand operates through RF signals from a base station or any other transmitter. The backscatter device phases and amplifies the received signal by selectively altering the reflecting characteristics of the surface. This can be done in several ways some of which include changing the load impedance characteristic of the device which in effect determines how the signal is echoed back to the receiver. The echoed signal includes all details of the backscatter gadget and allows the system to send information without an inner power source as well as without using standard ways of interacting.

1.2.1 Applications of Backscatter Communication

Backscatter communication has several promising applications:

a. **IoT Device Connectivity**: Backscatter Communication is well designed mainly for low-power IoT devices that need a small power for their operation. These devices include antennae reflecting and modulating similar to existing radio frequency (RF) signals and converse without creating their signals, thereby cutting down drastically on power usage. This is important for smart home devices, environment sensors, and wearable technology devices we have today.
b. **Smart Agriculture**: Specifically in agriculture BackCom may be used in establishing a network of several sensors that measure aspects such as soil moisture, temperature, and crop status (Saiam, 2023). Sustainable and easy to maintain,

these sensors can 'talk' to a central system and do not require batteries to relay collected data.

c. **Environmental Monitoring**: Thus, it is easy to utilize Backscatter Communication systems in environmental monitoring applications with sensors placed in remote areas. These sensors use atmospheric RF signals from active base stations in current networks to send vital information on climate, water levels, and wildlife tracking, among others, without needing to be connected to fixed power sources.
d. **Smart Cities**: In smart city-related works, BackCom can be employed in the fields of traffic management, waste disposal, public safety systems, and more. Due to the connectivity of the numerous low-power devices, efficient data acquisition and processing in real-time support urban planning and management and resource usage can be achieved.
e. **Non-Orthogonal Multiple Access (NOMA) Systems**: Backscatter Communication can be combined with NOMA approaches so that multiple devices occupy the same communication channel at once. This jointly improves the total network throughput and energy consumption and is suitable for high-density IoT practical application scenarios where many devices require bidirectional communication (Zhao, 2023).
f. **Ambient Backscatter Communication Systems (AmBCS)**: AmBCS uses virtually any available RF signals from multiple sources; for instance, Wi-Fi signals or even cellular signals. This is especially advantageous for areas where it is both hard or expensive to install specific communication infrastructure. This makes it possible for devices to work under extremely low power conditions but still be connected.
g. **RFID and Supply Chain Management**: Technological advancement of backscatter has developed from simple RFID uses in many complex identifying systems that allow two-way operation. These new enhancements enable inventory management in supply chains, thus enhancing operations and decreasing costs.

1.3 Challenges and Future Directions

Both RIS and Backscatter Communication face challenges that must be addressed for successful implementation in 6G networks:

a. **Complexity in Design and Deployment**: The integration of RIS into existing infrastructures needs to be planned and designed to cooperate cohesively and create the optimum impact.
b. **Interference Management**: Interference becomes increasingly important with the increased use of backscatter communication to make communications most effective.

c. **Standardization Efforts**: For these technologies to go mainstream, it was necessary to define standard norms for implementing them for use.

Reconfigurable Intelligent Surfaces and Backscatter Communication are novel additions to the progressions in wireless communications technology (Jian, 2022). Both of these companies' current and future abilities will arguably play a major role in addressing the problems arising from the rising demand for connection in the emerging 6G world.

2. PRINCIPLES OF RIS-ASSISTED BACKSCATTER COMMUNICATION

In wireless communication technology, changes in communications have provoked the concept of establishing new and enhanced methods of connectivity, reachability, and efficiency. Such Reconfigurable Intelligent Surfaces (RIS) and backscatter communication have been developed as some of the potential solutions. Backscatter is also applicable in that it gives RIS the ability to actively control the wireless signals while backscatter communication is a low-power approach through which devices can send data by simply reflecting other signals which might be broadcasted in their environment (Saiam, Chowdhury, & Hasan, 2023). They are analyzed and grouped according to this classification in this paper, together with the integration, benefits, drawbacks, and potential investigative areas of RIS-assisted backscatter communication principles. In Fig 2 Reconfigurable Intelligent Surfaces are thin panels made of many elements that can alter the EM wave transmission by reflection, refraction, or absorption. From the above discussion, it is found that RIS could be installed in urban environments or in indoor and outdoor zones to enhance the quality of wireless communication. These are the elements in an RIS with which one may alter the waves' phase and amplitude to control the waves that enter into it. RIS exploits methods of signal processing and control methods which from time to time may involve AI & ML (Xu, 2023).

Figure 2. RIS- Assisted backscatter communication

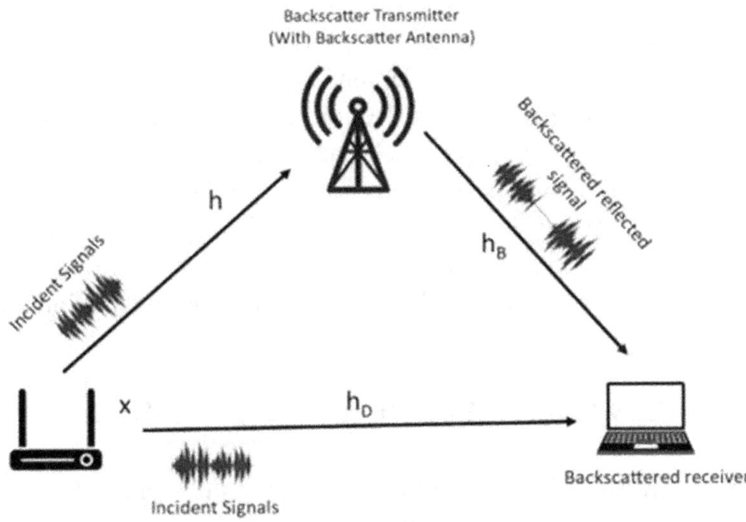

The integration of Reconfigurable Intelligent Surfaces and backscatter communication is a breakthrough in the advance of wireless communication (Wu, Guan, & Hua, n.d.). From the enhancement of signal quality, communication capacity, and energy efficiency, the backscatter communication system along with the RIS can provide fresh networking prospects for numberless applications such as the IoT and smart environment. The future evolution of these technologies depends fundamentally on the efficiency with which the problems of channel identification, interference mitigation, and standardization will be solved. It is therefore possible for the backscatter communication systems to become stronger, more dependable, and environmentally sensitive in the future as more research and development of the communication systems with assistance from RIS is achieved. It should also indicate that the studies should examine existing algorithms for new channel estimation, interference, and signal decoding to enhance the usefulness of RIS-assisted systems. AI and ML can enhance RIS and backscatter communication for creating intelligence and intelligent operative wireless networks (Sur & Bera, 2021). The experimental assessment for the purpose of verifying the RIS assisted backscatter communication systems shall provide an idea regarding the efficiency of the systems which can be utilized for real world applications. New application areas that can be discovered and explored more about the use RIS-aided backscatter communication include smart city, environment sensing, and health care because it will attract more attention and usage.

2.1 The Integration of RIS and Backscatter Communication

Incorporating RIS and backscatter communication has several advantages or benefits that complement each other. RIS enhances the surrounding setting of backscatter communication to ensure that reflections can reach a targeted recipient with enhanced quality and strength. When backscatter devices use RIS, communication can be done over long distances since physical barriers may not affect the line of sight (Le, 2021). This makes the spectrum opportunity size to be properly exploited thereby advocating for a high probability of total capacity for the communication system through the integration of RIS and backscatter communication.

3. CHALLENGES AND RESEARCH DIRECTIONS

The deployment of RIS to the framework of the upcoming 6G networks offers intriguing opportunities and profound challenges at the same time. With this technology growing, researchers are pointing out areas of concern in an attempt to harness the full potential of RIS for enabling backscatter communication for IoT use.

a. **Channel Estimation:** Channel state information (CSI) is important for optimal steering and directing signals, and indeed, of the three categories of methods under investigation, CSI is deemed the most important. The mobile nature of wireless settings hinders the attainment of accurate CSI and thus degrades the performance of RIS-assisted systems. CSI is accurate only under the best surface conditions; therefore, researchers need to enhance algorithms to deliver accurate CSI images under different scenarios.
b. **Regulatory Constraints:** The use of RIS technology triggers some concerns in regulation such as spectrum regulation as well as regulation of the installation of these technologies. Non-compliance must be discouraged on the other end while the freedom to innovate must be encouraged to encourage the take-up of RIS in 6G networks across the globe (Hamid & Begh, 2024). Multi-sectoral partnerships may be used to effectively manage these regulatory systems.
c. **Hardware Limitations:** The physical realization of RIS means the physical implementation that brings some problems connected with the size, areas of place, and, consequently, the stability. The placement and orientation of RIS elements also must be accurately defined to ensure that the technology has the best possible impact while the panels themselves are also designed to be highly durable and damage-proof against the outside world.

To address the aforementioned challenges, several research directions are emerging:

d. **Advanced Algorithms:** Selecting better algorithms for channel estimate and beamforming would be beneficial to establishing a RIS-assisted backscatter communication system. Thus, making use of such things as machine learning may provide efficient solutions to real-time adaptations.
e. **Integrated Network Architectures:** Interrogation of using network-controlled modes for the construction of RIS system architecture can enable convenience in incorporating with current communication structures. This ranges from exploring the synergies between RIS and Non-Orthogonal Multiple Access (NOMA) for resource management to enhanced user experience.
f. **True Time Delay Mechanisms:** Studying True Time Delay (TTD) structures within RIS improves signal processing capacity which in turn provides better control over signal reflecting and phase shift. This could result in dramatic enhancements in total system effectiveness and dependability. Through the above-generalized challenges, potential research directions that revolutionize and improve backscatter communication for 6G IoT networks can be realized through RIS making wireless communications more efficient and sustainable.

3.1 Channel Estimation and Optimization

This paper presents a detailed view into channel estimation and optimization as two keys to deploying Reconfigurable Intelligent Surfaces (RIS) in 6G networks, with a further focus on backscatter communication for IoT applications (Zhang, Huang, & Liu, 2024). Precise CSI is required for all signal processing for RIS to control its parameters and enhance the connected links.

a. **Dynamic Environments:** A wireless channel by its nature is a dynamic one depending on mobility, barriers, and variations in environment to name but a few. The latter is an important requirement for RIS operation, and for accurate estimation of CSI that is required for beamforming and phase shifting.
b. **Pilot Overhead:** It is well known that most of the conventional techniques for channel estimation exert huge power on pilot signals for obtaining precise CSI, thus causing high overhead constraints. These methods include the Least Squares (LS) method and Linear Minimum Mean Square Error (LMMSE) method which may be resource wasteful. This is true when the number of users and reflecting elements are high, the overhead consideration becomes a pilot.

c. **Complexity of Multi-User Scenarios:** In several user schemes, the interconnection between the different users' channels causes difficulties regarding the estimation. Sophisticated methods are necessary to capitalize on this symbiosis while keeping somewhat optimistic an approach based on the optimization of expectations of quantities that are less likely to be estimated inaccurately and use up a great deal of a pilot's time.

3.1.1 Research Directions for Optimization

a. **Advanced Estimation Algorithms:** The research is typically centered on designing increasingly complex methods for decreasing pilot cost while enhancing estimation precision. Compressive sensing is also besides machine leaner algorithms which include deep neural networks presently being used to improve CSI sampling in difficult situations.
b. **Optimization Techniques:** Optimization-based algorithms can solve challenging non-convex issues of phase shift settings and beamforming vectors in RIS-assisted systems. For instance, recursive algorithms for automatically adjusting the phase, according to real-time feedback, often enhance system performance of most systems.
c. **Integration with Emerging Technologies:** Integration of RIS with other innovations such as NOMA, can help in enhancing the techniques of resource management and boosting the general performance of the entire system. Another important area of research is the channel regulation and information modulation techniques necessary for improved performance during multifaceted operations.

3.2 Hardware Design and Implementation

This paper emphasizes the integration of Reconfigurable Intelligent Surfaces (RIS) into 6G networks where the hardware design and implementation are essential for the improvement of backscatter communication for IoT devices (Nguyen, Sur, & Do, 2022). If we analyze the future development of RIS technology, there are the following problems that arise during the design and deployment stages.

3.2.1 Challenges in Hardware Design

a. **Scalability and Size:** Wisdom has it that it is most appropriate to design RIS with a large number of reflecting elements all else being equal. But when the number of elements is increased it gives very large size and weight which becomes cumbersome during installation and deployment of various environments. The challenge now becomes being able to design small-size products and still be objective enough not to sacrifice functionality with compactness.

b. **Cost Efficiency:** Hence, although RIS technology can be considered cost-efficient since it utilizes a passive design, overall system cost may increase with more elements and intricate control circuitry. Overall, maintenance of the enhancements' performance is highly relevant in the excessively competitive area and, for the same reason, the affordability of all the proposed solutions is highly inadmissible in numerous IoT-constrained contexts.
c. **Durability and Environmental Resilience:** RIS units have to be expected on exposure to different conditions, such as temperature, humidity, and physical shocks. Keeping products both durable and lightweight is always an engineering problem.

3.2.2 Research Directions for Implementation

a. **Material Innovation:** Enhancing the performance of the material by improving its mechanical properties while at the same time reducing its weight and manufacturing cost is also the other research focus. These generalized superstates might improve the signal reflection properties of advanced materials in addition to energy efficiency.
b. **Integration with Energy Harvesting:** An inclusion of various energy harvesting functionalities to the RIS designs to improve their sustainability. With this feature implemented, RIS helps the device to recover energy from the surrounding RF signals or nearby base stations which makes them more appropriate for battery-less IoT gadgets (Ahmed, 2024).
c. **Advanced Control Mechanisms:** The sensing system must enable real-time control of the phase shifts and the beamforming gain. Extensive study into network-controlled modes may help to tailor integration to existing communication structures improving overall system functionality.
d. **Modular Designs:** Subsequent issues can be size and cost, which can be solved by finding possible modular solutions providing an opportunity for the further extension of the product. The RIS units could in principle be built modular and thus be incrementally deployed according to the required application.

By responding to these challenges through advanced RD, the design and hardware realization of the RIS can create more opportunities for backscatter communication within the 6G IoT networks, to effectively facilitate improved and sustainable wireless connection technology.

3.3 Security and Privacy Concerns

The combination of Reconfigurable Intelligent Surfaces (RIS) with IoT in the context of the developing 6G networks brings major improvements to wireless communications. However, this evolution has some fundamental questions and security and privacy issues that need to be solved for it to be secure and reliable.

3.3.1 Security Concerns

a. **Data Privacy and Integrity:** While RIS increases the capacity for backscatter communication, data transfer rates are increased proportionally thus raising the consequences associated with interception and manipulation of data. To third parties, weaknesses may present an opportunity to gain unauthorized access to personal user information. It is therefore very important to employ adequate encryption techniques, that will guarantee security during the data transfer for instance potential application of quantum encryption (da Costa, 2023).
b. **Network Vulnerabilities:** The introduction of new features and design elements of 6G networks, which involve the establishment of highly interconnected devices and the central role of artificial intelligence (AI) in their management, opens up new kinds of threats. AI vulnerabilities, opportunities for attackers to embed themselves, and possible associated risks of AI being exploited by malicious actors create smart cyber threats. These threats, therefore, require constant vigilance, and proactive security response.
c. **Advanced Persistent Threats (APTs):** Due to the projected higher data throughput in the 6G networks, they will become compelling attack targets for APTs – long-term sophisticated adversaries with the potential to deliver complex attacks. These threats can easily gain unauthorized access to computer networks which makes it necessary that the networks have extra threat detection systems that can e-alert an organization in cases of attacks.

3.3.2 Privacy Issues

a. **Identity Theft and Fraud:** Because 6G networks might allow for more effective connection in line with today's enhanced identity theft scams. The availability of user data increases the need for strong methods of user authentication to prevent fraud and unauthorized access.

b. **User Data Harvesting:** One major issue of using sensor nodes in the RIS environments is over concern of data snooping on user data by parties with malicious intentions. Such risks call for the proper formulation of privacy principles and rigorous protection and security provisions at the design phase.

To tackle these security and privacy challenges effectively, future research should focus on:

a. **Developing Comprehensive Security Frameworks:** It is possible to build an understanding of security measures that are recommended for RIS technology on an international level to strengthen protection against cyber threats.
b. **Leveraging AI for Security:** When the security prospects rely on AI-derived measures, they are capable of enhancing threat identification and combating new attacks.
c. **Enhancing Physical Layer Security (PLS):** The analysis of the possibilities of RIS ensuing to PLS may contribute to protecting the security of voice communications against various threats, including eavesdropping.

4. APPLICATIONS OF RIS-ASSISTED BACKSCATTER COMMUNICATION IN 6G IOT NETWORKS

The integration of Reconfigurable Intelligent Surfaces (RIS) can potentially transform backscatter communication in the next-generation IoT networks, namely the 6G. From the literature discussed above, it can be inferred that the special features of RIS, coupled with the ultra-low power-consuming backscatter communication, have positive synergistic impacts in resolving some of the problems associated with IoT applications (Chen, 2020).

a. **Enhanced Signal Quality:** Optimizing the propagation environment using RIS can enhance backscatter signals quality of signals by at least five times. Due to phase and amplitude shift control of reflected signals, RIS offers possibilities to solve several problems, such as multipath fading and interference, and SNR improvement. This is especially useful for the Internet of Things devices used in densely populated cities where conventional channels don't work well.
b. **Energy Efficiency:** The first of the benefits of the backscatter communication system is that it does not require substantial energy so devices only reflect existing RF signals. When integrated with RIS, which can scavenge energy from existing RF sources, the efficiency improves even more. This capability disseminates

flexibility through battery-operated IoT devices to function efficiently over a prolonged time without charging.

c. **Multi-User Communication:** RIS technology supports the ability to broadcast signals to many IoT devices by using beamforming capabilities. Specific beams in RIS can satisfy high Data Rates in interaction with different users and low latency which makes this concept acceptable for different applications such as Smart City and Industrial Automation.
d. **Improved Data Rates:** Some of the ideas, that can be implemented to achieve better spectral efficiency in RIS-assisted backscatter communication systems include higher order modulation schemes like the Quadrature Amplitude Modulation or QAM. Most research work has shown that employing methods such as polarization diversity and MIMO formation increases the amount of throughput but at the same time, reduces the error rates.
e. **Robustness Against Interference:** The characteristic of RIS to control the wireless channel while interacting with devices increases the immunity of the hybrid network to interferences from other devices. This ability is especially important, especially in high device density scenarios to facilitate reliable communication in other important applications like health monitoring and environmental measurement.

In fig 4 Self-interference-cancellation in the backscatter mode of operation in 6G IoT networks gives rise to revolutionary opportunities that improve signal quality, and energy efficiency, and support multiple users simultaneously. While the research challenges persist in the areas of hardware design, channel estimation, and security concerns, the vision of RIS as an enabler of efficient and sustainable wireless communication for IoT devices will remain a reality in future IoT systems (Kumaravelu, 2022).

Figure 3. Applications of RIS-Assisted backscatter communication

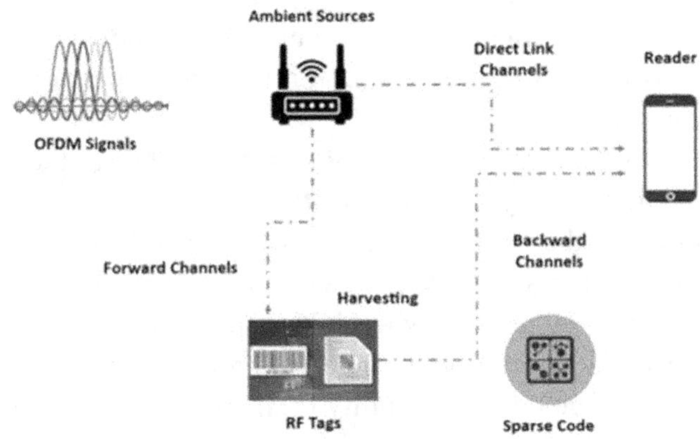

4.1 Smart Cities and Infrastructure

The addition of RIS into smart cities is an important breakthrough in the category development of 6G networks, especially in backscatter communication for IoT customers. The implementation of new RIS technologies means that advanced capabilities to revolutionize the structures of urban society can be fashioned.

4.1.1 Applications in Smart Cities

a. **Urban Environment Monitoring:** RIS can help monitor real-time conditions of urban areas by improving the communication channels between the multitude of sensors installed across the city. To this extent, this capability makes it possible to gather reliable data on the quality of air in a city, traffic patterns, and energy consumption, which can be effectively used in decision-making processes regarding the further development of cities.
b. **Smart Transportation Systems:** Through enhancing scatted signals quality and coverage, RIS could foster better interchange of information between automobiles and structures in smart transportation. This improvement is beneficial for the usage of V2X communication, an emerging key technology for self-driving cars and smart traffic (Muhammed, 2022). The flexibility of signal paths means that important information gets to vehicles in time, increasing safety and efficiency.

c. **Energy Management in Buildings:** Emphasizing that RIS can improve the Internet of Things connectivity implemented in energy management systems of smart buildings. Thus, through the managing of communication between employed energy sensors and control systems, RIS has the capability of assisting in giving improved demand response control and precise real-time analysis on the utilization of energy.
d. **Public Safety and Security:** The improvement of RIS in the regions where people are residing densely provides a scaffolding to public safety by improving communication for the EMSs. Increased signal quality and range guarantee that emergency agencies can always stay in touch while executing dangerous operations, specifically in areas with high populace or a lot of interference.
e. **Sustainable Urban Development:** The power-saving feature of RIS technology is a suitable match for sustainability in smart cities. Therefore, RIS has the potential of reducing the wired connections, particularly in densely populous areas and extracting energy from the ambient RF signals, which would assist in the formulation of greener urban development.

Smart cities are expected to receive a boost from the implementation of RIS-assisted backscatter communication. Promoting greater data speed and data consistency across multiple applications, such as transportation and energy distribution or consumption, RIS technology has the potential to architect and therefore develop 'smart and sustainable' city environments. This paper identified that there are problems associated with the implementation concept and security issues that researchers have yet to solve when it comes to the use of RIS in the 6G IoT network to support future city demands; thus, it is only when the above problems are solved that RIS's full potential can be effectively utilized to provide solutions to new challenges.

4.2 Industrial IoT and Automation

With the integration of RIS into IoT, and automation, effectiveness and communication stability in the execution of functions across the different domains shall be greatly impacted. This paper examines how RIS technology benefits provide values that are attainable by satisfying the industry 4.0 demands and offered by backscatter communication progression.

4.2.1 Applications in Industrial IoT and Automation

a. **Enhanced Machine-to-Machine Communication:** In practical applications like manufacturing industries, the need for better channels between machinery can be provided by RIS which tends to minimize stray paths. This has to be enhanced specifically for the operational real-time data exchange for the various automated systems that are hardwired into manufacturing processes.
b. **Energy Efficiency:** Backscatter communication and Integrating Reflecting Indoor Environment Illuminating Systems result in low power so the IoT devices are sustainable. Unlike other devices that can cause reflection and modulation of the same signals or even the generation of entirely new RF signals, these can lead to a substantial reduction in energy use. This capability is especially valuable for situations when a large number of devices is used, for example in a factory or a warehouse.
c. **Real-Time Monitoring and Control:** By enhancing the dependability of sensor information transfer, RIS may enable continuous tracking of industrial processes. Higher signal quality guarantees that important information concerning the capability of the machinery and environmental conditions together with operational parameters reaches the control room instantaneously, thus facilitating interventions.
d. **Support for Advanced Automation Technologies:** The integration of RIS with some of the most significant automation technologies of the current era robotics and artificial intelligence can improve operations. For instance, RIS can help achieve accurate localization and navigation of autonomous robots in confined industrial spaces depending on the control and stability of links established.
e. **Scalability and Flexibility:** RIS is designed in a modular nature that ensures it may be deployed incrementally in these industrial settings. There are extra componentry bits that can be added as a business evolves or demand evolves which does not require significant modifications to the structures.

The addition of RIS-assisted backscatter communication within Industrial IoT networks provides a significant opportunity to improve overall automation and operationality (Ahad, 2024). If the difficulties in energy usage, communication dependability, and speed of real-time data processing are solved, RIS technology will help make progress in smart manufacturing and other industry fields. With ongoing improvements in these technologies through research, options for revolutionizing current industry practices will also advance for creating smarter and more adaptive industrial areas.

4.3 Healthcare and Wearables

Reconfigurable Intelligent Surfaces (RIS) have tremendous potential in healthcare and wearable technology inside 6G networks, especially in improving backscatter communication for IoT devices. These combinations can enhance patient surveillance and data sharing and, in general, the whole process of healthcare services.

4.3.1 Applications in Healthcare and Wearables

a. **Remote Patient Monitoring:** By improving the connectivity of wearable devices, RIS-enabled backscatter communication can help with health status real-time monitoring. Such devices can capture and send essential data like the patient's heart rate, blood pressure, and glucose levels as a continuous process; The increased reliability of signals provided by RIS means that significant data about patients' health condition is delivered without delay, which may be critical for proper treatment.
b. **Energy Harvesting for Wearables:** Besides, most wearable devices contain small batteries and are used for a short time as their battery life runs out quickly. Through reflection of ambient RF signals to these devices, RIS can help in the energy harvesting process and the frequent charging of these devices could be avoided. This capability is highly useful for long-term health monitoring applications where sustained device functioning is inevitable.
c. **Enhanced Data Transmission Rates:** This work also demonstrated that with RIS, there is a possibility of directing signal paths, thus enhancing the chances of improved data transfer rates between wearables and healthcare systems. This enhancement is critical for bandwidth-intensive applications, e.g., video consultations or the transfer of detailed radiological images.
d. **Improved Signal Coverage in Healthcare Facilities:** Wireless communication is also difficult in hospitals and clinics because of the tight technical environment. RIS can enhance signal coverage and eliminate dead spots within these facilities, so that wearable devices stay connected with central monitoring systems irrespective of harsh surroundings.
e. **Integration with Artificial Intelligence:** Depending on the RIS and integrated AI algorithms, health data gathered from wearables might be further improved. For example, current centralized machine learning models can handle incoming data streams better if the communication lines are reliable to support predictive analytics that would anticipate diseases before they occur.

Such applications as RIS-assisted backscatter communication in healthcare and wearable technologies in 6G networks will enhance patient outcomes, and wearables monitoring efficiency. As much as it has the potential to enhance energy efficiency, data rates, and connectivity, RIS technology can serve as an important enabler to emergent smart health systems. With further researches to answer questions concerning with the hardware designing and security the capabilities of this technology will be unleashed, further enabling the deployment of client healthcare solutions that enhance patient care and organizational functions.

5. FUTURE TRENDS AND OUTLOOK

The insights based on trends and prospects of Reconfigurable Intelligent Surfaces for 6G networks will help drive wireless communications into new paradigms with special emphasis on backscatter communication for IoT. Due to the ever-increasing need for saturated data rates, low latency and efficiency in channel utilization, the RIS technology is coming out as a key solution.

a. **Integration with Advanced Technologies:** We anticipate that endowing RIS with AI, machine learning, and edge computing capabilities will improve their performance. AI can be used to fine-tune the amount of control of the RIS elements in real-time and with respect of the environment and/or users. This integration will provide the capability of intelligent network management thereby increasing the performance of the system.
b. **Expansion into New Frequency Bands:** Since 6G will be considered using terahertz (THz) communications, a system such as RIS will be critical in regulating the propagation profiles of such elevated frequencies. THz networks may be deployed since RIS has the potential of controlling electromagnetic waves more efficiently than increasing path loss and improving the signal coverage.
c. **Sustainability and Energy Efficiency:** As concerns about the usage of energy in telecommunications become persistent, RIS provides the direction to green networks. Through the concept of energy harvesting from idle Radio Frequency signals and usage of passive reflection to manage power consumption, RIS can greatly help in lowering the carbon footprint of wireless communications.
d. **Enhanced Security Features:** RIS can enhance the security at the physical layer since it allows direction control of signals and also reduces interceptions. Developing sound security approaches utilizing RIS effective capabilities on security wireless transferred data would probably form the center of future research.

e. **Market Growth and Commercialization**: The worldwide market of RIS is expected to grow rapidly due to increase adoption of RIS in various industries and sectors such as telecommunication, smart city, healthcare, and industrial IoT. It will be expected that companies will devote many resources into research to develop solutions that are cost effective and easily scalable for implementation in RIS.

It is expected that in the future RIS will remain an important part of developing wireless networks, including new generation 6G networks. Despite this, this paper presents evidence that as research advances, and practical executions surface, RIS is poised to support unprecedented interconnectivity and effectiveness for IoT solutions. Higher data rates, better energy consumption, and strong securities make RIS a promising enabler for attaining the vision of 6G networks. The future trend surrounding RIS proves the notion that RIS will be critical in defining the new generation of wireless communication. Solving present challenges and encouraging new solutions in an expanding digital environment make RIS technology one of the most promising tools for improving people's ability to communicate and interact with each other.

6. CONCLUSION

The incorporation of Reconfigurable Intelligent Surfaces (RIS) in 6G provides a step forward in wireless technology, especially for boosting backscatter communication in IoT devices. When the need for connectivity arises with the strains of modern-day connectivity, RIS has radical solutions to many issues of today such as coverage, power consumption, and data delivery reliability. RIS technology is in the right place and at the right time to remake almost all possible fields of human activity ranging from smart city and industrial Internet to the Internet of Health and other parts of the Internet of Things. Furthermore, by changing the frequency characteristics of electromagnetic waves, RIS can optimize the signal quality, and coverage range and interfere less with other networks leading to optimal network performance. The opportunity to scavenge energy from environmental RF sources also contributes to the sustainability of IoT devices making the nodes more practical for long-term deployment. However, the accomplishment of RIS in 6G networks will encounter some challenges such as hardware design challenges, security issues, and policy problems at this level. Further improvements on such systems and designs are essential to enhance them to meet the challenge that near-future applications present. Moving forward, the integration of the RIS with advanced technologies like AI and edge computing will be key to opening up ever previously imagined application possibilities in the area of intelligent network management and appropriate resource allocation. While discovering these innovations, RIS would leave its mark as a key enabler of the future wireless communications environment to support higher connectivity, efficiency, and sustainability. The uses are unlimited and as we engage this technology it is projected that we will even enhance our interaction with our digital environment.

REFERENCES

Ahad, A., Jiangbina, Z., Tahir, M., Shayea, I., Sheikh, M. A., & Rasheed, F. (2024). 6G and Intelligent Healthcare: Taxonomy, technologies, open issues and future research directions. *Internet of Things : Engineering Cyber Physical Human Systems*, 25, 101068. DOI: 10.1016/j.iot.2024.101068

Ahmed, M., Raza, S., Soofi, A. A., Khan, F., Khan, W. U., Xu, F., Chatzinotas, S., Dobre, O. A., & Han, Z. (2024). A survey on reconfigurable intelligent surfaces assisted multi-access edge computing networks: State of the art and future challenges. *Computer Science Review*, 54, 100668. DOI: 10.1016/j.cosrev.2024.100668

Basharat, S., Hassan, S. A., Pervaiz, H., Mahmood, A., Ding, Z., & Gidlund, M. (2021). Reconfigurable intelligent surfaces: Potentials, applications, and challenges for 6G wireless networks. *IEEE Wireless Communications*, 28(6), 184–191. DOI: 10.1109/MWC.011.2100016

Chen, Y., Liu, W., Niu, Z., Feng, Z., Hu, Q., & Jiang, T. (2020). Pervasive intelligent endogenous 6G wireless systems: Prospects, theories and key technologies. *Digital Communications and Networks*, 6(3), 312–320. DOI: 10.1016/j.dcan.2020.07.002

da Costa, Daniel Benevides, "6G: Vision, Applications, and Challenges." *Fundamentals of 6G Communications and Networking* (2023): 15-69. DOI: 10.1016/j.dcan.2020.07.002

Hamid, H., & Begh, G. R. (2024). IRS assisted UAV communications for 6G networks: A systematic literature review. *Wireless Networks*, 1–29. DOI: 10.1007/s11276-024-03798-y

Ihsana Muhammed, P. (2022). Reconfigurable Intelligent Surface (RIS)-Assisted UAV Cellular Communication. In *Unmanned Aerial Vehicle Cellular Communications* (pp. 115–135). Springer International Publishing.

Jian, M., Alexandropoulos, G. C., Basar, E., Huang, C., Liu, R., Liu, Y., & Yuen, C. (2022). Reconfigurable intelligent surfaces for wireless communications: Overview of hardware designs, channel models, and estimation techniques. *Intelligent and Converged Networks*, 3(1), 1–32. DOI: 10.23919/ICN.2022.0005

Kumaravelu, V. B. (2022). Unmanned Aerial Vehicle-Assisted Reconfigurable Intelligent Surface for Energy Efficient and Reliable Communication. In *Unmanned Aerial Vehicle Cellular Communications* (pp. 173–201). Springer International Publishing.

Le, C.-B., Do, D.-T., Li, X., Huang, Y.-F., Chen, H.-C., & Voznak, M. (2021). Enabling NOMA in backscatter reconfigurable intelligent surfaces-aided systems. *IEEE Access : Practical Innovations, Open Solutions*, 9, 33782–33795. DOI: 10.1109/ACCESS.2021.3061429

Liang, Y.-C., Zhang, Q., Wang, J., Long, R., Zhou, H., & Yang, G. (2022). Backscatter communication assisted by reconfigurable intelligent surfaces. *Proceedings of the IEEE*, 110(9), 1339–1357. DOI: 10.1109/JPROC.2022.3169622

Liaskos, C., Tsioliaridou, A., Ioannidis, S., Pitsillides, A., & Akyildiz, I. F. (2021). Realizing ambient backscatter communications with intelligent surfaces in 6G wireless systems. *IEEE Wireless Communications*, 29(1), 178–185. DOI: 10.1109/MWC.101.2100233

Nguyen, M.-S. V., Sur, S. N., & Do, D.-T. (2022). *A Comprehensive Review on Physical Layer Design for Smart Cities*. IoT and IoE Driven Smart Cities. DOI: 10.1007/978-3-030-82715-1_1

Saiam, M., Chowdhury, M. Z., & Hasan, S. R. (2023). "Crypto Holographic Reconfigurable Intelligent Surface-Assisted Backscatter Communication." *The Fourth Industrial Revolution and Beyond: Select Proceedings of IC4IR+*. Singapore: Springer. *Nature in Singapore*, •••, 149–156.

Saiam, M., Chowdhury, M. Z., Hasan, S. R., & Jang, Y. M. (2023). Reconfigurable intelligent surface assisted BackCom: An overview, analysis, and future research directions. *ICT Express*, 9(5), 927–940. DOI: 10.1016/j.icte.2023.07.004

Sur, S. N., & Bera, R. (2021). Intelligent reflecting surface assisted MIMO communication system: A review. *Physical Communication*, 47, 101386. DOI: 10.1016/j.phycom.2021.101386

Wu, Qingqing, Guan, Xinrong, & Hua, Meng. "Intelligent Reflecting Surface For B5G/6G Wireless Networks." DOI: 10.1016/j.phycom.2021.101386

Xu, F., Hussain, T., Ahmed, M., Ali, K., Mirza, M. A., Khan, W. U., Ihsan, A., & Han, Z. (2023). The state of AI-empowered backscatter communications: A comprehensive survey. *IEEE Internet of Things Journal*, 10(24), 21763–21786. DOI: 10.1109/JIOT.2023.3299210

Zhang, H. (2021). *Reconfigurable Intelligent Surface-Empowered 6G*. Springer. DOI: 10.1007/978-3-030-73499-2

Zhang, S., Huang, W., & Liu, Y. (2024). A systematic survey on physical layer security-oriented to reconfigurable intelligent surface empowered 6G. *Computers & Security*, •••, 104100.

Zhao, Y. (2023). Reconfigurable intelligent surfaces for 6G: Applications, challenges, and solutions. *Frontiers of Information Technology & Electronic Engineering*, 24(12), 1669–1688. DOI: 10.1631/FITEE.2200666

Chapter 6
AI-Powered Reconfigurable Intelligent Surfaces:
A Review of Functioning, Research Challenge, and Future Directions in Wireless Communication

Sangeetha Arumugam
https://orcid.org/0009-0001-4668-3797
Dr. G.R. Damodaran College of Science, India

S. Kalaivani
Dr. G.R. Damodaran College of Science, India

V. Bakyalakshmi
Dr. NGP Arts and Science College, India

ABSTRACT

Reconfigurable Intelligent Surfaces and Artificial Intelligence combined offer a major breakthrough in wireless communication technology. These are a promising technology that has the potential to significantly improve network performance and signal propagation in wireless communication. RIS is capable of optimizing wireless channels, resulting in increased signal strength, decreased interference, and expanded coverage by dynamically modifying the electromagnetic parameters of the surrounding environment. This study provides an extensive overview of the current state of AI-powered RIS-assisted networks, and investigates the functioning processes, deployment approaches, and possible applications of RIS technology.

DOI: 10.4018/979-8-3693-8099-4.ch006

The limitations of hardware design, computational complexity, interoperability, and security issues are highlighted as major research problems. This review paper explore as a foundational reference for research challenges and providing insights into the transformative potential of AI-RIS and guiding future innovations in this rapidly evolving field.

1. INTRODUCTION

The future information society's transition from "connected things" to "connected intelligence" is heavily reliant on wireless connectivity. Future wireless networks will require ultra-high data speeds, ultra-high dependability, ultra-low latency, and super-massive connectivity to facilitate data sensing, collection, transmission, and processing. The main enabling wireless technologies include massive multiple-input multiple-output (MIMO), millimeter wave (mm Wave), ultra dense networks, AI-powered wireless networks, and edge computing, caching, learning, and network slicing (Basar *et al.,* 2019).

Modern wireless systems development operates on the principle that radio environments are externally established and beyond control. The primary goal in designing systems, protocols, and algorithms has been to minimize the adverse impacts of radio environments. To unlock the full potential of future wireless networks, it is crucial to explore the possibility of creating a radio environment that is programmable and controllable. This would enable system optimization to leverage the radio environment as a variable (Di Renzo *et al.,* 2020). Intelligent reflecting surfaces (IRSs) and large intelligent metasurfaces (LIMs) are other names for reconfigurable intelligent surfaces (RISs), which are designed to artificially reconfigure the electromagnetic wave propagation environment in order to reduce energy consumption and improve the spectral efficiency of wireless networks. RISs are specifically built utilizing planar (or even conformal) artificial metasurfaces that can be programmed by an intelligent controller and have numerous reflection amplitude/phase changes. Hence, the phases and/or amplitudes of the incident signals can be controlled by the RIS-enabled smart radio environments, increasing the required signals and reducing the interfering signals (Wu.Q *et al.,* 2019).

As such, RISs have the potential to significantly alter the design of wireless networks, especially when paired and integrated with other cutting-edge wireless technologies like AI-powered wireless networks, edge intelligence, terahertz communications, and ultra-massive MIMO. Consequently, the goal of this essay is to support and raise awareness of this new line of inquiry. By adjusting the radio propagation environment using RIS-powered smart radios, next-generation wireless communications will be able to counteract unfavorable propagation situations like deep fading. 5G and beyond wireless communications are expected to be made possible via millimeter wave and terahertz communications. As radio frequencies rise, tens or perhaps hundreds of antenna components will be needed for base stations and even portable devices (Huang *et al.,* 2020).

In order to address the exponentially increasing demand for wireless data services, massive MIMO can offer previously unheard-of capacity improvements. Antenna arrays are deployed on device terminals as well as base stations. Buildings and other structures are more likely to block electromagnetic radiation when radio frequency increases because they decrease the diffraction and scattering impact. As such, it is challenging to guarantee a ubiquitous coverage of wireless services in 5G and beyond using traditional cellular technologies. The recently developed RIS technology offers an innovative approach to the issue by artificially modifying the electromagnetic wave propagation environment.

RISs typically use a reconfigurable array of inexpensive, energy-efficient reflecting components, which can be made to reflect incoming electromagnetic waves with a programmable phase shift with the help of an ingenious controller. Reconfigurable intelligent surfaces (RIS) can reduce the negative propagation characteristics of wireless communication systems by carefully positioning and utilizing reflect/passive beam shaping to generate an additional high-quality channel link (Renzo *et al.,* 2020).

Reconfigurable intelligent surfaces (RIS) in conjunction with artificial intelligence (AI) are setting the standard for next-generation wireless communication technologies. AI has revolutionized many industries. Reconfigurable Intelligent Surfaces (RIS) are emerging as a game-changing technology in wireless communication because of their unparalleled capacity to alter the electromagnetic wave propagation environment. Reconfigurable intelligent surfaces (RIS) use artificial intelligence (AI) to increase their versatility and efficiency. Because of this, RIS systems are crucial for developing artificial surfaces with electromagnetic wave control and next-generation networks. These surfaces consist of a multitude of active or passive elements that can be dynamically adjusted to change incident signals in the proper manner (Tang *et al.,* 2020).

Engineered surfaces, or RIS, are made up of many active or passive components that can be used to individually modify the amplitude, phase, or polarization of incident electromagnetic waves. By making these changes, the surface can optimize communication performance by steering, reflecting, or focusing signals in desirable directions. RIS are adaptable tools for managing wireless environments since they may be installed in a variety of ways, including on walls, ceilings, or even as isolated panels. Unprecedented breakthroughs in wireless communications are anticipated as a result of the convergence of AI and RIS technologies, which will improve capacity, coverage, and signal quality while lowering expenses and energy usage (Björnson *et al.,* 2020).

Reconfigurable intelligent surfaces has the remarkable ability to shape and control the environment in which electromagnetic waves propagate. RIS-enabled systems offer a method that is more flexible and adaptable than standard wireless communication systems, which rely on static infrastructure and fixed propagation parameters. Transmissions in hitherto unreachable locations can be dynamically altered to increase signal strength, decrease interference, and lead toward desired directions with the use of RIS. This groundbreaking potential is crucial for addressing the growing demand for consistent, quick, and ubiquitous wireless access.

By using complex algorithms and machine learning techniques to optimize surface designs instantly, artificial intelligence (AI) is used in RIS to further improve its capabilities. RIS systems can adapt effectively to changing network requirements, user demands, and ambient conditions thanks to artificial intelligence (AI). Machine learning methods, particularly deep learning and reinforcement learning, enable autonomous decision-making to improve system performance, as well as the prediction of optimal configurations and learning from historical data. Artificial intelligence is key to RIS advanced signal processing techniques. AI-driven beam creation, for instance, offers precise control over the direction and focus of electromagnetic waves to enhance signal intensity and minimize interference (Li *et al.,* 2021).

AI algorithms can also enhance multi-user communication by dynamic resource allocation and RIS configuration modifications to handle multiple users simultaneously. The combination of AI with RIS enables self-optimizing and self-healing networks that can provide seamless connectivity and enhanced user experiences.

1.1 System Architecture

Typically, the architecture of RIS powered by AI comprises of the following elements:

- **RIS Panel:** Several reconfigurable parts, or metasurfaces, make up the surface itself, which interacts with incident waves. The way the waves are reflected or refracted can be changed by adjusting these components in response to control inputs.
- **Controller:** The primary control unit of the RIS system interfaces with the metasurfaces. This controller transmits commands to every component of the RIS after processing data from external sensors, such as user positions or wireless channel estimations. This control unit incorporates AI algorithms to facilitate intelligent decision-making and flexible responses.

- **Sensing and Estimation Unit:** In order to obtain current information about the wireless environment, AI-powered systems require sensors or external data gathering devices. This unit gathers and feeds information to the controller for processing, including channel state information (CSI), user locations, mobility patterns, and interference levels (Ozdogan *et al.,* 2020)
- **Wireless Communication System:** The RIS communicates with base stations, user gear, and other network infrastructure as part of the communication system. It aids in the optimization of signal routes to raise performance indicators including latency, capacity, and signal-to-noise ratio (SNR).

1.2 Operation and Functionality

The primary functions of an AI-powered RIS can be decomposed into multiple operational phases:

1. **Environmental Sensing:** The RIS system monitors the wireless environment continually, collecting information on user locations, obstructions, and sources of interference. The system uses this real-time data to comprehend the channel's current condition.
2. **AI-Driven Decision Making:** To forecast how the environment will change, AI models analyse sensory data. Machine learning systems, for instance, could forecast human movement or the presence of impediments. The AI recommends RIS configurations that can improve communication quality the most based on these assumptions.
3. **Surface Reconfiguration:** The control unit modifies the phase shifts or amplitude settings for every metasurface element after identifying the ideal configuration. The technology can create nulls, focus signals, or redirect beams in unwanted directions thanks to this real-time reconfiguration.
4. **Performance Monitoring:** The system keeps an eye on the communication performance following surface reconfiguration to see whether the altered configuration produced the intended outcomes. Based on feedback, AI systems may adjust the parameters, continuously learning and adjusting to new circumstances (Zhang *et al.,* 2020).

1.3 Benefits of AI-Powered RIS

Over traditional wireless systems, AI-powered RIS has a number of benefits, such as:

- **Energy Efficiency:** RIS uses a lot less power than base stations or conventional relay systems because it doesn't need to generate or amplify signals. Utilizing AI further optimizes energy use by ensuring that resources are distributed effectively.
- **Expanded Coverage:** Artificial intelligence-driven reconfigurable intelligent surfaces (RIS) can increase the reach of wireless networks by bouncing signals into places where reception is poor, such indoor spaces or urban canyons. RIS can sustain strong signal strength and coverage by constantly adapting to changes in the surrounding environment.
- **Enhanced Capacity and Throughput:** Managing signal routes and reducing interference contributes to enhancing the network's overall capacity. AI makes sure that RIS adjusts to variations in demand or network congestion, which results in increased throughput and decreased latency.
- **Low Complexity and Cost**: Passive, low-cost metasurfaces that are rather simple to build are frequently used in the construction of RIS. When compared to conventional active antenna arrays or relay stations, this lowers the complexity and cost of establishing large-scale RIS systems (Wu *et al.,* 2020).

1.4 AI-Powered RIS Offer Several Advantages

1. **Enhanced Efficiency:** Automation of repetitive operations, including as data input, report creation and image processing, expedites workflows and lessens the administrative strain on staff members and radiologists.
2. **Better Diagnostic Accuracy:** Artificial intelligence algorithms help understand complicated images by seeing irregularities and subtle patterns that the human eye can overlook, which results in diagnoses that are more accurate.
3. **Better Image Quality:** By lowering noise and artifacts, AI can improve image clarity and improve the display of important features.
4. **Predictive analytics:** AI examines both past and current data to forecast patient outcomes and the course of a disease, enabling early intervention and individualized treatment regimens.
5. **Workflow Optimization:** AI helps radiology operations run more smoothly by controlling scheduling, giving priority to urgent patients, and making the most use of the equipment, all of which increase productivity.
6. **Decision Support:** AI offers insightful analysis and suggestions to help radiologists make well-informed decisions, especially in complicated situations.
7. **Decreased Human Error:** AI-powered automation and support reduce human error and variability in data management and image interpretation.
8. **Cost Savings:** AI can assist in lowering the running expenses of radiology departments by increasing productivity and decreasing errors.

9. **Scalability:** AI-powered solutions are appropriate for both small clinics and major hospitals since they can readily manage growing data volumes and imaging investigations.
10. **Improved Patient Experience:** Timely therapies follow quicker and more accurate diagnoses, which can enhance patient outcomes and satisfaction (Huang *et al.,* 2021).

2. FUNCTIONING OF AI-POWERED RIS

2.1 Basic Functions of RIS

The Reconfigurable Intelligent Surface (RIS) is a technology that uses many small, movable components and is organized on a flat, planar surface. Each of these elements can be changed separately to regulate electromagnetic waves, including radio signals. Modern wireless communication relies on Reconfigurable Intelligent Surfaces (RIS), which alter electromagnetic signals using latest technologies. The core functions of RIS as follows:

Electromagnetic Wave Control and Reflection:

RIS maximises the transmission channel by reflecting incoming signals in various directions towards a specific area. This involves manipulating the phase and amplitude of the signals to optimize their path and enhance reception at the intended location. Essentially, steering helps improve signal quality and connectivity by ensuring that the energy is focused where it is needed most, reducing interference and enhancing overall performance.

Beamforming:

Beamforming is a wireless communication technology that enhances signal delivery by directing stronger signals towards specific devices, improving clarity, reducing interference, and adapting to changing conditions like user movement or obstacles for optimal performance. Digital beamforming involves the use of software for more complex changes, while analogue beamforming involves the use of hardware to manage signals. The dependability and efficiency of wireless communication is greatly enhanced by beamforming. This targeted approach allows for better utilization of the available spectrum.

Noise Mitigation:

RIS manage and reduce interference using intelligent redirection of signals away from problematic areas such as congested zones or obstacles, and direct signal to specific users which significantly reduce background noise. RIS enhances network performance by providing higher-quality signals, optimizing network efficiency, allowing higher data rates, lower latency, minimizing disruptions and more reliable connections.

Environment Adaptation:

Reconfigurable Intelligent Surfaces (RIS) are designed to dynamically respond to changing user demands and circumstances by changing their attributes in real-time. Due to its adaptability, RIS is capable of optimizing network performance in response to changing circumstances, ensuring reliable connectivity even when users move or traffic patterns change. By directing signals efficiently, RIS maintains strong connections and high data speeds while minimizing disturbances. By minimizing interference and effectively distributing bandwidth, it also cleverly controls resources, improving network efficiency as a whole. At the end of the day, RIS greatly improves the standard of service, resulting in consumers having a more reliable and seamless communication experience (Zhang *et al.,* 2021).

2.2 Integration of AI with RIS

Integrating AI with Reconfigurable Intelligent Surfaces (RIS) can significantly enhance their performance and adaptability in wireless communication systems.

Dynamic Configuration and Optimization

Artificial Intelligence is used by Reconfigurable Intelligent Surfaces (RIS) to proactively adapt to changing circumstances. While AI algorithms are employed to evaluate the RIS's reflection properties, machine learning models use past data to forecast best configurations. To provide optimal signal transmission in a real-time data, for instance, it dynamically modifies the RIS's reflection parameters based on the analysis of user locations and signal quality. Consequently, timely updates and proactive configurations offer dependable communication events in a dynamic environment while significantly improving network speed, reliability, and user experience overall. It enhances network performance through the use of real-time adaptation and predictive modeling.

User Behaviour Prediction

User behavior prediction in Reconfigurable Intelligent Surfaces refers to use AI to find patterns of user movement and behavior to forecast future requirements. As a result, the RIS is able to optimize its configurations proactively. For each unique user, the RIS improves Quality of Service (QOS) by efficiently allocating resources depending on their needs. It results in a more adaptive and user-centric

Channel State Information (CSI) Estimation

Reconfigurable Intelligent Surfaces (RIS) employ Artificial Intelligence (AI) and machine learning approaches to improve the Channel State Information (CSI) estimation. By using this estimating method, the CSI's accuracy is improved, which is vital for maximizing the RIS's reflection properties and improving communication. CSI fluctuations can be predicted using machine learning-based predictive models that take into account user movement and environmental conditions. The RIS may make configuration adjustments ahead of time according to this proactive approach, which guarantees more dependable and effective wireless connection.

Interference Management

Reconfigurable Intelligent Surfaces (RIS) use artificial intelligence (AI) for real-time interference detection, which efficiently locates and categorizes interference sources. Because of this flexibility, the RIS may dynamically modify its reflection properties, reducing the detrimental effects of interference on signal quality and network performance as a whole. Moreover, the deployment of collaborative learning facilitates the exchange of knowledge and information among various RIS units, leading to a network-wide optimization that improves overall performance and lowers interference. This dual strategy guarantees a smoother and more effective user experience throughout the network in addition to bolstering the dependability of wireless communication.

Resource Allocation and Management

Reconfigurable Intelligent Surfaces (RIS) use AI-powered optimized resource allocation to efficiently manage power and spectrum resources. AI makes sure that users get the best service possible by allocating these resources based on network conditions and real-time demand. AI also examines traffic patterns to help with load balancing among several RIS units. This strategy encourages effective use of network resources by preventing any one unit from getting overloaded. When combined, these tactics raise network performance as a whole, boost user satisfaction, and preserve dependable connectivity. In the end, optimizing wireless communication systems' efficacy and efficiency depends heavily on AI-driven resource management.

Edge Computing Integration

Reconfigurable Intelligent Surfaces (RIS) use local data processing, which integrates AI at the edge to enable real-time data analysis and decision-making. Being close to the data sources lowers latency dramatically, improving performance and responsiveness. RIS can instantly adjust to changing conditions and user demands by processing information locally. Furthermore, edge devices can cooperate with each other due to collaborative AI, exchanging knowledge and experiences to enhance learning. The development of more potent RIS optimization techniques is the result of this collaboration. The technology can thereby improve user experiences and resource management. In general, the development of effective and responsive wireless communication environments greatly depends on collaborative AI and local data processing.

Simulation and Modelling

Artificial Intelligence (AI) is used in digital twin technology in Reconfigurable Intelligent Surfaces (RIS) to generate virtual versions of the real world, enabling the simulation of different setups and situations. This capability offers insights into possible performance outcomes, which helps with the planning and optimization of RIS deployment. Furthermore, scenario analysis powered by AI is used to model various deployment techniques and analyze the potential effects of different RIS configurations on overall network performance. Through the employment of these technologies, decision-makers may improve user experiences and maximize resource consumption while also increasing the efficacy and efficiency of wireless communication systems.

Security Enhancements

Reconfigurable Intelligent Surfaces (RIS) employ artificial intelligence (AI) for anomaly detection, which uses AI to continuously monitor network behavior, detect possible security risks, and enable quick reactions to abnormalities or assaults. This proactive strategy improves the network's overall security. Furthermore, AI is essential to data privacy management because it applies security measures that are customized based on user behavior patterns that it has learnt. AI strengthens user confidence in the system by ensuring that user data is adequately protected. When combined, these methods protect the network and help users communicate in a safe and dependable environment.

Enhanced Decision-Making Algorithms

AI is able to optimize configurations through trial-and-error interactions with the environment in Reconfigurable Intelligent Surfaces (RIS) through reinforcement learning. With the help of this method, the system can eventually figure out what settings work best for certain scenarios. Furthermore, AI can use multi-agent systems to coordinate actions utilizing decentralized decision-making techniques in scenarios involving numerous RIS units. By coordinating efforts across RIS, overall network performance and efficiency are improved since immediate responses and efficient resource allocation are guaranteed. When combined, these tactics greatly increase wireless communication systems' efficiency and adaptability.

Adaptive Channel Access

Artificial Intelligence (AI) has the potential to allow RIS to function as cognitive radios, adapting their operation dynamically to disturbances and spectrum availability. To reduce interference and increase data speeds, RIS can use AI to select the best frequencies.

Network Slicing and Virtualization

AI is capable of managing network slicing for a range of applications (such as AR/VR and IoT) by dynamically modifying RIS configurations to satisfy the unique needs of each slice. AI can make sure that RIS resources are distributed effectively by examining the requirements of various slices.

Environmental Awareness

By combining AI algorithms with environmental sensor integration, the RIS is able to learn about external factors (such as traffic or weather) and modify its settings accordingly to ensure peak performance. Better reflection techniques can be made possible by AI's assistance in producing real-time maps of the surroundings, complete with barriers and user locations.

Quality of Experience (QoE) Monitoring

AI can use feedback from users about their experiences to continuously improve the performance of the RIS. This can involve adjusting settings based on user satisfaction metrics. Machine learning can be employed to predict QoE based on historical usage patterns, enabling proactive adjustments to the RIS.

Integration with 5G/6G Technologies

The integration of Reconfigurable Intelligent Surfaces (RIS) with 5G/6G technologies is improved by AI, optimizing performance alongside Massive MIMO (Multiple Input Multiple Output) systems to enhance capacity and coverage. It adapts to real-time conditions to efficiently manage network resources and intelligently reflects signals to enhance coverage in complex environments. Furthermore, AI optimizes RIS operations and reduces power consumption, leading to increased energy efficiency. This integration ensures better overall network performance, sustainability, and seamless connectivity.

Data-Driven Insights for Deployment

Data from past deployments can be analyzed by AI to guide future RIS placements, pinpointing areas that will result in the greatest performance improvements. Through machine learning, it is possible to predict potential breakdowns or maintenance requirements for RIS components, decreasing downtime and enhancing dependability.

User-Centric Design

The use of AI can help create user interfaces that adjust to each user's preferences and actions, simplifying the interaction and customization of RIS. AI can enhance the overall performance of RIS systems by analyzing data from multiple users and incorporating collective experiences and usage patterns (Nadeem *et al.,* 2020).

3. RESEARCH CHALLENGES IN AI-POWERED RECONFIGURABLE INTELLIGENT SURFACES (RIS)

3.1 Technical Challenges

RIS (Reconfigurable Intelligent Surfaces) that are powered by AI encounter numerous fundamental technical obstacles associated with their intricate interactions with wireless signals and environmental changes.

Channel Estimation and Sensing Accuracy:

Accurate channel state information (CSI) is crucial for adjusting RIS surface elements to achieve optimal performance. Yet, obtaining precise CSI in real-time is challenging due to environmental changes, user mobility, and the involvement of numerous elements. The development of AI algorithms capable of effectively estimating the channel, even with uncertain or limited data, continues to pose a major challenge.

Scalability and Complexity:

The complexity of optimization increases exponentially with the number of RIS elements and network size. This greater scale must be handled by AI models without sacrificing performance or needing unnecessarily high processing power. One of the main research challenges is optimizing large-scale RIS deployments and real-time integrating them into current network architectures.

Energy Efficiency:

Enhancing spectral efficiency through RIS is beneficial, but effectively controlling power usage for both the surface and AI algorithms presents a crucial challenge. Optimizing AI models for minimal energy consumption is essential, particularly in environments with limited resources, while still maintaining high performance.

Model Training and Real-Time Adaptation:

Large amounts of data are frequently needed for AI models' training and optimization. It is challenging to guarantee that an AI-powered RIS can instantly adjust to shifting circumstances, such as user mobility, interference, and fluctuating traffic loads. It is critical to develop AI methods that enable self-optimization, rapid adaption, and continuous learning in a dynamic wireless environment (Yuan *et al.*, 2020).

3.2 Practical Challenges

Practical difficulties arising from physical deployment and system integration impede the application of AI-powered RIS in real-world environments:

Hardware Limitations

Deploying AI-powered RIS poses a major challenge primarily due to the hardware involved. RIS is usually made up of a vast array of passive elements that have the capability to alter the incoming electromagnetic waves. Nevertheless, creating and producing these surfaces with the required precision and control presents considerable obstacles.

Component Precision: High-precision beam formation demands sophisticated hardware with low error phase shift manipulation. It's possible that the granularity offered by current materials and circuitry designs is insufficient for precise control.
Durability and Wear: These surfaces must perform dependably across a range of environmental factors, such as dust, humidity, and temperature swings. One of the biggest challenges is creating RIS hardware that is durable and can sustain performance over time.
Cost and Scalability: RIS systems are still too expensive to produce in large quantities and at a price point that allows for widespread use (Mishra *et al.*, 2021).

Energy Efficiency

AI-powered RIS systems, especially for applications such as smart cities or rural communication infrastructures, need to function well in energy-constrained contexts. The RIS devices are often inactive; but, when they are combined with AI systems, the computing demands on them may increase.

Power Consumption of AI Models: AI algorithms frequently call for a significant amount of processing power, especially those that use deep learning and real-time optimization. If the power consumption of the AI components is not tuned, this could compromise the low-energy benefits that RIS provide.
Energy Harvesting and Management: Although integrating energy-harvesting technology into RIS may provide a solution, existing approaches (such as solar or RF energy harvesting) have limited output and require improvement to enable efficient RIS operations (Abeywickrama *et al.*, 2021).

Real-time Adaptability

AI-powered RIS must react fast to changes in the propagation conditions (such as obstructions and user mobility) due to the dynamic nature of wireless settings. There are multiple obstacles in ensuring real-time adaptability of RIS.

Latency in Control Systems: Detecting a change in the environment and adjusting the RIS configuration can be hindered by the time delay, which can impede real-time adaptability. Both efficient algorithms and high-speed communication between the RIS and the controlling AI system are necessary to achieve low-latency performance.

Algorithmic Complexity: It can be computationally demanding to optimize the RIS configuration in real time across a variety of devices and frequencies while taking user mobility, network capacity, and interference into account. It is imperative that research be done on simplifying these algorithms without compromising performance (Tang *et al.*, 2020).

Deployment and Integration

Incorporating RIS into current wireless communication systems isn't simple. Implementing it practically requires coordination with existing infrastructure, guaranteeing compatibility, and addressing logistical hurdles in positioning RIS panels in ideal locations.

Site selection and Installation: The positioning of RIS is critical to their efficacy and should take the surrounding environment, the planned coverage area, and possible sources of interference into consideration. It is a continuous research problem to determine the optimal deployment strategy for optimum advantage.

Seamless Integration with Current Networks: AI-powered RIS must communicate with the base stations, routers, and other wireless devices that make up the current network infrastructure. Practical adoption requires ensuring interoperability with a wide range of technologies (e.g., 5G, Wi-Fi 6, and IoT protocols).

Control and Coordination of Multiple RIS

Coordination between multiple deployed RIS in an environment is crucial to prevent interference and optimize their combined effectiveness.

Decentralized Control: It is possible to decrease latency and boost failure resilience by coordinating several RIS panels decentralized from a central control unit. Decentralizing control does, however, provide difficulties for RIS panels in terms of coordination, communication, and decision-making.

Scalability Issues: Managing a growing number of RIS in an environment becomes significantly more complex. There is a need for research to create control algorithms that can efficiently handle large quantities of RIS.

Security and Privacy

RIS powered by AI can bring about fresh security and privacy issues in wireless communication networks.

Vulnerability to Attacks: The manipulation of signals by RIS could make them vulnerable to malicious actors aiming to disrupt communication. It is crucial to focus on researching ways to secure RIS from such attacks while still preserving their flexibility.

Privacy Concerns: RIS have the ability to modify signal propagation, which means that they might be used to track users or collect data on their conversations and movements. One of the main challenges is creating privacy-preserving algorithms that reduce these hazards.

3.3 Regulatory and Standardization Challenges

A number of regulatory and standardization issues surface as AI-powered Reconfigurable Intelligent Surfaces (RIS) gain traction in the wireless networking and telecommunications industries. To overcome these obstacles, more research is needed to integrate RIS technology with current systems. The following are some of the primary research questions in the area of standards and regulations pertaining to AI-powered RIS:

Lack of Unified Standards: At present, there is no consistent set of regulations overseeing the utilization and implementation of RIS technologies. Various areas might embrace distinct technical criteria for frequency ranges, signal manipulation, and data protection, posing challenges in guaranteeing the compatibility of RIS systems across diverse networks. Setting universal standards is crucial to ensure seamless incorporation into 5G/6G and future networks (You *et al.,* 2021).

Spectrum Allocation: The modification of radio signals by RIS raises concerns about spectrum interference and coexistence with current communication systems. The implementation of AI-powered RIS presents novel approaches to spectrum utilization, potentially necessitating regulatory entities to reconsider conventional spectrum allocation and interference management policies.

Data Privacy and Security: The integration of AI algorithms into RIS leads to a higher volume of data moving between the network, devices, and the environment. Data privacy concerns, especially in sensitive environments where user data could be leveraged to enhance network performance, must be addressed by regulatory frameworks. Furthermore, the implementation of RIS could create new security weaknesses, necessitating the development of suitable guidelines by regulatory bodies to counter potential threats.

Compliance and Certification: It is crucial to make sure that AI-powered RIS technologies adhere to the current electromagnetic emission and safety standards. The challenge lies in certifying devices to comply with regulatory standards for electromagnetic interference, exposure limits, and safety thresholds, especially because RIS has an impact on signal propagation in uncontrolled environments (ElMossallamy et al., 2020).

4. FUTURE DIRECTIONS

Reconfigurable Intelligent Surfaces (RIS) powered by AI have the potential to greatly improve network efficiency in upcoming wireless communication systems. Through the utilization of AI algorithms, RIS can intelligently adjust signal reflection, resulting in improved spectrum usage and energy efficiency. Future progressions might revolve around integrating deep learning and reinforcement learning methods to dynamically customize RIS for different real-time environments. This would optimize transmission paths, decrease interference, and ensure smooth connectivity, particularly in densely populated urban areas and high-mobility situations. In the progression towards 6G and beyond, the focus will be on the scalability of RIS within wireless networks. Research in the future will likely investigate how AI can facilitate the seamless integration of RIS into extensive network architectures containing thousands of RIS elements. This will involve the management of intricate network topologies and the optimization of interactions between multiple RIS to achieve consistent and reliable communication. AI will be vital in addressing the challenges of large-scale deployment, automating decision-making processes to ensure cooperative functioning of RIS elements without excessively burdening network resources.

The future holds potential for the development of fully autonomous RIS systems with AI capabilities for self-deployment and self-maintenance. Through the use of AI predictive analytics, RIS could independently recognize and resolve signal disruptions, environmental changes, or hardware malfunctions without the need for human intervention. In the coming years, AI will play a role in predicting and addressing dynamic communication needs, leading to reduced operational costs and consistent quality of service (QoS) for end users in diverse settings, ranging from smart cities to remote locations.

Ensuring strong security and privacy will be crucial with the growing usage of AI-powered RIS. Future developments might include using AI to actively monitor and address potential cyber security threats in real-time by identifying irregularities in signal behaviour or patterns that may suggest malicious activities like eavesdropping or jamming. AI-driven RIS could adjust signal properties to guarantee secure transmissions, possibly integrating block-chain technologies or advanced cryptographic methods to enhance the security and reliability of wireless communications in an increasingly interconnected world.

5. CONCLUSION

The landscape of wireless communication is undergoing a transformative shift with the emergence of AI-Powered Reconfigurable Intelligent Surfaces (RIS). The integration of artificial intelligence with intelligent surface technology has the potential to bring about unprecedented advancements in network performance, efficiency, and flexibility. This review has comprehensively outlined how AI-Powered RIS function, emphasizing their capabilities in dynamically controlling wireless environments and improving signal propagation.

The optimization of RIS operation is mostly dependent on AI algorithms. These systems can adjust to changing channel conditions, user requests, and environmental elements by utilizing machine learning and deep learning algorithms. Coverage, data rates, and energy efficiency can all be significantly increased with the real-time reconfiguration of the surface features. As this evaluation has made clear in several places, the combination of AI and RIS improves user experience while also optimizing resource allocation to meet the increasing need for high-capacity wireless networks.

In conclusion, AI-Powered Reconfigurable Intelligent Surfaces have great potential for the future of wireless communication. With the assistance of AI, we can develop flexible, efficient, and smart wireless networks that can meet the needs of an increasingly interconnected world. Moving forward requires a cooperative approach to tackle existing challenges and take advantage of the opportunities presented by this ground-breaking technology. As we are on the cusp of a new era in wireless communication, it is crucial to encourage innovation, standardization, and practical implementation to fully realize the potential of AI-Powered RIS.

REFERENCES

Abeywickrama, S., Zhang, R., & Cheng, J. (2021). Intelligent reflecting surface-aided wireless communication: A physical layer security perspective. *IEEE Wireless Communications Letters*, 10(5), 974–978. DOI: 10.1109/LWC.2021.3066027

Basar, E., Di Renzo, M., De Rosny, J., Debbah, M., Alouini, M.-S., & Zhang, R. (2019). Wireless communications through reconfigurable intelligent surfaces. *IEEE Access : Practical Innovations, Open Solutions*, 7, 116753–116773. DOI: 10.1109/ACCESS.2019.2935192

Björnson, E., Özdogan, Ö., & Larsson, E. G. (2020). Intelligent reflecting surface versus decode-and-forward: How large surfaces are needed to beat relaying? *IEEE Wireless Communications Letters*, 9(2), 244–248. DOI: 10.1109/LWC.2019.2950624

Di Renzo, M., Zappone, A., Debbah, M., Alouini, M.-S., Yuen, C., & Tretyakov, S. A. (2020). Smart radio environments empowered by AI reconfigurable meta-surfaces: An idea whose time has come. *IEEE Journal on Selected Areas in Communications*, 38(11), 2450–2525. DOI: 10.1109/JSAC.2020.3007211

ElMossallamy, M. A., Zhang, H., & Han, Z. (2020). Reconfigurable intelligent surfaces for wireless communications: Principles, challenges, and opportunities. *IEEE Transactions on Wireless Communications*, 20(2), 1096–1110. DOI: 10.1109/TWC.2020.3036433

Huang, C., Zappone, A., Alexandropoulos, G. C., Debbah, M., & Yuen, C. (2020). Reconfigurable intelligent surfaces for energy efficiency in wireless communication. *IEEE Transactions on Wireless Communications*, 19(8), 5153–5169. DOI: 10.1109/TWC.2020.2974743

Huang, Y., Song, H., & Wang, W. (2021). AI-enabled intelligent reflecting surfaces for secure communication in IoT networks. *IEEE Internet of Things Journal*, 8(5), 3622–3633. DOI: 10.1109/JIOT.2020.3038689

Li, X., Mao, J., Huang, Y., & Zhang, R. (2021). Intelligent reflecting surface enhanced wireless networks: Two-timescale beamforming optimization. *IEEE Transactions on Wireless Communications*, 20(12), 8092–8107. DOI: 10.1109/TWC.2021.3074935

Mishra, D. K., Joham, M., & Debbah, M. (2021). Learning to reflect and transmit: A novel intelligent reflecting surface with relay hybrid architecture. *IEEE Transactions on Wireless Communications*, 20(3), 1453–1469. DOI: 10.1109/TWC.2021.3051381

Nadeem, Q.-U.-A., Kammoun, A., Chaaban, A., Debbah, M., & Alouini, M.-S. (2020). Intelligent reflecting surface-assisted wireless communication: A discrete phase-shift model. *IEEE Transactions on Communications*, 68(3), 2114–2127. DOI: 10.1109/TCOMM.2019.2963665

Ozdogan, O., Bjornson, E., & Larsson, E. G. (2020). Using intelligent reflecting surfaces for rank improvement in MIMO communications. *IEEE Wireless Communications Letters*, 9(1), 60–64. DOI: 10.1109/LWC.2019.2941232

Renzo, M. D., Debbah, M., & Phan, V. (2020). Reconfigurable intelligent surfaces vs. relaying: Differences, similarities, and performance comparison. *IEEE Open Journal of the Communications Society*, 1, 798–807. DOI: 10.1109/OJCOMS.2020.3002955

Tang, W., Chen, M. Z., Wong, K.-K., & Wang, S. (2020). Path loss modeling and measurements for reconfigurable intelligent surfaces in indoor environment. *IEEE Transactions on Wireless Communications*, 19(5), 3961–3970. DOI: 10.1109/TWC.2020.2969473

Tang, W., Dai, J. Y., & Wong, K.-K. (2020). Wireless communications with reconfigurable intelligent surfaces: Path loss modeling and experimental validation. *IEEE Transactions on Wireless Communications*, 20(1), 421–434. DOI: 10.1109/TWC.2020.3024887

Wu, Q., & Zhang, R. (2019). Towards smart and reconfigurable environment: Intelligent reflecting surface aided wireless network. *IEEE Communications Magazine*, 58(1), 106–112. DOI: 10.1109/MCOM.001.1900107

Wu, Q., Zhang, R., & Larsson, E. G. (2020). Intelligent reflecting surface-aided wireless communications: A tutorial. *IEEE Transactions on Communications*, 69(5), 3433–3465. DOI: 10.1109/TCOMM.2020.3036433

You, C., & Zhang, R. (2021). Wireless communication aided by intelligent reflecting surfaces: Active or passive? *IEEE Transactions on Wireless Communications*, 20(11), 7270–7284. DOI: 10.1109/TWC.2021.3097401

Yuan, X., Shi, Y., & Debbah, M. (2020). Indoor intelligent reflecting surface enhanced wireless local area network: A case study. *IEEE Wireless Communications Letters*, 9(2), 199–203. DOI: 10.1109/LWC.2019.2955559

Zhang, L., Liang, Y.-C., & Peng, L. (2020). Intelligent reflecting surface aided multi-user MIMO: Exploiting deep reinforcement learning for beamforming. *IEEE Journal on Selected Areas in Communications*, 38(8), 1809–1822. DOI: 10.1109/JSAC.2020.3001081

Zhang, S., Huang, C., & Debbah, M. (2021). Machine learning methods for reconfigurable intelligent surfaces: State of the art and future prospects. *IEEE Communications Surveys and Tutorials*, 23(4), 2688–2721. DOI: 10.1109/COMST.2021.3085383

Chapter 7
Machine Learning for Design Solutions of Reconfigurable Intelligent Surface Applications:
Intelligent Surface Applications and Future Trends

S. Sudha
V.S.B. Engineering College, India

Kiran Shrimant Kakade
Symbiosis International University, Pune, India

N. Ashokkumar
https://orcid.org/0000-0002-4034-484X
Mohan Babu University, India

Sheshang Degadwala
Sigma Institute of Engineering, India

ABSTRACT

They have made the conversation much better because so many people use them. One example is the intelligent surface (RIS), which can be changed to help wireless devices work better. Some studies say that ML-optimized (RIS)--assisted wireless communication systems might be a good choice since they help the signal stay strong in the wireless frequency band. That being said, we can't say which method or ways of doing things are the best because there are so many of them. More than one study

DOI: 10.4018/979-8-3693-8099-4.ch007

has looked at these two ideas as ways to improve radio networks. As part of the study, they want to make a list of all of them. The project's goal is to find the best ways to use machine learning to make a system that works with RIS. There are different groups for each type of speed boost, database, machine learning, and hard code.

1. INTRODUCTION

To set up the fifth and sixth types, you should know a lot about how to do it right. It is possible to talk to your car and do things like IIoT and Industry 4.0 over 5G and 6G networks (Mannweiler, C.et al. 2017). To work, they need to be ready, not cost too much, and use little power. Tools can connect to 5G networks better than 4G networks, which makes apps work better. A lot of people are interested in 6G wi-fi networks this week, though. They're safer, and you can quickly give and receive data. Everything is tied and tracked at the same time, so it's easy to find things. You can see where they are, what shape they're in, and how fast they're going. 6G networks will be able to stay safe, plan ahead, and meet their own needs (Andrews, J.G. et al. 2024), but they won't be able to do these things. We can now talk to each other in more ways to help us reach our goals. A RIS is a smart surface that can be changed. A lot of people believe they are an important part of 6G.

If RIS helps wireless systems change how they send and receive data, "smart radio" will be in the area. A RIS is also called an intelligent reflection surface (IRS) or a big intelligent surface (LIS). From this point of view, a flat surface is made up of cheap, unimportant pieces that are simple to move around. While they are being sent electronically, anything can be changed at any time (Gutierrez, C.A, et al. 2021). One of the best things about RIS is this. RISs are used for many things because of this. Some ways to do this are to link the Internet of Things and sensor networks, keep the physical layer safe, let unmanned aerial vehicles (UAVs) talk to each other, and send and receive power. The RIS is now a well-known mm Wave tool in the world of 6G (Wild, T, et al. 2021). It gets information and looks for trends that can help it guess what will happen. After that, rules or steps can be made from these designs. This type of AI is called machine learning. Based on these trends, you can make smart, flexible choices that will help you guess values quickly.

Word, picture, and sign artists have done a lot of great work with this tool. Since it can do many things, many people think it is a good way to see how well Wi-fi networks work. Learn what a real talk is to get this done fast. A lot of people now study both machine learning and radio broadcast systems at the same time. You could use this to move on to the next set of networks. As networks get better, they will need more work because they will be able to set up, run, and handle faster data rates on their own. It's easy and doesn't take much work to make radio ads that use machine learning. A lot of experts are interested in how portable gadgets use machine learning at the moment. It is stronger, works better, and is easy to change because people learn quickly. Also, this is very true for tools that let people use predictions right away to help them choose better. At that point, most of the work, lessons (Tan, D.K.P, et al. 2021), and tests were about what RIS is and how it works. On the other hand, one study looks at how machine learning can be used with RIS networks. The people who worked on the study were mostly interested in how to use DL to help RIS talk to each other better.

It's easy to understand how machine learning RIS systems work because they talk about how they are put together. Fused learning (FL), watched learning (SL), and reinforcement learning (RL) are a few of these. There are a number of different points of view and a list of the most important changes that are made with each new version of the study system. A great deal of machine learning is already being used to make RIS simple to use (Pechac, P. et al. 2012). In this work, they are talked about and given names. RIS-based machine learning apps can be compared by how many users they have, what the paper is for, the system model that is given based on the inputs and outputs, and the method and model design that is used. Putting AI and RIS together took a lot of work. This book is mostly about how to deal with signs. The AI-RIS study is made up of different parts. It can read the world, get channels, make beamforming designs, and plan resources, among other things.

The literature review talks about both the AI program that was used and the place where it was used. It looks into how to make Wi Fi networks that are built on RIS work better (Di Renzo, M. et al. 2020). Some of the machine learning methods used in the RIS radio link study have already been talked about. They use new ideas like switch learning, hierarchy learning, and graph learning. The main point of the study is to quickly show how machine learning (ML) can be used in RIS messages. You need to figure out the best way to send cases, change results, add rates, and free up space. You should also figure out the best way to send CSI. You can answer this question in a lot of different ways. Some people also say that you can only use the computer tools, the RIS phase, and the SINR (signal-to-interference noise ratio) for a certain amount of time. They go over what the new device can and can't do. It also says what it can be used for (Liang, Y.C, et al. 2019).

(Liu, Y et al. 2021) A lot of the early writings were about things that were already known. This study, on the other hand, looks at new ways to use AI and machine learning in wireless systems that are built on RIS. The study also looks at how RIS could make better and faster phone calls. AI and RIS can work together to make things better, too. This will help people who build Wi-fi systems make the best ones over time. This study adds the following important things: This is the most up-to-date list of RIS study guides for books. They learn in simple ways, in depth, and with machines. You can use the data sources on this list to learn machine learning skills that will help the RIS-assisted language system work better. (Khalid, W, et al. 2021).

2. TECHNICAL BACKGROUND

Radio links need to be able to send and receive EMF waves in order to work. People who listen send and receive these waves most of the time. You couldn't send data without these waves. Not every gadget can work with every level and band. Ads come in many shapes and sizes. (Zhang, J, et al. 2020) Electromagnetic waves are used to send data over power lines, and then the data is sent back over power lines. A radio broadcast system works best when its parts are put together in the right order. The antennas are what set them apart. RISs can send. This means that when EMF waves pass through them, their phase and strength can be changed. All of these can change the path that energy waves take. (Rahmatov, N, et al. 2023) RISs can change the shape and direction of these parts to send, receive, or send back electromagnetic waves. RISs can do many things at the same time because of this. This is the first thing that comes to mind when you think about what RISs are all about. It's made up of many small pieces that together make up a line.

Any one of these parts can be changed to change the electromagnetic waves that go through them (Chu, Z, et al. 2022). You can make these parts out of metals, dielectrics, semiconductors, and other things. When it comes to electricity, these can be very different. You can put them together in a lot of different ways. One of the many things that metasurfaces can do is act as antennas, filters, polarizers, and beam formers (Kumaravelu, V.B, et al. 2023). In RISs, the pieces can always change shape and where they are placed. This goes one step further than meta surfaces. Putting those two thoughts together will help you do this. This is how RISs can change how they look and work with the wi-fi system. There are antennas that work better and can be bent more. In many ways, RISs are better than regular antennas. They can change form and arrange themselves in a certain way. A lot of these spots can shrink, but antennas can't. These fit better on phones with little space (Huang, C, et al. 2020).

There it is, the sixteenth one. These days, a lot of people want smart surfaces that can be changed. That's why they're here: to make Wi-fi groups work better. RISs can change form and be active or not active when they are active. They can also move their electric waves around and change where, when, and how strong they are. (Han, X, et al. 2020) The RIS knows how to make waves hit it and then go back to where they came from. This lets the RIS be changed so that it can work with wireless feeds. When you do this, Wi-Fi solutions work better.

We have learned a lot about RISs for three important reasons:

- How well the rainbow does its job: (Di Renzo, M, et al. 2020) A lot of noise between transmission lines can be cut down by paying attention to how the phase changes of the RIS parts. This makes the range work better in the end.
- Smart use of energy: signal energy can be sent directly to the monitor that needs it in wireless systems, which can help save energy (Huang, Y, et al. 2020).
- It should be easy for people to give and receive signs. Other people should find it hard to hear. The real layer will be safer this way. One way to do this is to make the setting hard for real people to use.

This page tells you everything you need to know: Table 1. There is a lot to learn and do in the area of radio communication systems that use RIS.

Table 1. Shows that RISs can be used to study a lot of different things.

Research Challenges	Research Opportunities
Making models of channels	Spectrum efficiency
Design of a protocol	Using energy well
Complexity (Ding, Z, et al. 2020)	Layer of physical security
Time synchronization	Multiple RIS systems working together and cooperating
Energy efficiency	THz Waves and Millimeter Waves could be used to send information (Attaullah, H.M, et al. 2021).
THz Waves and Millimeter Waves could be used to send information. This is possible with THz Waves and Millimeter Waves (Alexandropoulos, G.C, et al. 2019).	It works with other systems like MIMO and NOMA
Setting up and maintaining	The ability to improve reach and capacity both inside and outside
Handling of interference	Integration is possible in 5G and other wireless networks

More research needs to be done on RIS-aided radio transmission systems to fully understand what they can do and to find good ways to use them. (Elkashlan, M, et al. 2021) This is why these steps are still being changed. No. 23 No. 23 People are looking for new ways to use RISs and setting up big test beds to see how well radio communication systems that use RISs work. This is where useful ways to make and use RISs are formed. It has everything. There are lots of things to learn because it's still very new. But there are some issues that need to be fixed first (Yuan, X, et al. 2021, Rehman, F.U, et al. 2023). They are very good at everything they do in radio communication systems where they are used.

3. ML APPLICATIONS FOR RIS-ASSISTED COMMUNICATIONS

With AI, this part talks about faster ways to talk on the radio. We look at a model's source code, network design, database (if it's important), and machine learning method to see how good it is. Systems that use RIS and machine learning can have problems like the ones in Picture 12. You can find the best phase-shift with a lot of free tools. To make these tools, they use beamforming and tool predictions. To look for signs, you can group the modulation, find the spectrum, or demodulate the signal that is being sent. It is possible to do all of these things. The last thing we'll look at is this: Figure 1. Through the RIS, they share data from the UE.

Figure 1. Uses of machine learning to improve radio systems that use RIS.

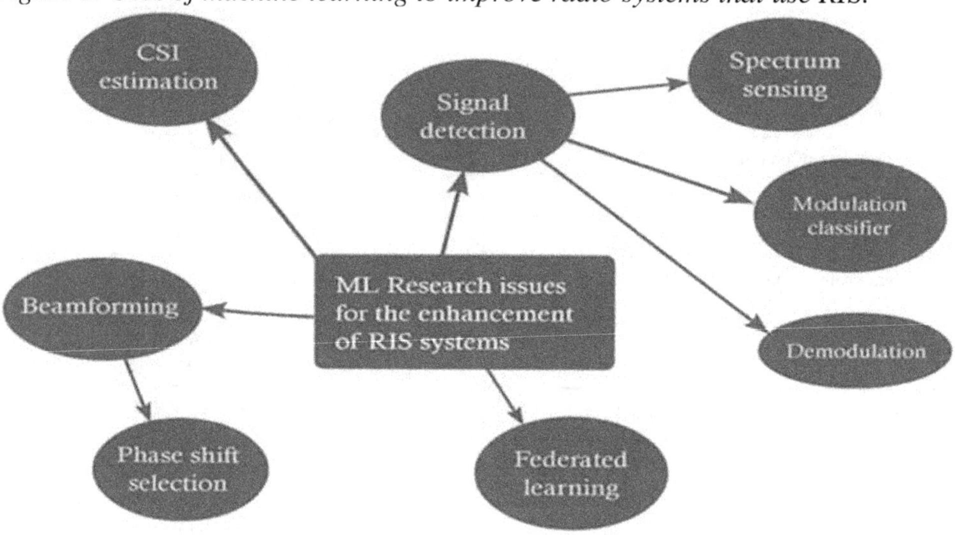

This part is mostly about mmWave, RIS, and MIMO gear. These places are also good for businesspeople to find data for their computer work. As a last job, we talked about the main kinds of radio broadcast systems that use RIS.

3.1. Resources for Generating Databases to Train ML on Wireless Communications

This type of machine learning tries to find trends in data that computers can use to figure out what to do on their own. A lot has been written about the machine learning method used to study radio communication systems, mostly the Rayleigh, Rician, and Saleh–Valenzuela models. You need to find the Monte Carlo Wi-Fi link to do this. Either they split the real and imaginary parts of the channel number or they check how big each part is. Nerve networks can only work in the real world, but channel realizations can deal with things like large numbers. This is the best way to use machine learning to make RIS-accelerated messages better, as shown in Table 2. With MIMO millimeter-wave gadgets, some people can learn things that other people can't. It has a ray-tracing model and a traffic game for cars that are both made to work in a 5G world where people can move around.

Table 2. That can be used to teach machine learning techniques that are used to make RIS-based communication systems better.

Type of Datasource	Description	Publicly Available?
Synthetic	You will use a model for ray-tracing and traffic to make maps based on what 5G can do.	No
Synthetic	One way to make a dataset that works with machine learning is to use the Remcom ray-tracing tool.	Yes
Physical	A list of channel readings for a number of different types of antennas and methods for getting them	Yes
Physical	They were taken on RIS in the 6 GHz band and OFDM transceivers and are in this set.	Yes

Anyone can use it for machine learning jobs in MIMO radio systems, it doesn't matter what kind of set it is. Even when they don't know much about the modulation scheme, AMR and other deep learning methods help them figure it out. When mmWave MIMO is turned on, the records are what pick the beam.

3.2. Estimation of CSI

For RIS-based communication tools, machine learning is used to figure out CSI. Next, we need to teach a computer program how to guess the CSI channel vectors that will be sent. This might take place in real life or online. This study has been put down and is mostly DL-based. This is the case because mazes show how the paths in MU systems work. Because channel guess and transmission failure are linked, the CSI is one of the hardest things to figure out when RIS is used to help communications. This means that the phase-shift pick can be made better at the RIS after a good guess of the channel has been made.

Figure 2. Diagram of ML-based CSI estimation for RIS communication systems.

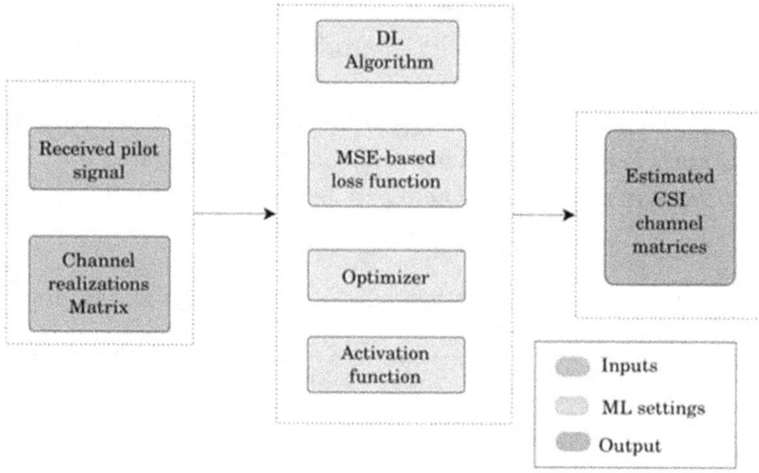

AI was used to figure out the channel, as shown in Find 13. This is how DL can figure out what went wrong if something does go wrong. The above CSI frames change the loss function, as shown by the mean square error (MSE). Little mean square error (MMSE) is one thing that sets these systems apart. Another change is the least mean square error (LMSE). The schedule is one more thing to think about. In this step, the model is made, and the firing functions that the neurons will use are chosen. Table 3 shows a number of ways to guess CSI on a computer or phone. These are some ways that RIS and deep learning are used. Here are the things they added, along with notes for each building.

Table 3. ML applications for CSI estimation on RIS-aided communications.

Contributions	Remarks
A structure based on a convolutional neural network (CNN) to determine CSI	There is no need for any more training for the model if the user's position changes by up to four degrees.
In order to find the CSI, two different types of sensors were mixed: active reflected and compression.	Deep learning doesn't need to know anything about the RIS array's shape ahead of time, but it does need a large sample.
AI-based way to find symbols in RIS wireless protocols and figure out CSI	Based on the signals it has seen, the suggested model tries to guess the CSI and phase angles.
We can still use DL to guess channels from user devices to the radio access system (RIS) even when the SNR and amount of multipaths change	Learn how to use deep learning to guess channels from user devices to radio access system (RIS) and see how well it works with various sample rates (SNRs) and multipath counts.
The low-rank sparsity of channels is used to look at two deep learning models for CSI estimates.	If the number of sensing devices is raised, the idea might work better.

One of the first steps in this direction was to set up a deep learning system for CSI in RISs. The test data were read by the UE and then sent to the DL model. It took some care to turn on and off the RIS parts in order to get the training data. Setting up a set of inputs and outputs for each channel is one way to do this. It had to be changed into channel vectors before the DNN could use it. It could get data from both straight channels and channels that were stacked on top of each other. There was no need for the person to learn in that way if their position moved by up to four degrees. It was made with an SGD planner, dropout, and a small set of 128 data points.

This helped the system work better. A 15-layer Convolutional Neural Network (CNN) showed it. Each of these 64 3x3x64 filters had 64 on it. Adam used a ReLU tool to make a plan and get things going. If you watch CNN, you might understand the RIS in CSI better. CNN models were used to make a deep learning system by the people who wrote this. "ESDR" stands for "single-enhanced deep super-resolution neural network." That was one of them. For a single-ray case, it might work well as CSI. To figure out what CSI stood for, the multi-scale deep super-resolution neural network (MSDR) was used.

It was thought by many that CSI would be better if the MU streams were already set up. It was shown how to make a network that is reflective of RIS. It was made up of Adam-optimized linear layers that had the sigmoid function turned on. Users' gadgets could still talk to the main station, even though there was no line of sight (LOS). The DL network was used to find the CSI at the BS in this case. It had a few more return signs and pilot controls.

The models that were looked at are shown in Table 3. The given DL characteristics, the style of the network, and the machine learning method were checked to see if they were unique. Figure 13 shows this. It also describes the kinds of data that were created with the help of the tools we talked about in Section 4.1. By giving away the source code, they let other people use their method and get the same results as them. This is shown in the last column of Table 4. The GitHub library has a lot of apps. This page also has the code that was used.

Table 4. Shows different machine learning frameworks, methods, and data sources that can be used to estimate CSI in RIS-based communication systems.

ML Algorithm	Architecture and ML Methods	Database Used	Available Source Code?
DL	There are dropout, SGD optimizer, minibatch, and two nine-layer CNNs.	Synthetic	MATLAB R2018b does work.
DL	NN that has a number that can be changed and a Way to Start ReLU	MIMO Deep	MATLAB R2018b does work.
DL	NN with steps that can be changed, the Adam planner, and the tanh activation function	The fake	Not
DL	CNN has 15 convolutional layers, 64 filters that are 3x3x643x3x64 pixels in size, the ReLU activation function, and the Adam planner.	Synthetic	Yes, Python and MATLAB R2018b
DL	Plans for EDSR, ReLU activation, and MDSR based on CNN	Synthetic	Of course, Python
DL	We used an FFDNet that is based on CNN, 3x33x3 filters, conv2D layers, ReLU activation, and group normalization.	Synthetic	Of course, Python
DL	32 filters and a CNN-based network were turned on.	[56] DeepMIMO	Yes

3.3. Beamforming Applications

When RIS is used to help wireless systems, test signals or channel realizations are used to show beamforming based on machine learning what to do. Like the ways we talked about in Section 4.2 for CSI. But most of the new study these days is about how to use a DL plan to raise the sum rate. Figure 3 shows that beams for systems that use RIS can be made with deep learning. Figure 3 shows a different way to learn, but it looks like the goal is the same as in Figure 3. The sum-rate cap helps you get the best result in the first case. An organization called RIS sends out beamforming and phase-shift vectors.

Figure 3. Scheme for DL-based beamforming for RIS-assisted communications systems.

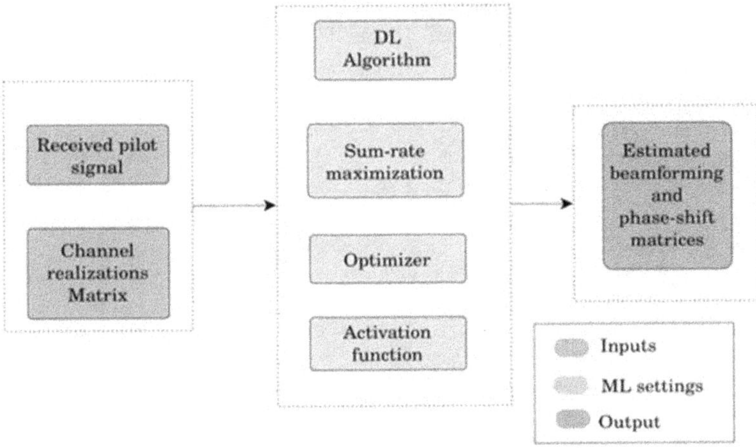

DRL, or differential reflection, has been used in a number of different ways instead of deep learning to shape beams. They can do this on their own, so you don't need to teach them. In this case, channel matrices could be thought of as an RL setting that creates the state space. The best ones that can be used to make a channel are these phase shift and beamforming matrices. You can also give the sum rate as a quick gift. Figure 15 shows a common way to use the DRL loop to send radio messages with the help of RIS. The motion is made up of beamforming matrices and phase changes as of a certain time t. A place called $S(t)$ holds the channel matrices from Equation (1) and the agent's most recent activity. The agent wants to win as many prizes as possible. Some people choose the sum-rate measure because it is the best for beamforming, but it takes a lot of work. If the goal stays the same, the steps are always changed to use a DRL method like DDPG, TD3, or SAC. This is important for making the model. Figure 4 shows a closed loop that will keep the beamforming and phase-shift matrices up to date and make them better if the sum-rate bonuses are high enough.

Figure 4. General approach of DRL for beamforming on RIS-aided communication systems.

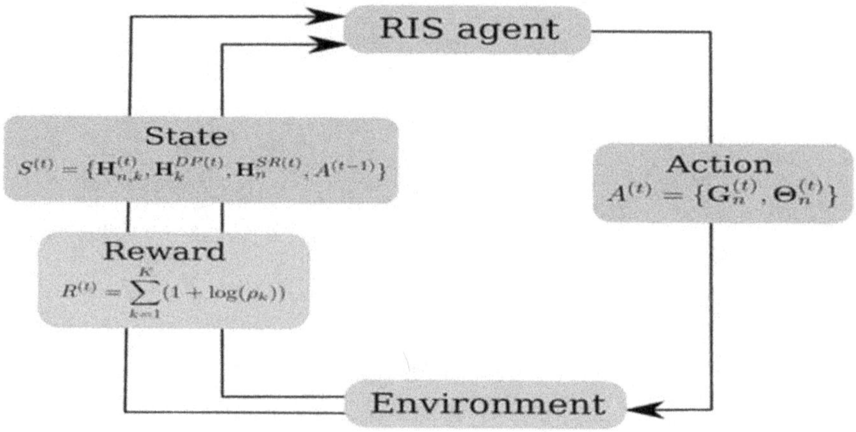

Deep learning has been used for more than just beamforming. Because of this, they don't need a skill set. You can think of channel matrices as an RL setting that helps make the state space in this case. For each channel, it finds the best phase shift and beamforming models, among other things. The general rate is also used and abused as a cheap gift. Figure 4 shows an easy way to use the DRL loop to make RIS radio calls. At time t, the phase changes and beamforming matrices are what make the work possible. This room is what action h(t) is made of. In (1). Most of the time, the sum-rate measure is picked as the best beamforming choice. Because the output always happens, a DRL method like DDPG, TD3, or SAC is used to switch the steps around. This is done so that a model can be made. This step needs to be taken in order to build the plan. Figure 4 displays a closed loop that can quickly modify the beamforming and phase-shift matrices to make them better when the sum-rate effects are at their strongest.

Table 5. Shows the updated techniques using machine learning to shape beams in RIS-based transmission systems.

Contributions	Remarks
Based on deep learning, this method lets the RIS figure out how to best handle the data that comes in; only channels at the busy parts are shown.	No previous understanding of the RIS array's shape is needed for this method to achieve data rates that are almost perfect.
Using the area for local spread and a DL-based method for phase change at the RIS	It doesn't need as much training and does a better job than the usual least-squares guess.
To guess the phase change of RIS mirrored parts using an unsupervised deep learning method	The model can set up shifts in real time while having a good rate.
This is the best way to make beamformers at the BS and reflection factors at the RIS work their best.	After a few tries, the plan can figure out how to speed up or slow down.
Active beamforming might work better from the base station to users if the method is based on DL, while inactive beamforming might work better for the RIS.	A standard RIS system can't match the suggested answer when it comes to BER performance.
an ML way to make the weighted sum rate better in RIS-based systems based on the traits of the product	With this plan, things work out better than with the BCD method.

continued on following page

Table 5. Continued

Contributions	Remarks
It was based on the BCD method to come up with a shared design solution that had beamforming at the entry point and phase vector.	Measurements that use 100 RIS parts show that the plan makes speed much better.
In a MU-MISO setting, the DRL method is used to plan broadcast beamforming and phase changes at the RIS.	This plan can work even if there are physical problems with RIS-aided wireless systems.

Another idea is to learn how to beamform without being seen. A method learned off-line is used in the study to make sure that the online predictions of how the RIS phase will change are always correct and up to date. The suggested plan includes a batch normalization system, an Adam optimizer system, and a ReLU activation system. An ANN has five fully linked layers with neurons that can be changed. First, plans were made for how to get the data that the DL method would need. After that, those plans were made. An AI-based method can be used to improve both the RIS reflection factors and the BS beamformers. As many people as possible used smart phones for this project. They used this to figure out how to speed up the RIS parts.

It would be better if RIS signal processing could learn. It was tuned and the beam was shaped at the same time. One DNN was made for the device being used, and the other was made for the computer.

This is true because RIS passive beamforming uses less power than UAV active beamforming. The plan was built on TD3, which had two players, each with three networks. MLP ideas were used to put together every network. The prize SEE helped stop the UAV in the play area. People needed to know which way things were going. It was put together from the UAVs in the state area. In both Section 4.2 and Table 6, the machine learning methods and beamforming factors were looked at next to each other. We give you a list of the source databases that were used for each move along with the source code.

Table 6. Shows the ML settings, datasets, methods, and source code that can be used for the new beamforming jobs on RIS-based wireless systems.

ML Algorithm	Architecture and ML Methods	Database Used	Available Source Code?
DL	You can change the fully linked layers of a perceptron, activate RELU, and lose MSE.	Deep MIMO	No
DL	There are two NNs with fully linked layers, an Adam controller, and ReLU activation.	Synthetic	No
DL	The ANN has 5 levels, and the amount of neurons, the Adam planner, and the ReLU activation can all be changed.	Synthetic	No
DL	You can use Python and MATLAB R2018b.	Synthetic	Yes, Python and MATLAB R2018b
RL	Reviewer and player networks in the DDPG algorithm, as well as a browning affect	n/a	Yes, Python (works as written)
RL	The Adam and Xavier optimization, the tanh activation function, and the SAC method all have three MLPs.	n/a	Yes, Python
RL	A loss function called MSE, an Adam optimizer, and a way based on Q-learning with a DQN agent are some of the things that make It up.	n/a	Yes, Python

3.4. Federated Learning Applications

Figure 11 is a flow chart that shows how each piece of user equipment (UE) has its own model. What the model is made of is sent to the base station (BS) by the RIS. With the global aggregation method, you can change every part of a global machine learning model. The people who worked on this study say that a FL model made on AirCom should be used. This model helps the group of models grow by following the rules set by MSE. Next, you need to figure out a set of MSE problems and pick out the best device. Getting rid of as many problems as possible is one way to do this. Once that is done, a different method is used to make a good plan for how to connect the beamformers. They talk about a way to make the RIS FL plan even better when it's not live with AirComp in. Using look-ahead data and a Lyapunov structure, the FL method can be broken down into its separate parts.

The BCD method is used to tune for phase shift. At the speaker, the sounds are then split up. A simple method based on changing parts one after the other can be used to make a RIS that needs a certain shift. The point of the study is to find faster ways to share tools and talk about them using FL-RIS to help communication. The writers suggest a different way to get better that doesn't involve the factors. This method is called block coordinate descent (BCD) in full. The MM method is then used to pick the best users and send power in the best way. Finally, we can find the RIS phase shift with the help of SDR and Gaussian reduction.

3.5. ML Applications for Signal Decoding

It was written that for a CNN-based RIS MIMO system, an OFDM signal needs to be demodulated. This would make the system work better by lowering the amount of data that is lost. That is, the user equipment (UE) screen was the best place to work with the info. The model was tested with data from the border station (BS) after being shown how to make fake channel realizations. The Saleh–Vanenzuela model was looked at in order to make the Rayleigh fading channels. As planned, different CNNs with two hidden layers, the Adam optimization, and a ReLU activation function were looked at. Also, 32 screens that are three by three were used. What the results were were decided by a function called softmax that was used as an activation function. It was also checked to see how well the BER and SER worked.

They used a CNN-GRU plan that everyone knew to find signs in RIS-based communication systems. Since the new network wasn't up and running yet, the OFDM data that was sent over the channel was copied and used to train it. The BER and SER numbers were also checked to make sure they were right for that type. CNN says it had 64 3x3-pixel screens and a surface for exchanging information. A sigmoid activation was used on the GRU network's output layer. It was also used in the deeper layers with two sigmoid shape activations. The GRU network was made up of four levels. The output layer had a sign that showed the person's most likely class. An Adam-based loss and a loss that isn't based on Adam were both used.

3.6. ML-Based Applications for RIS-Assisted Communications Modeling

Make tools that look like systems that use RIS to help them talk to networks. This will help researchers learn more about how RIS surfaces talk to networks. Open-source channel modeling tools can be used inside or outside of work. These ways explain how wireless data really move when there is a radio interference system (RIS).

The RIS needs to know a lot about what's going on around it in order to make deals between the real and digital worlds. No matter when the app is made, this is a very important step that must be taken. There are many ways to reach the above goals that you can learn about in the books.

These Are Some Things That Can Be Done

You will need a source, a receiver, a Reconfigurable Intelligent Surface (RIS), and an area you want to feel in order to use this method. It bounces off of a few things in the target area before it gets to the listener. The RIS's job is to change the signal where you want to hit it. When the reader looks over the information it has, it can get data that makes sense. It's possible that these messages are mirrored links or lines of sight (LoS). Someone, a radio system with many inputs and outputs (MIMO), and a radio interface system (RIS) are what they use.

In all of these ways, RIS's features help make radar, wireless sense, location, and maps better. One smart way to deal with spectral clutter is to use a device that sends data and watches it. These two things will make the range work better. You'll need a smart surface that can be moved around and has a tracking unit built in for this to work. To manage and direct the RIS in real time, you need to add an LCD screen, a computer unit (MCU), and/or a Field Programmable Gate Array (FPGA). There are more ways to look at how the RIS-based sense system works, as shown in Figure 6. One way is to use a block model.

Figure 5. The RIS can also do a lot of other things with electromagnetic waves. It can send and receive any number of single, dual, or quad beams and change directions. These jobs are done with buttons on the RIS that can be changed.

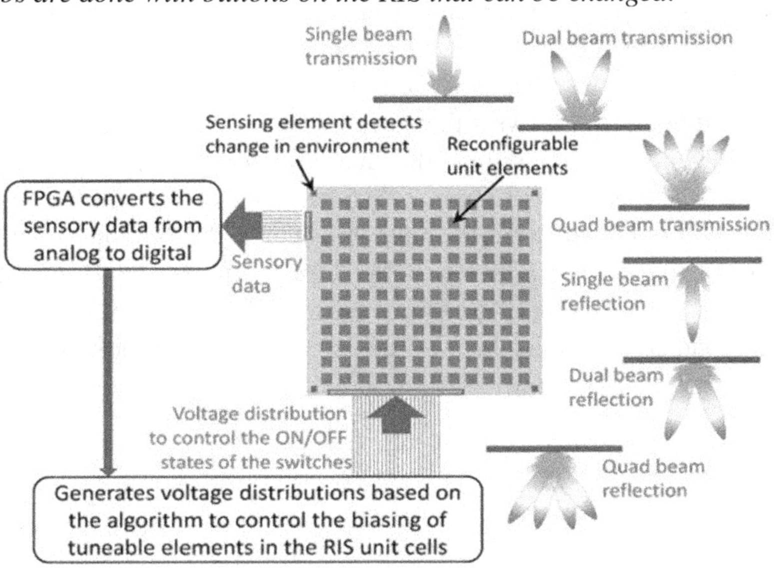

What the sensor unit learns from changes in the world around it is sent to a computer as a stream. The data is then sent to a processor, which could be an FPGA or an MCU. Through the digital code method, the voltage spread is made. This is how the FPGA is meant to work. Some of the parts of the RIS that can be pulled are avatar diodes, MEMS, and NEMS. This is possible because of how the electricity is spread out. In real time, you can also change the way the RIS spreads out. Two kinds of parts that can be changed are P-i-n diodes and varactor diodes.

4. UNIT CELL DESIGN

There are more ways to look at how the RIS-based sense system works, as shown in Figure 1. One way is to use a block model. This study shows a smart surface that can be changed. This surface can send out one beam, two beams, or four beams with a phase change of 90°. It can also pick up one, two, or four beams that have the same polarization as the electric signal that hits it. The switches must be set up in a certain way for something to be possible. We learned how to make this kind of adjustable RIS with a 2-bit unit element that can be changed for the transmission and reflection phases from the picture in Figure 7. Two p-i-n diodes are inside its five metal layers and three shielding layers. It was said that this unit cell might have a frequency-selective surface that turns based on time and polarization. Metal gratings are on the top and bottom of the unit cell, right next to each other. Because of these gratings, waves that are sent change direction. The loss slope for the FR4 resin base is 0.02, and the relative permittivity is 4.4.

This is used in all of the layers that keep things warm in the process. Between the two base layers, the letter g shows that there is space for air. The rate of speed is 500,000 hertz, and the force is 15 mm. It's 0.5 mm wide on both the top and bottom pieces. The main layer, which has the number 1 on it, is the thinnest. It's only 0.3\m thick. There is a p-i-n diode built into both of the solid layers at the middle base's top and bottom. An Infineon BAR63-02V p-i-n diode is used to set it up. So that they could be called unit cells, the p-i-n diodes were made to look like other RLC circuits. Figure 7f and 2g show this.

Figure 6. Each of the three insulator plates in this model of the RIS's unit cell is separated by air. These pictures show the air gap from different angles: (a) from the side; (b) from the top; (c) from the bottom; (d) from the middle; and (e) from the middle. Infineon BAR63-02V p-i-n diode model: (f) ON and (g) OFF states.

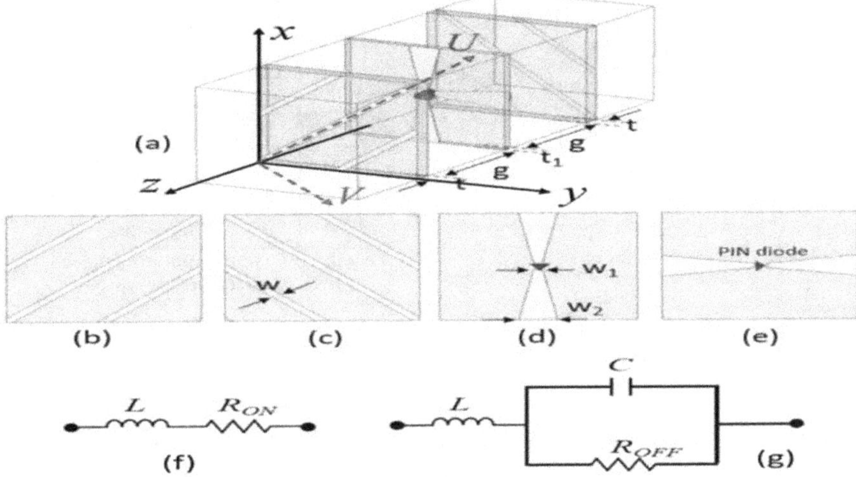

4.1. Unit Cell Simulations

To make full-wave electromagnetic models, a professional electromagnetic (EM) device known as CST Microwave Studio is used. This is because the inductance of the p-i-n diodes stays at -0.6 nH while they are working. However, the resistance changes as the voltage does. This is an RLC circuit since the p-i-n diode is "ON" when the bias voltage is greater than the cutoff voltage. It has a minimum resistance of R = 1 and an inductance of 0.6 nH on the inside. If you turn the p-i-n diode "OFF," it can also be thought of as a parallel RLC circuit with R = 5 kΩ, 0.25 pF of capacitance, and 0.6 nH of inductance.

The p-i-n diodes are set up on both sides of the main base layer inside the unit cell, as shown in Figure 8a. There is a letter D1 on top of the middle layer diode and a letter D2 on the bottom. There are four different ways to connect these two p-i-n diodes to make pairs that can switch states, as shown in Figure 8b. There are different ways to set things up. To make a good model of the unit cell, you need to use both Floquet port stimulation and uneven border conditions. In other words, the Floquet modes that hit the screen are always polarized at 45 degrees, no matter what the scan angle is. To make the first mode, the TE(00) and TM(00) modes are connected in a straight line. In CST MWS, this setting is set to face 45 degrees to the Floquet port's u-axis.

Figure 7. We can see the direction of the p-i-n diode on both sides of (a) if we break up the middle layer of the unit cell. As shown in the table, the p-i-n diodes can switch states, and when they do, the unit cell spreads out.

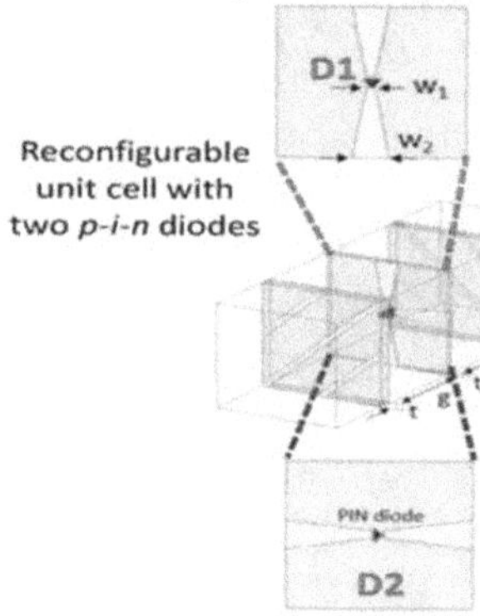

(a)

Diode ON	Diode OFF
1	0

D1	D2	Unit cell is highly:
1	0	Transmitting
0	1	Transmitting
1	1	Reflecting
0	0	Reflecting

Colour codes, corresponding digital codes and scattering behaviour.

(b)

There is a lot to say about how the unit cells reflect light whether both diodes are on or off. When one diode is on and the other is off, the unit cells can send and receive important notes. People have also seen that the polarization of the electric field being sent is different from the polarization of the electric field being made. When an electric field hits something or hits something else, it doesn't change its direction. The PIN diodes in the main layer control how this unit cell's direction changes.

The switchable polarization-rotating unit cell works better when you look at things in two different ways, XYZ and UVZ. The UV-axis is 45 degrees away from the XY-axis in this shape, as shown in Figure 2a. On the top of a unit cell, copper strips are lined up along the U-axis. It has a V-shaped bottom with copper strips lined up. PIN diodes are set up in metal strips along the X-axis at the top and the Y-axis at the bottom of the middle switching layer. The cloth has lines on both sides. Their names are 01 and 10. They both show how this unit cell's polarization spin works. To learn more about how the unit cell works, different models were run for each layer. Waves with a voltage of V can pass through the top layer, but waves with a voltage of U can't. It looks like a grid of wires.

These two numbers for the unit cell are not the same, as shown in Figure 9. Figure 9a and 4b show three different ways that D1 and D2 can be turned off. These show the amounts and steps of the transmission when D1 is off and D2 is on. It shows how much and how many steps are in each pair of signs. Between 1.35 GHz and 2.1 GHz, the broadcast level is more than -3 dB for both shift states. This can be picked up by all radio bands. If you change the states from 01 to 10, or from D1 = OFF and D2 = ON to D1 = ON and D2 = OFF, this won't happen. That's why the transmission sizes don't change. The electric field only changes direction when the diode is on. One state (01) is 180° away from the other state (10), which is 0.5° away. The electric field now moves in a different direction because of the change. This change is shown in picture 4c.

Figure 8. In different swapping states, this picture shows how the suggested unit cell spreads out. When one of the p-i-n diodes is turned on, these are the (a) transmission strengths and (b) transmission phases of the cross-polar part. This is how the two unit cells send different bits of data. There are two types of co-polar components, which can be ON or OFF. The forms and phases of these components are shown in (d) and (e). When two unit cells copy each other, they each go through a different set of steps.

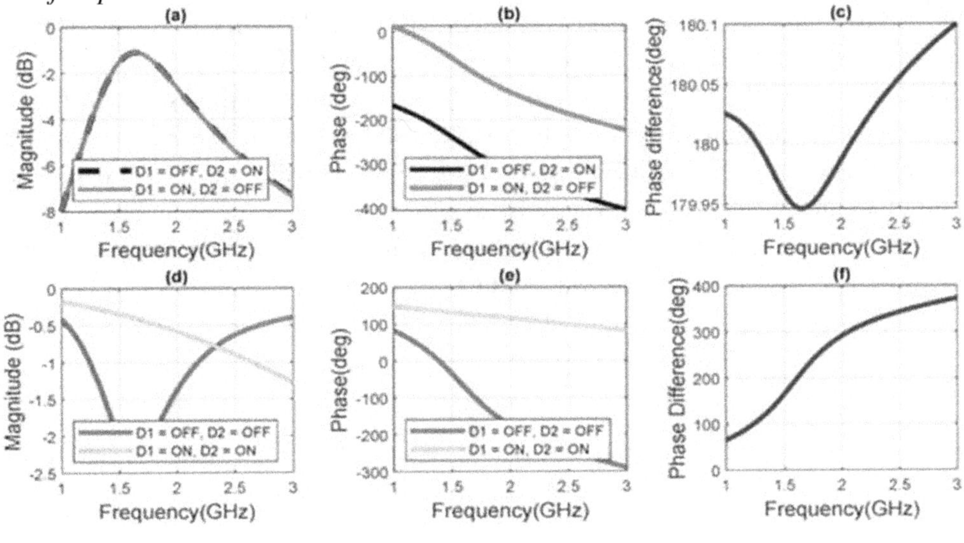

If D1 and D2 are both off (Figure 4d), the changing states are different. The same is true if D1 and D2 are both on (Figure 4e). In both the 1 GHz and 3 GHz switching states, there are echoes of more than -2.2 dB. When you go from 1 GHz to 3 GHz, the space between the mirrored steps does get bigger. If you look at a phase range of 180°∘±30°, as shown in Figure 4f, you can still get a narrow band function between 1.44 GHz and 1.64 GHz.

4.2. Multi-Functional RIS Design and Simulations

We made a 20x20 multipurpose RIS that can always be set to act in a different way. It covers 640x640 mm2. The study that looked at how the unit cell could spread led to it. In the track, there are 800 p-i-n diodes. When the number 0 is present, the p-i-n diode is ON. It is off when the number 1 is there. For ease of use, we split the area into two-by-two subregions, each with ten-by-ten unit cells. In Section 3, we talked about how to bias, and we believe that each sub-region can be biased on its own. In this case, changes can also be made to the way it spreads. On the other hand, because each part of the RIS can be done separately, it can be handled more better. All the p-i-n diodes in the same column could share the bias line. They could use the same DC power and be set up the same way.

$$\theta = \sin^{-1}\left(\frac{\lambda}{nP}\right) \tag{1}$$

It's shown by a sign in the middle. This graph displays the number of units in each time era. For each measure, the letter P tells you how big the span is. (1) can be used to figure out how much the beam bends after it's been cut in half. This is the RIS that was thought of. In both transmission and reflection modes, it let us split the beam and let specular transmission happen. Figure 10 shows that the transmission states are more direct than the reflected states. This drop might be because the suggested unit cell has a transmission magnitude greater than −1.5 dB and a return magnitude less than −1.6 dB. Just so you know, this is shown in Figure. This kind of reflected information is weaker because it spreads out in a way that wasn't planned. This is how the mirror was set up.

Figure 9. The 3-D radiation patterns that RIS sets send are also talked about. for each of the four sub-regions, with (D1 = ON; D2 = OFF). The pattern of radiation in three dimensions is the same pattern of radiation in four dimensions. (c) RIS with two sub-regions next to each other having (D1 = ON; D2 = OFF) and the other two sub-regions next to each other having (D1 = OFF; D2 = ON). For each of the sub-regions, this is how the radiation looks in three dimensions: (e) RIS with (D1 = ON; D2 = OFF) for some sub-regions and (D1 = OFF; D2 = ON) for the other two sub-regions that change.

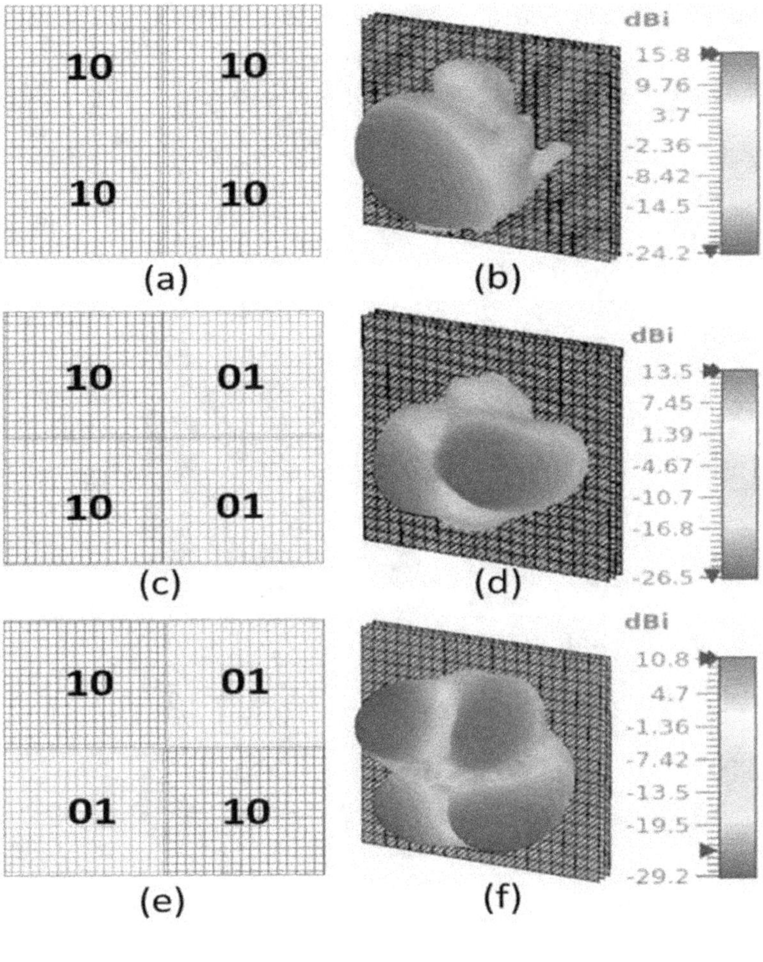

Figure 10. This type of radiation works for the RIS systems because it is three-dimensional and bistatic. The RIS with (D1 = ON; D2 = ON) for each of the four sub-regions and the three-dimensional B-type radiation pattern that goes with it (b). The two-dimensional, three-dimensional radiation pattern that goes with (c) RIS is shown in (d). (c) (D1 = ON; D2 = ON) means that two subregions next to each other are on and two subregions next to each other are off. What you see in (f) is the matching two-dimensional three-dimensional radiation pattern. Set D1 and D2 to OFF to turn off two sub-regions that switch off. Set D1 and D2 to ON to turn on two more sub-regions that will switch between them.

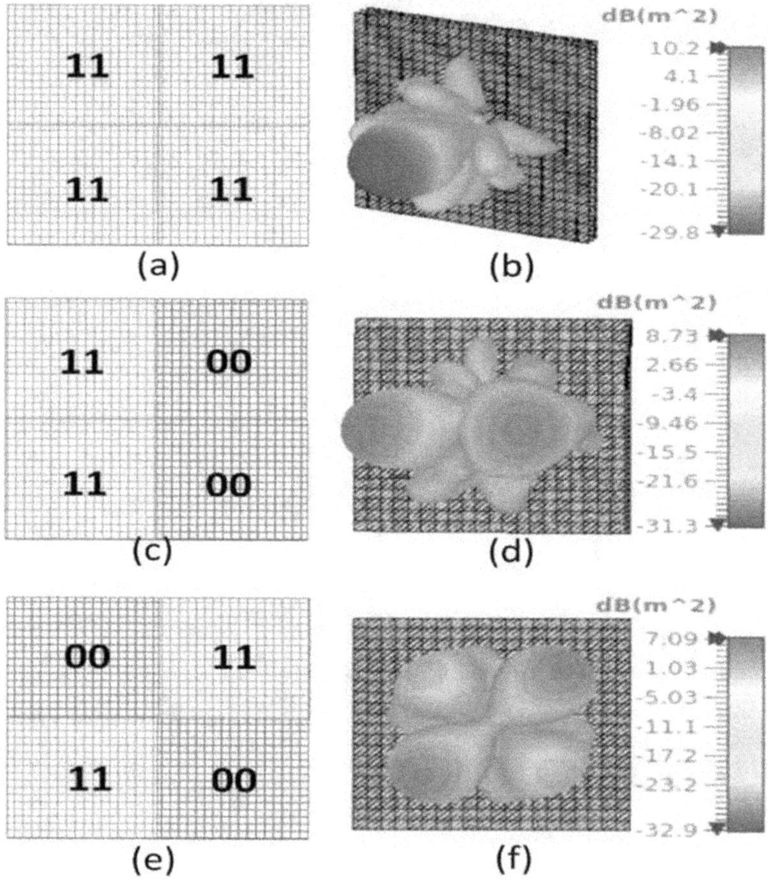

We made six different ones to show that the suggested RIS could do many things. This is how full RIS models are made in the CST microwave lab: with open boundary conditions and waveguide ports. The code sequence is shown in Figure 10, and the right three-dimensional radiation patterns for sending RIS energy are shown in Figure 11. They are also in the right order in Figure 10. The FPGA will handle the code process during the real run.

5. FUTURE RESEARCH TRENDS FOR ML-BASED IRS-ASSISTED WIRELESS COMMUNICATION

Models that are based on machine learning can help you get better radio service. Figure 12 shows the various methods an IRS-based radio transmission system could be put to use. Someone can't see anything because something is in the way, as shown in Figure 12a. The IRS might be able to help them talk about this. As you can see in Figure 12b, the IRS could be used to make very large machines. As shown in Figure 12c, controlled beamforming and IRS can be used together to make the physical layer safer. From what we can see in Figure 10d, it looks like mmWave huge transfer could work well with IRS. In Figure 10e, you can see how IRS could be used to send power straight to home Internet of Things networks. We will, of course. As we move forward, we will talk about some areas of research that may use both taxonomy and machine learning.

Figure 11. IRS-aided wireless communication applications.

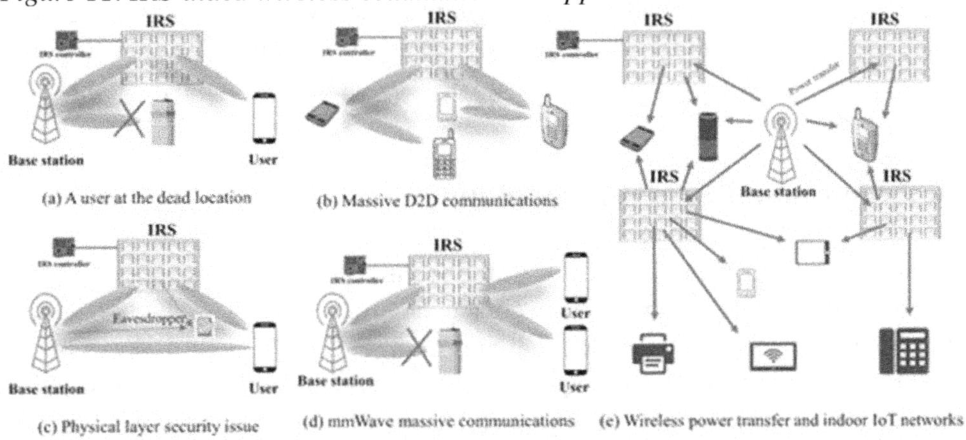

5.1. Optimal Placement of IRS

It's too bad that the mirror surface can't be moved or changed once it's set up. Because of this, you should know a lot about the IRS before you use it in real life. To get the best results, you might want to think about IRS gadget jobs that depend on how well they work. This is taken into account when the IRS is set up in and out of the country. Machine learning models can help you think about certain things in the best way. New models could help figure out how well something will work before it is installed because each one is likely to be different.

5.2. Dynamic Hybrid Beamforming

The BS signal can be bounced around by taxi devices to control beamforming. Because beamforming rests on parts of the IRS device that are not in use, it can make shapes that are unique to the user's tools. It might be possible to figure out how to guess the beamform for the user equipment (UE) service with the help of machine learning methods. Feedback from the user equipment (UE) may also be part of the dynamic beamforming method for mobile UEs. To get the best results, you need to try out new training data and models.

5.3. This Getting Information and Showing Models

When making ML models, it's important to gather data, because the data tells you how accurate the model is. Skewed data can lead to imperfect models, which is why it's hard to collect data for IRS-based communication methods that use machine learning. The main things that change the guess are how well a signal is recognized, its CE, and how the receiver's beamforming design works. Data collection tools can make or break the use of machine learning-based methods. This might be something to study in the future.

5.4. Constrained System Modeling

There are tools in cars that can change the beamforming by moving the BS signal around. Beamforming can make forms that can only be made with the user's tools when it's used on parts of the IRS that aren't being used. It might be possible to guess the beamform for the user equipment (UE) service with the help of machine learning. It is possible for users' equipment (UE) to send input as part of the dynamic beamforming method for mobile UEs. To get the best results, you should use new training data and models.

5.5. Channel State Characterization

Several models have been made to help classify communication that the IRS has helped with. A physics model, an impedance network-based model, an independent diffusive scatter-based model, and a framework based on tiles and codebooks are some of these models. Machine learning is now being used to study CE, and different methods have been used to guess how channel modeling will turn out. For instance, most of the methods for growth that are available are hard to use with computers. In the future, researchers may look into new ways to refine things.

5.6. IRS for IoT Network

Things that connect to the internet are being made all the time because people want them so much. With the help of 6G and the Internet of Things (IoT), the fourth industrial revolution is going faster. With infrared radio connection, more Internet of Things devices can be joined together both inside and outside. With all the Internet of Things devices spread out, it's tough to say how well the wi-fi will work. Machine learning systems will be able to copy how the world works better after these changes. A lot of the things that are connected to the internet are sensor nodes, which log data about different parts of the world. We can learn more about how the IRS changes the links between things that are online if we do more research that uses machine learning.

5.7. Protection Against Eavesdropping

There is a lot of talk about physical layer security (PLS) these days because security has become important in communication systems. PLS has a number of advantages over the coding method. It will be harder to keep track of all the keys for all the devices in a 6G network because PLS is easier to compute and the networks are not centralized. IRS made beamforming so that sending data could be done safely. With machine learning, it is possible to predict different nature dangers and then give the right reactions to those models. In a future study, the usefulness of this method could be checked.

5.8. MmWave Communication

The technology known as millimeter wave (mmWave) could help make the speed needed for 6G communication faster. There have been a few studies that use IRS-assisted mmWave transmission, but we need more studies to fully understand how it works. Different ways to modify signals, antenna arrays, plan methods, signal patterns, and using machine learning techniques for the distributed IRS are all things that could be looked into further.

5.9. EDGE Intelligence

"Edge computing" means moving operations away from the computer and to the edge of a network. This lets the source that is closest to the network do the work. There are many perks to using edge computing to connect user equipment (UE) to the server, such as low latency, low cost, privacy, and high data usage. Modern computers that use machine learning could be a good way to handle network tasks. IRS and ML should be able to improve edge computing in the long run, even though a lot more study needs to be done.

5.10. Hybrid Communication Implementation

To raise the RF frequency, you can also use a different method called visible light communication (VLC). Through IRS-based transmission, a new device that combines VLC and radio frequency (RF) could open up new study areas. For instance, people could use a mixed communication gadget to talk to each other while they are underwater. RF communication can be used to send data to a computer close to the surface. Visual communication can be used to send data farther below the surface. IRS works with both radio waves and light waves because it is based on wave reflection.

6. CONCLUSION

The purpose of this study was to find out how machine learning can be used in communication systems that use RIS. Making a training database, figuring out what kind of machine learning method was used, and talking about the problems that came up in each study setting were the major parts of the work. This paper talks about where RIS-assisted networks might be useful in the future and what machine learning tools need to be worked on. Researchers need to look into what machine learning techniques can do in order to make building wireless communication systems better and easier on computers. It's because the subject is very interesting and has been looked into a great deal. This paper is meant to help experts, students, and developers come up with fresh and useful ideas for the next generation of radio systems.

Andrews, J. G., Buzzi, S., Choi, W., Hanly, S. V., Lozano, A., Soong, A. C., & Zhang, J. C. (2014). What will 5G be? *IEEE Journal on Selected Areas in Communications*, 32(6), 1065–1082. DOI: 10.1109/JSAC.2014.2328098

Attaullah, H. M., & Javed, I. A. (2021). Wireless Robotic car control through human interface using eye movement. *Interdiscip. J. Appl. Basic Subj.*, 1, 1–6.

Di Renzo, M., Ntontin, K., Song, J., Danufane, F. H., Qian, X., Lazarakis, F., De Rosny, J., Phan-Huy, D. T., Simeone, O., Zhang, R., Debbah, M., Lerosey, G., Fink, M., Tretyakov, S., & Shamai, S. (2020). Reconfigurable intelligent surfaces vs. relaying: Differences, similarities, and performance comparison. *IEEE Open Journal of the Communications Society*, 1, 798–807. DOI: 10.1109/OJCOMS.2020.3002955

Ding, Z., & Vincent Poor, H. (2020). A simple design of IRS-NOMA transmission. *IEEE Communications Letters*, 24(5), 1119–1123. DOI: 10.1109/LCOMM.2020.2974196

Guo, H., Liang, Y. C., Chen, J., & Larsson, E. G. Weighted sum-rate maximization for intelligent reflecting surface enhanced wireless networks. In *Proceedings of the 2019 IEEE Global Communications Conference (GLOBECOM)*, Waikoloa, HI, USA, 9–13 December 2019. DOI: 10.1109/GLOBECOM38437.2019.9013288

Gutierrez, C. A., Caicedo, O., & Campos-Delgado, D. U. (2021). 5G and beyond: Past, present and future of the mobile communications. *Revista IEEE América Latina*, 19(10), 1702–1736. DOI: 10.1109/TLA.2021.9477273

Han, Z., Li, H., Li, H., & Han, X. Wireless power transfer using reconfigurable intelligent surfaces: Challenges and opportunities. IEEE Access 2020, 8, 136843–136859. DOI: 10.1109/TLA.2021.9477273

Hu, M., Li, Y., & Di Renzo, M. Reconfigurable intelligent surface-based wireless energy transfer: A survey. IEEE Access 2020, 8, 91838–91851. DOI: 10.1109/TLA.2021.9477273

Huang, C., Alexandropoulos, G. C., Zappone, A., Debbah, M., & Yuen, C. (2019). Reconfigurable intelligent surfaces for energy efficiency in wireless communication. *IEEE Transactions on Wireless Communications*, 18(8), 4157–4170. DOI: 10.1109/TWC.2019.2922609

Huang, C., Zappone, A., Alexandropoulos, G.C., Debbah, M., & Yuen, C. Holographic MIMO surfaces for 6G wireless networks: Opportunities, challenges, and trends. IEEE Access 2020, 8, 205729–205747. DOI: 10.1109/TLA.2021.9477273

Khalid, W., Yu, H., Do, D.T., Kaleem, Z., & Noh, S. RIS-aided physical layer security with full-duplex jamming in underlay D2D networks. IEEE Access 2021, 9, 99667–99679. DOI: 10.1109/TLA.2021.9477273

Kumaravelu, V. B., Imoize, A. L., Soria, F. R. C., Velmurugan, P. G. S., Thiruvengadam, S. J., Do, D. T., & Murugadass, A. (2023). RIS-Assisted Fixed NOMA: Outage Probability Analysis and Transmit Power Optimization. *Future Internet*, 15(8), 249. DOI: 10.3390/fi15080249

Liu, Y., Huang, Y., & Zhang, J. Intelligent reflecting surface-aided beamforming for multiuser communication systems. IEEE Access 2020, 8, 43238–43253. DOI: 10.1109/COMST.2021.3077737

Liu, Y., Liu, X., Mu, X., Hou, T., Xu, J., Di Renzo, M., & Al-Dhahir, N. (2021). Reconfigurable intelligent surfaces: Principles and opportunities. *IEEE Communications Surveys and Tutorials*, 23(3), 1546–1577. DOI: 10.1109/COMST.2021.3077737

Niu, H., Lin, Z., Chu, Z., Zhu, Z., Xiao, P., Nguyen, H. X., Lee, I., & Al-Dhahir, N. (2022). Joint beamforming design for secure RIS-assisted IoT networks. *IEEE Internet of Things Journal*, 10(2), 1628–1641. DOI: 10.1109/JIOT.2022.3210115

Pan, C., Ren, H., Wang, K., Kolb, J. F., Elkashlan, M., Chen, M., Di Renzo, M., Hao, Y., Wang, J., Swindlehurst, A. L., You, X., & Hanzo, L. (2021). Reconfigurable intelligent surfaces for 6G systems: Principles, applications, and research directions. *IEEE Communications Magazine*, 59(6), 14–20. DOI: 10.1109/MCOM.001.2001076

Rahmatov, N., & Baek, H. (2023). RIS-carried UAV communication: Current research, challenges, and future trends. *ICT Express*, 9(5), 961–973. DOI: 10.1016/j.icte.2023.03.004

Rehman, F. U., Attaullah, H. M., Ahmed, F., & Ali, S. (2023). Data Defense: Examining Fintech's Security and Privacy Strategies. *Engineering Proceedings*, 32, 3.

Rost, P., Mannweiler, C., Michalopoulos, D. S., Sartori, C., Sciancalepore, V., Sastry, N., Holland, O., Tayade, S., Han, B., Bega, D., Aziz, D., & Bakker, H. (2017). Network slicing to enable scalability and flexibility in 5G mobile networks. *IEEE Communications Magazine*, 55(5), 72–79. DOI: 10.1109/MCOM.2017.1600920

Subrt, L., & Pechac, P. (2012). Intelligent walls as autonomous parts of smart indoor environments. *IET Communications*, 6(8), 1004–1010. DOI: 10.1049/iet-com.2010.0544

Tan, D. K. P., He, J., Li, Y., Bayesteh, A., Chen, Y., Zhu, P., & Tong, W. Integrated sensing and communication in 6G: Motivations, use cases, requirements, challenges and future directions. In Proceedings of the 2021 1st IEEE International Online Symposium on Joint Communications & Sensing (JC&S), Dresden, Germany, 23–24 February 2021; pp. 1–6.

Tang, Z., Hou, T., Liu, Y., Zhang, J., & Zhong, C. (2021). A novel design of RIS for enhancing the physical layer security for RIS-aided NOMA networks. *IEEE Wireless Communications Letters*, 10(11), 2398–2401. DOI: 10.1109/LWC.2021.3101806

Wild, T., Braun, V., & Viswanathan, H. Joint design of communication and sensing for beyond 5G and 6G systems. IEEE Access 2021, 9, 30845–30857. DOI: 10.1109/LWC.2021.3101806

Yang, L., Meng, F., Zhang, J., Hasna, M. O., & Di Renzo, M. (2020). On the performance of RIS-assisted dual-hop UAV communication systems. *IEEE Transactions on Vehicular Technology*, 69(9), 10385–10390. DOI: 10.1109/TVT.2020.3004598

Yuan, X., Zhang, Y. J. A., Shi, Y., Yan, W., & Liu, H. (2021). Reconfigurable-intelligent-surface empowered wireless communications: Challenges and opportunities. *IEEE Wireless Communications*, 28(2), 136–143. DOI: 10.1109/MWC.001.2000256

Chapter 8

Design Challenges and Applications of Wireless Powered Unmanned Aerial Vehicles Assisted by Reconfigurable Intelligent Surfaces

Nitisha Waghela
https://orcid.org/0009-0003-7002-5377
Sage University, Indore, India

Swapnil Waghela
https://orcid.org/0000-0003-0917-8932
Shivajirao Kadam Institute of Technology and Management, Indore, India

ABSTRACT

Reconfigurable intelligent surfaces (RIS) and wirelessly piloted unmanned aerial vehicles (UAVs) are a revolutionary step in wireless communication and energy transfer. This chapter focuses on the integration of RIS technology and wirelessly piloted UAVs, including various design challenges and potential applications. RIS made of programmable metamaterials can dynamically control electromagnetic waves, thereby improving signal strength, coverage, and efficiency. When combined with wireless energy transfer (WPT) technologies, UAVs can operate for long periods of time without requiring frequent battery replacement or landing.The major challenges faced in this integration include optimizing RIS configurations for effective WPT, ensuring uninterrupted communication in dynamic environments,

DOI: 10.4018/979-8-3693-8099-4.ch008

& maintaining energy efficiency. In addition, this chapter investigates the implications of RIS-assisted WPT for UAVs in various fields such as disaster management, agricultural monitoring, & logistics.

8.1 INTRODUCTION TO RIS AND WIRELESS POWERED UAVS

Reconfigurable Intelligent Surface (RIS) is a new and advanced technology that helps control electromagnetic waves and improve the performance of communication systems. RIS is a type of smart surface primarily made of programmable metamaterials. (Dajer et al., 2022) These metamaterials are materials whose electromagnetic properties can be controlled in a very specific way as shown in Figure 1.

Figure 1. Reconfigurable Intelligent Surface (RIS)

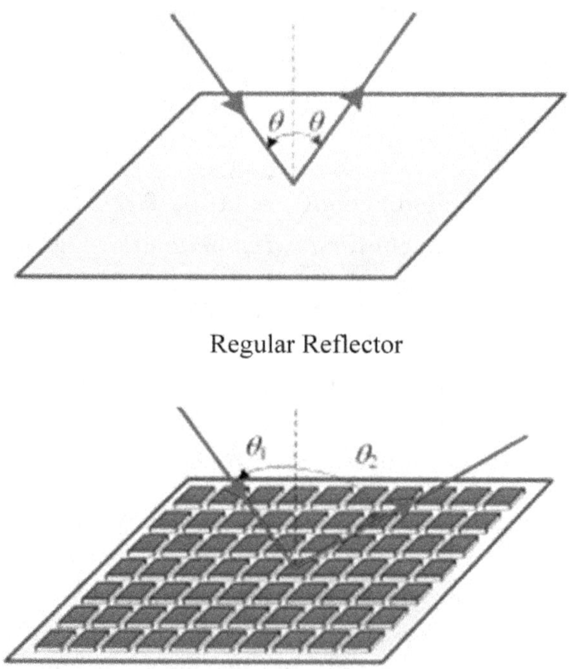

The main function of RIS is to help change the path of electromagnetic waves, bend their direction, and reflect or absorb them. It can enhance the signal strength in wireless communication networks, improve signal coverage, and overall increase the efficiency of the communication network.

For example, if a mobile phone or wireless device is in an area with a weak signal, RIS can reset its surfaces to improve the signal strength and provide better connectivity to the user. In this way, RIS can make communication networks more stable and reliable, especially in places where traditional base stations or signal towers cannot reach.

Introduction to Wireless Powered UAVs and Their Importance in Modern Applications

Wireless powered unmanned aerial vehicles (UAVs) are drones that are powered by wireless power transfer (WPT) instead of batteries. The main advantage of these UAVs is that they usually do not require frequent battery replacement or landing, as they are powered while in the air.

Figure 2. General block diagram of Wireless Power Transfer (WPT).

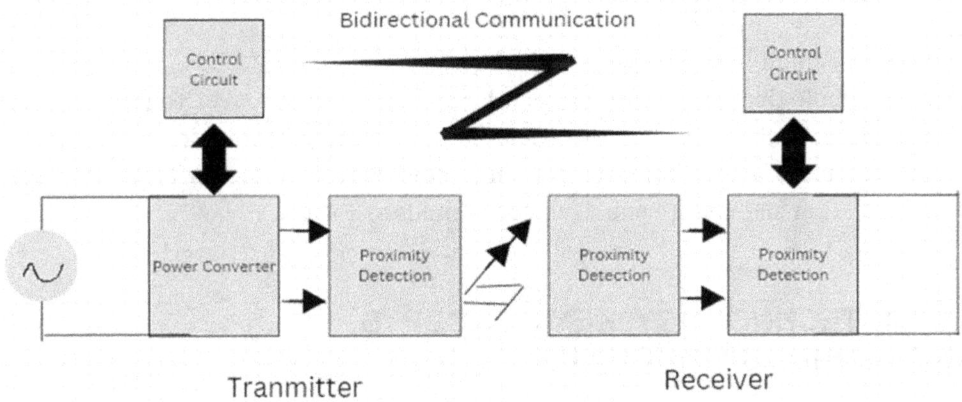

A typical wireless power transfer (WPT) system consists of a transmitter that is supplied with high frequency AC (alternating current), generated by a high frequency power converter. This transmitter is connected to a receiver, which is connected either through an electric, magnetic, or electromagnetic field. This induces a voltage in the receiver. The power converter is used to transmit the voltage generated in the receiver to the load, which adjusts the voltage according to the load specifications. (Korakianitis & Vokas, 2019) Figure 8.2 shows the block diagram of a typical wireless power transfer (WPT) system.

Wireless power transfer means that UAVs are powered without any physical connection. It involves sending energy from a power source (such as a station or another UAV) to another UAV via power waves. This technology gives UAVs the ability to fly longer distances, which is especially useful in areas where long flight times are required, such as disaster relief, agricultural monitoring, or surveying in remote areas.

These UAVs are important in many modern applications:

Disaster Management: During disasters such as earthquakes or floods, UAVs can be used to conduct aerial surveys of the affected areas and deliver relief materials to needy places. Through wireless energy, these UAVs can fly for long periods of time, making it possible to deliver help on time.

Agricultural Monitoring: UAVs are used to monitor the condition of crops in agricultural fields. With the help of these UAVs, farmers can better track the health and growth of their crops, so that measures can be taken to increase the yield.

Logistics and Delivery: The use of UAVs is increasing in e-commerce and delivery of goods. Wireless power transfer enables UAVs to fly continuously without stopping, allowing them to deliver goods faster and over longer distances.

Thus, the combination of RIS and wirelessly operated UAVs is very powerful from a technological point of view, as it helps in providing longer flight duration, better communication, and energy efficiency. These two technologies together provide better and effective solutions, opening new directions for aerial monitoring, communication and energy management in modern society.

8.2 RIS TECHNOLOGY: KEY CONCEPTS AND WORKING PRINCIPLE

Reconfigurable Intelligent Surfaces (RIS) are a new and advanced technology used to improve signals in communication systems. RIS uses programmable metamaterials. Metamaterials are special types of materials whose electromagnetic properties differ from normal materials. In RIS, these metamaterials are specially designed so that they can control and change the path of electromagnetic waves.

RIS consists of millions of tiny elements called “reflectors” or “interferers”. These elements interact with electromagnetic waves and bend them in such a way that the quality and strength of the signal can be enhanced. These metamaterials are designed to reflect, absorb or refract waves, thereby improving the performance of communication networks.

The key point is that these elements are programmable, which means that their function can be easily controlled by changing them. (Hong, 2023) For example, when a signal is weak or obstructed, the elements of a RIS can be reset in such a way that they can direct that signal and make it stronger.

How RIS can improve signal strength, coverage and energy efficiency

There are several major improvements that can be made to communication systems through RIS, such as improving signal strength, coverage and energy efficiency:

Increasing Signal Strength

The RIS can change the direction of the signal so that it reaches the receiver directly, thereby increasing the signal strength. When the signal is weakened due to an obstruction (such as a building or tree), the RIS can bend the signal in such a way that it can reach the receiver by bypassing the obstruction. This improves communication and provides better connectivity to the user.

Improved Coverage

The signal can be spread even in areas where communication was previously difficult by using RIS. For example, in a remote or dense area where traditional signal towers cannot reach, a RIS can help spread the signal to these areas. This increases the coverage capacity of the communication network and allows more people or devices to be connected.

Improved Energy Efficiency

An important advantage of a RIS is that it increases the energy efficiency of the communication network. Using a RIS allows the signal to be sent directly and effectively, reducing unnecessary energy losses. In addition, devices connected to a RIS generally require less energy, as they only modulate the signal, not generate it. This means that the communication system can do more with less energy. (Li et al., 2023)

For example, in a RIS-enabled network, if a mobile phone is in an area with a low signal, the RIS can divert the signal in such a way that the mobile phone receives a strong and stable signal, saving its battery life.

Benefits of Using RIS

1. Improved Signal Quality

The biggest advantage of RIS is that it can significantly improve signal quality. Generally, signals in communication networks can be weak, especially in areas where radio waves encounter obstacles (such as buildings, trees or other obstructions). In this case, the technology of RIS modulates the signal in such a way that it can overcome these obstacles and receive a better quality signal.

The programmable metamaterials of RIS are designed in such a way that they can reflect, absorb and direct electromagnetic waves. When a signal is weak, RIS directs it and bends it and sends it in the direction where the receiver is located. Through this process, the signal strength increases and the connectivity of the network becomes more stable and faster. (Sharma, Chehri, & Fortier, 2021)

This means that when we are transferring data from a mobile phone or any other wireless device, the improvement in signal quality by RIS transfers the data faster and without any interruption. Also, if there is a problem in the network, RIS helps to fix it, thereby improving connectivity.

2. Low Cost

RIS technology can be implemented at a lower cost than traditional signal towers and base stations. Generally, signal towers and other infrastructure need to cover a large area, which are expensive and are also difficult to build and maintain. In addition, land, materials and labor are required to install these structures, which significantly increases the overall cost.

With RIS, new structures do not have to be built or expensive towers constructed. RIS can be used on existing infrastructure such as walls, roofs or other surfaces, which helps to improve the wireless signal. By installing the programmable elements of RIS on these surfaces, signal enhancement can be implemented. This means that the cost of the network is significantly reduced, making it an affordable solution, especially in places where it was previously difficult to build expensive signal towers.

In addition, improving signal quality through RIS also reduces the load on signal towers, thereby increasing their efficiency and reducing their maintenance costs.

3. Long Distance Communication

Another important advantage of RIS is that it helps transmit communication over long distances. RIS technology can increase the coverage and distance of the signal compared to traditional networks. Usually, a lot of signal towers are required to send signals over long distances, but with the use of RIS, the signals can be controlled and redirected in new ways, increasing the reach of the signal.

This is especially important in remote areas where it is difficult to extend communication towers or other infrastructure. RIS can modulate the signal so that it can reach areas where signals were not able to reach before. For example, in hilly areas, dense forests or rural areas where network coverage is poor, RIS can divert the signal in a direction that provides good connectivity even in these areas.

In addition, the use of RIS makes communication networks more flexible. It gives networks the ability to work in areas where previously communication was problematic, such as underground or inside buildings, ensuring communication even in remote and off-grid locations.

RIS technology offers the benefits of improved signal quality, lower network costs and longer communication distances. It is especially effective in locations where traditional network solutions are expensive or impossible. With RIS, communication systems can become more stable, faster and cost-effective, making networks more flexible and effective for the future (Hasan, Sabuj, Hamamura, & Hossain, 2024).

8.3 WIRELESS POWER TRANSFER (WPT) AND ITS IMPORTANCE FOR UAVS

Wireless Power Transfer (WPT) is a technology that transmits energy wirelessly through the air. This technology uses electromagnetic waves or magnetic fields to transmit energy from one place to another. This method is especially beneficial for devices that move at high speeds or do not require regular battery replacement or charging.

Figure 3. UAV-WPT overview with single UAV. (Ojha, Raptis, Passarella, & Conti, 2023)

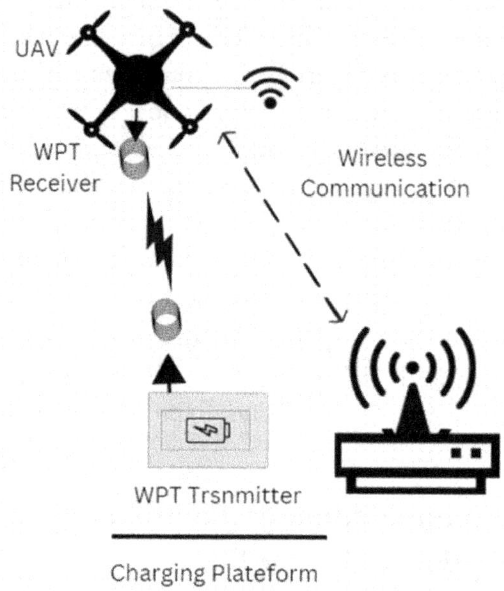

UAVs (unmanned aerial vehicles), or unmanned aerial vehicles, such as drones, are usually powered by batteries, which degrade over time and need to be recharged. This hinders their efficiency as the UAV has to land frequently to charge, reducing flight time. This is why wireless power transfer can become a revolutionary solution for UAVs. In Fig. 8.3, we depict a simple WPT framework for recharging of the UAVs. The recharging process of the UAVs are aided by a charging platform at the ground. The UAV can detect the charging platform using different means such as wireless communication or image processing, and aligns its WPT receiver with the WPT transmitter embedded in the charging platform. On the other hand, the charging platform can also employ sensor-based systems to better align the WPT transmitter and receiver for better efficiency of the recharging process. Depending on the application and requirements, there can be multiple UAVs employed in operation and multiple charging platforms deployed at the ground for facilitating the WPT. (Ojha, Raptis, Passarella, & Conti, 2023)

With WPT, UAVs can be charged in the air without any physical contact. WPT involves the transfer of energy using electromagnetic waves or magnetic fields. This energy is sent from a sending station to a receiving device (such as a UAV). While the UAV is flying, it receives continuous power from a nearby WPT station, so its battery never runs out and the UAV can operate without stopping.

Benefits of WPT for UAVs

Longer Life of UAVs: WPT technology increases the battery life of UAVs. Usually, when a UAV runs out of battery, it has to land on the ground to charge. But with WPT, the UAV can charge continuously while in the air, putting less strain on its battery and increasing battery life. This means that the UAV does not need to stop to charge as often, increasing its efficiency and flight time. (Nguyen et al., 2020)

Reduction in Maintenance Costs: When a UAV has to land frequently to charge, it can reduce the maximum capacity of the battery and require early battery replacement. WPT allows the UAV to receive energy for a longer period of time, which puts less strain on the battery and increases its life. This reduces the cost of battery replacement, and also reduces the overall maintenance cost of the UAV. Moreover, using WPT technology eliminates the need for UAVs to stop at charging stations frequently, making it a cost-effective solution. (Gong, Yoon, & Noh, 2024)

Increased Operating Time: One of the major advantages is that the flight time of the UAV is significantly increased due to the energy transfer by WPT. UAVs have limited batteries, which get exhausted after some time, but through WPT it can be ensured that the UAV continues to get energy to fly. This means that UAVs get the ability to fly for many hours without interruption, which is especially helpful in tasks that require long-distance and time-consuming surveillance or data collection. (Yan, Shi, & Zhang, 2020)

Long-Range Operation: WPT gives UAVs the ability to operate even in situations where traditional battery-based systems cannot. Through WPT, UAVs get continuous power, allowing them to operate for long periods of time even in remote areas. For example, during disaster relief operations, UAVs may need to operate for long periods of time to send help to remote areas, which can be accomplished through WPT. (Nguyen et al., 2020)

Continuous and Automatic Power Supply: WPT provides an automatic power supply system. This means that UAVs never have to worry about running out of power. As soon as the UAV reaches the power transfer area, it gets power automatically. This process happens without any human intervention and thus the operation and control of UAVs becomes more efficient.

Wireless Power Transfer (WPT) is a significant technological improvement for UAVs, which not only increases their efficiency but also reduces their operating costs. This technology gives UAVs the ability to fly for longer periods of time without stopping, as well as increase its battery life and reduce maintenance costs. Moreover, WPT enables UAVs to operate even in remote and difficult terrains, making it highly useful in various fields such as disaster management, surveillance, and supply distribution. (Wang & Ma, 2016)

8.4 DESIGN CHALLENGES IN RIS INTEGRATION WITH UAVS

The combination of RIS and UAVs can be helpful in creating an advanced communication and energy transfer system. However, some critical design challenges have to be faced to implement this technology in an effective manner. Major among these challenges include optimizing the configuration of RIS, ensuring robust communication in dynamic environments, and balancing flight time with energy efficiency.

1. Optimizing RIS Configuration for Effective Wireless Power Transfer (WPT)

Setting the right configuration to optimize wireless power transfer (WPT) using RIS is a critical design challenge. In the WPT process, energy has to be transmitted to the UAV through RIS without any wires. RIS consists of many small elements that are capable of directing electromagnetic waves. But it is necessary to set these elements in such a way that they are able to transmit energy to the UAV effectively.

Optimizing the RIS configuration involves arranging these elements in such a way that they send waves in the direction of the UAV with minimal energy loss. This requires taking into account that the position of the UAV can change continuously, and hence the configuration of the RIS must also be dynamic so that the UAV can receive energy at all times. This is a challenging task, as it involves constantly estimating the position of the UAV and resetting the RIS accordingly.

2. Ensuring Robust Communication in Dynamic and Challenging Environments

The integration of RIS and UAV cannot be effective unless the communication between the two is robust and continuous. UAVs usually fly in the air and their position keeps changing. Also, they may operate in different environments, such as between tall buildings, in mountainous terrain or in dense forests. Communication in these environments can be challenging as obstacles can affect the signal.

To address this challenge, it is necessary to design communication using RIS in such a way that the UAV maintains a robust and stable communication link with the RIS. This requires ensuring that the surfaces of the RIS can properly reflect or direct the signal. In addition, as the UAV's state changes, the RIS must adjust to this change so that there is no disruption in communication. This process is technically complex and requires a highly efficient RIS design and a fast response system.

3. Balance Energy Requirements With Energy Efficiency and Flight Time

The use of RIS and WPT can help maintain UAVs in uninterrupted flight, but proper energy management is essential. UAVs require energy during flight, and if energy efficiency is not handled properly, the operation of the UAV may be affected.

The RIS must be designed to transfer energy efficiently and provide sufficient energy to the UAV. But the flight and operation of the UAV also requires additional energy, such as sensors, cameras, and other equipment mounted on the UAV. Therefore, it is important to ensure that the energy from the RIS is sufficient so that the UAV can perform all tasks without interruption.

Maintaining energy balance between RIS and UAV is one of the major design objectives. This requires that RIS can transmit maximum energy efficiently and UAV can also use energy efficiently. If the right energy balance is not achieved, the flight time of UAV may be reduced, and it may need to land. Therefore, to address this challenge, special attention is paid to energy efficiency in the design of both RIS and UAV.

The combination of RIS and UAV can revolutionize wireless communication and energy transfer, but its design involves many challenges. Some of the major challenges are to optimize RIS for effective WPT, ensure robust communication in dynamic environments, and balance flight time with energy efficiency. After overcoming these design challenges, the integration of RIS and UAV can become more effective and efficient, which can prove to be extremely helpful in various practical applications (Abdalla, Rahman, & Marojevic, 2020).

8.5 APPLICATIONS OF RIS-ENABLED UAVS IN THE REAL WORLD

Unmanned aerial vehicles (UAVs) using Reconfigurable Intelligent Surface (RIS) technology can be used in many important areas. In a large and diverse country like India, this technology can make a significant contribution in many areas such as disaster management, agricultural monitoring and logistics. The use of RIS-enabled UAVs in these areas is described in detail below.

1. Disaster Management: Improving the Performance of UAVs in Search and Rescue Missions

India is frequently hit by natural disasters such as floods, earthquakes, and cyclones, causing huge loss of lives and property. Quick delivery of aid during such disasters is of utmost importance. Unmanned aerial vehicles (UAVs) can be used for search and rescue missions in these disasters. However, the performance of these UAVs can be limited in difficult conditions, especially when communications are disrupted. This is where Reconfigurable Intelligent Surface (RIS) technology can help UAVs. (Bushnaq, Mishra, Natalizio, & Akyildiz, 2022)

RIS technology has the ability to direct the signal in a specific direction, thereby increasing the range of communication. For example, in flood-affected areas where communications are disrupted, RIS can amplify the signal so that UAVs can send video feeds and sensor data even in remote areas. Thus, rescue teams get information about the areas where people are stranded, and they can carry out relief operations quickly and safely.

Example: The 2018 Kerala floods were a devastating natural disaster that left thousands of people stranded and infrastructure completely disrupted. During this time, UAVs were used for relief operations to get information about the condition of stranded people and to evacuate them safely. However, the communication range of UAVs is limited in difficult conditions and their performance may be affected. If RIS-enabled UAVs were available at that time, they could have maintained communication in flood-affected areas. For example, in areas where signals did not reach or communication was lost, with the help of RIS, UAVs could send their signals strongly and could be successful in sending vital information to the rescue team. Thus, RIS-enabled UAVs can be more effective in identifying people stranded in remote areas, knowing their location and sending help immediately. How does RIS help UAVs? RIS technology helps in directing the signals of UAVs. After a flood or earthquake, communication in the affected areas gets disrupted and traditional communication devices are not effective. Through RIS technology, UAVs can enhance their communication capability and send videos and data to the rescue center without any interruption. This gives the rescue team information about the affected areas in real time and they can take appropriate steps according to the situation.

The use of RIS-enabled UAVs not only improves communication but also saves time, which is very important in any disaster situation. This brings speed and accuracy in disaster management and can significantly contribute to saving lives.

Thus, RIS-enabled UAVs can become very helpful during natural disasters like floods, earthquakes and cyclones in India. This technology can make rescue operations faster and more effective by increasing the communication range of UAVs in rescue missions, allowing people to be evacuated quickly and lives to be saved.

2. Agricultural Monitoring: Improving UAV Capabilities for Precision Agriculture

Agriculture is a vital sector in India, and farmers need new technologies to monitor their crops. The use of unmanned aerial vehicles (UAVs) can be very helpful in precision agriculture, as they help in detecting the condition of crops, soil moisture, and any diseases in plants. With Reconfigurable Intelligent Surface (RIS) technology, the capabilities of these UAVs can be further improved, allowing farmers to manage their farms better.

How do RIS-Enabled UAVs Improve Precision Agriculture?

RIS-enabled UAVs have the ability to receive continuous power, allowing them to stay in flight for longer periods of time. With the help of RIS, UAVs can obtain more clear and accurate data from different parts. Using this data, farmers can better understand the needs of their crops, such as determining the right amount of water, fertilizers, and pesticides. This increases yield and reduces the cost of production. (Velusamy et al., 2022)

For example, if a crop is short of water or is showing signs of disease, UAVs can quickly detect the condition of that crop and send information to the farmer. This way farmers can pay attention to those crops in time and take appropriate measures.

Example: Use of RIS-Enabled UAVs in Punjab and Haryana

In states like Punjab and Haryana, where large-scale farming is done, RIS-enabled UAVs can be used to monitor the fields. These UAVs can collect data from every corner of the fields, which lets the farmer know what is the problem in which part of the field and what needs to be done.

For example, if there is a paddy crop in Punjab and there is a shortage of water in a particular area, then RIS-enabled UAVs can immediately provide information about that place. Farmers can deliver water to that particular area at the right time, which benefits the entire crop. Similarly, if any part of the wheat crop in Haryana is infested with pests, UAVs can help the farmer spray pesticides on time by giving information about that area.

Benefits of RIS: More Precise and Efficient Farming

With the help of RIS, farmers can farm in a more precise and efficient way using UAVs. This means that they can not only make maximum use of their resources but also increase their crop production. Using water, fertilizers and pesticides in the right amount and at the right time not only increases production but also has less negative impact on the environment.

With the use of RIS-enabled UAVs, farmers can get detailed information about their fields, allowing them to take better care of their crops. In an agricultural country like India, this technology can prove to be very helpful for farmers, as it will increase their production efficiency and reduce production costs. This combination of RIS technology and UAVs can take precision agriculture to a whole new level in India.

3. Logistics: RIS-enabled UAVs improve delivery services

E-commerce and logistics are expanding rapidly in India, and the demand for delivery services is increasing even in remote areas. However, traditional delivery services in remote areas are often difficult and time-consuming. This process can be made more efficient and faster by using RIS-enabled UAVs.

How do RIS-Enabled UAVs Improve Logistics?

RIS technology allows UAVs to be kept in flight for longer distances, allowing them to reach remote locations. In addition, the quality of communication also improves with the help of RIS, allowing UAVs to accurately follow their route and send the necessary data without interruption. Thus, RIS-enabled UAVs can be used for fast, accurate and safe delivery of goods. (Nguyen Thi Khanh Chi, Le Thai Phong, & Nguyen Thi Hanh, 2023)

Example: How can Companies like Amazon and Flipkart use RIS-Enabled UAVs?

E-commerce companies like Amazon and Flipkart are growing rapidly in India and want to reach their customers even in remote villages and small towns. For these companies, the use of RIS-enabled UAVs can be an ideal way to operate delivery services quickly and efficiently.

For example, transportation may be difficult for traditional delivery services in the desert areas of Rajasthan or the hilly areas of the Northeast. In such a situation, RIS-enabled UAVs can overcome these difficulties. With the help of RIS, UAVs can be powered continuously, allowing them to travel long distances and reach remote locations without frequent stops. In this way, these UAVs can quickly deliver medicines, food items, and other essential materials to areas where traditional delivery services have difficulty reaching.

Real-Life Example

Suppose a small village in the hilly areas of Northeast India needs essential medicines. The roads to reach there are difficult and may take time. Using RIS-enabled UAVs, these medicines can be delivered to villages in a timely and safe manner. Similarly, in the desert areas of Rajasthan, where delivery by road takes time, food items and essential materials can be delivered faster.

RIS-enabled UAVs can give a new direction to logistics in India. This technology is not only useful in disaster management and agricultural monitoring but can also provide fast and accurate delivery services in remote areas in logistics. This combination of RIS and UAVs can open doors to new possibilities in India, improve communication, energy efficiency and service delivery and help meet people's needs faster.RIS-enabled UAVs will not only improve delivery services but can also improve the quality of life in remote areas of India.

8.6 TECHNICAL BARRIERS AND SOLUTIONS

The integration of RIS (Reconfigurable Intelligent Surface) technology and UAVs (Unmanned Aerial Vehicles) is a new revolutionary technology that can improve communication and energy transfer. However, there are some technical barriers to successful adoption of this technology in India. It is necessary to understand these barriers and find solutions to overcome them, so that RIS-enabled UAVs can be efficiently used in various sectors in India.

1. Technical Challenges in Integration of RIS and UAVs

Control of Signal Range and Quality: It can be difficult for RIS and UAVs to maintain the range and quality of the signal in different parts of India such as rural and hilly areas. RIS technology is used to strengthen the signal, but it is challenging to make it work effectively in different geographical areas. (Abdalla, Rahman, & Marojevic, 2020)

Energy Efficiency and Flight Duration of UAVs: It is a challenge to keep UAVs in flight for a long time, especially when they maintain continuous communication. The battery of the UAV can drain quickly, and there is also difficulty in implementing wireless power transfer (WPT) through RIS, as this technology is not yet fully developed. (Goh, Leow, & Nordin, 2023)

Maintaining stability in performance: In a country like India, where weather conditions keep changing (e.g., heavy rains in monsoon or high temperatures in summer), maintaining stable communication between UAVs and RIS can be challenging. The performance of the RIS can be affected in such changing conditions, which can reduce the communication capability of the UAV.

2. Solutions to These Challenges and Research Directions

Advanced Signal Management System: The signal management system can be improved to make the RIS technology more robust. For example, multi-beam RIS can be used, which can send signals in multiple directions at the same time. This will provide good communication to UAVs even in remote areas and they will be able to transmit data effectively. (Rahmatov & Baek, 2023)
Advanced Wireless Power Transfer (WPT) Technology: Research is being done to improve WPT technology so that UAVs can be powered for a longer period of time. For example, better energy harvesting systems and low energy consumption devices can be used. Scientists and researchers in India are working on WPT technology so that this technology can be more effective in the Indian environment and UAVs can be kept in flight for a longer period of time. (Liu, Li, Li, & Mi, 2024)
Weather Resistant Materials and Structures: Materials can be used in the construction of UAVs and RIS that provide stability in weather conditions. For example, if RIS and UAVs face extreme heat, cold or rain, there is no disruption in their communication. For this, researchers are developing cheap and strong materials that are more durable in Indian weather. (Sziroczak, Rohacs, & Rohacs, 2022)
Use of Machine Learning and AI: Communication can be made even more accurate by using machine learning and artificial intelligence (AI) in RIS and UAVs. AI-based algorithms can better guide signals, and adjust the routes and communications of UAVs according to changing conditions. This research is rapidly emerging in India, and the use of AI in RIS technology can make it more efficient. (Bithas, Michailidis, Nomikos, Vouyioukas, & Kanatas, 2019)

3. Impact of These Innovations in India

Efforts to solve technical barriers for RIS and UAVs can make a big difference in disaster management, agriculture, logistics, and other sectors in India. As development and research on these technologies continue, RIS-enabled UAVs will be used more efficiently in India and services can be delivered even in rural and remote areas.

There are technical barriers to the integration of RIS and UAVs in India, but various research and innovations are ongoing to overcome these barriers. Advanced solutions are being developed to deal with signal control, energy efficiency, and weather-related challenges. Technological improvements such as AI and machine learning will make these UAVs even better. This combination of RIS and UAV can bring great changes in communication, disaster management and other essential services in India and start efficient services in different areas of the country.

8.7 FUTURE DIRECTION OF RIS-ENABLED UAVS

The combination of RIS (Reconfigurable Intelligent Surface) and UAV (Unmanned Aerial Vehicle) can give a new direction to communication and energy transfer in India. In the future, there is scope for further advancement in the technology of RIS and UAV, which will improve energy efficiency, stability of communication and wider applications. These innovations will make it possible to provide efficient services in various sectors of India, be it disaster management, agriculture or logistics.

1. Future Advancement in RIS Technology

Better Signal Processing and Coverage: In the future, RIS technology can become more sophisticated, which will process and control the signal in a better way. For example, multi-beam technology can be developed in RIS, which will allow sending signals in different directions at the same time. This means that UAVs can maintain communication even over long distances and in difficult terrains.

Advanced Use of Artificial Intelligence (AI) and Machine Learning: RIS can be made even more autonomous by using AI and machine learning. In the future, the route of the UAV will be automatically determined through AI, and RIS will be able to automatically change the direction of the signal, so that there is no hindrance in communication. This can prove to be extremely useful in a vast country like India, where there are different geographical conditions. (Zhang, Li, Jian, Zhao, & Gao, 2021)

2. Potential Improvement in Energy Efficiency

Improved Wireless Power Transfer (WPT): In the future, the combination of RIS and WPT technology can be further improved, which will provide energy to the UAV for a longer time. For example, high-frequency WPT technology can be developed, which will make it possible to transfer energy over longer distances. This will allow UAVs to stay in flight for a longer time without stopping, which will help in better operation of logistics and other services in remote areas.

Greater Use of Solar Energy: In India, where solar energy is available in abundance, the combination of solar panels and RIS technology can make UAVs more energy efficient. Solar panels mounted on UAVs can store energy through RIS and recover excess energy using WPT. This will increase the flight duration of UAVs and reduce their energy requirement, allowing them to operate for longer periods of time. (Nguyen, Amri, Park, Kim, & Choi, 2022)

3. Improve Communication Stability

RIS Networks to Enhance Connectivity: In future, RIS networks can be set up in rural and remote areas of India, making the communication of UAVs more stable. As the technology of internet and communication is improving, it may be possible to set up small networks of RIS in different areas, which will never let the communication of UAVs get interrupted. (Pogaku, Do, Lee, & Nguyen, 2022)

Integration of 5G and 6G Technology with RIS: 5G has been introduced in India and 6G will also be available in the future. Integrating RIS with 5G and 6G can further improve the speed and stability of communication. This will allow UAVs to transfer data without any interruption and communicate faster even in difficult terrains.

4. Future Prospects for Wider Applications

Effective Use in Disaster Management: RIS-enabled UAVs will be used more effectively for disaster management in the future. Disasters such as floods, earthquakes and cyclones occur frequently in India. In such a situation, RIS-enabled UAVs can reach the disaster area quickly and deliver relief materials, medicines and essential goods.

Precision Management in Agriculture: Agriculture is a major sector in India and RIS-enabled UAVs can be used to bring precision in agriculture. In the future, these UAVs will be able to give farmers detailed information about their crops, so that they can use water, fertilizers and pesticides correctly. This will increase the yield and reduce the cost of farming.

Speed and Accuracy in Logistics: The use of RIS-enabled UAVs in logistics can become even more widespread in the future. In remote villages where it is difficult to deliver goods, UAVs can make deliveries more efficiently with the help of RIS technology.

The future of RIS-enabled UAVs in India is bright. Combining UAVs with RIS technology can improve many areas such as communication, energy efficiency and logistics. As the technology of RIS and UAVs continues to improve, their use is likely to increase in India as well. The development of these technologies will not only contribute to the development of the country but will also help in delivering services to remote areas, thereby strengthening India's communication and service network.

8.8 CONCLUSION

In this chapter, we have learned in detail about the combination of RIS (Reconfigurable Intelligent Surface) technology and UAVs (Unmanned Aerial Vehicles). Various aspects of how this technology can be used to improve communication, energy transfer and various critical services in various regions of India were discussed. Let us look at the summary of the key findings and conclusions of this chapter and learn how RIS-enabled UAVs can be used in India in the future.

Introduction to RIS and UAV: RIS technology is based on surfaces that can electronically control the direction and strength of the signal, thereby improving communication. UAVs can be used in disaster management, agriculture and logistics in various regions of India. With RIS technology, these UAVs become more energy efficient and can perform well even in difficult conditions.

Technical Challenges and Solutions: There are several technical hurdles in the integration of RIS and UAV, such as signal quality, energy efficiency and weather challenges. Several new solutions such as multi-beam RIS, use of solar energy, and AI have been suggested to use RIS and UAVs more efficiently in a country like India with diverse geography and climate.

Potential Applications in India: RIS-enabled UAVs can be used to give quick response in disaster management, bring precision in agriculture, and in logistics to deliver goods in remote areas. These applications can provide essential services in remote and rural areas of India.

Improving Energy Efficiency and Communication Stability: The combination of Wireless Power Transfer (WPT) and solar energy using RIS has been studied to increase the flight duration of UAVs. Also, the integration of RIS technology with 5G and future 6G networks for better communication has been discussed.

Future Uses and Perspectives of RIS-Enabled UAVs

In the future, RIS-enabled UAVs can be used to strengthen the communication network in India and make the services even more efficient. The possible approaches are as follows:

Durable and High Performance Communication Network: As the technology of RIS and UAVs develops, it will be possible to provide durable and high quality communication network in remote areas of India. This will improve the quality of internet and mobile network even in rural and backward areas, thereby helping in realizing the dream of Digital India.

Safe and Efficient Disaster Management: Disaster management can be made safer and faster with RIS-enabled UAVs. During natural disasters like floods, earthquakes and cyclones, these UAVs can go to vulnerable areas and collect information and deliver relief material, which will help in saving lives.

Precision and Smart Farming in Agriculture: The use of RIS-enabled UAVs in agriculture will enable farmers to keep a close watch on the status of their crops, allowing them to take timely decisions. This will not only increase the yield but also reduce the cost of farming. This will be a significant advancement for an agriculture-dominated country like India.

Improvement in Logistics and Access in Rural Areas: In the future, logistics networks can be further strengthened through RIS-enabled UAVs. Medicines, food items and other essential commodities can be easily delivered to areas that are usually difficult to reach by road. This will also provide better services to remote villages in India.

Advancement in Energy Efficiency and Environmental Protection: Improvement in energy efficiency through RIS technology will increase the chances of UAVs lasting longer, which will reduce energy consumption and also reduce pressure on the environment. This technology will be helpful in meeting India's energy saving and environmental protection goals.

RIS-enabled UAVs can bring a new revolution in communication, disaster management, agriculture and logistics in India. As this technology continues to improve, India's communication capacity, energy efficiency and service delivery will also improve. RIS-enabled UAVs will not only help in delivering essential services to rural and remote areas but will also contribute to sustainable development and environmental protection. This convergence of RIS and UAV will give a new dimension to India's digital and technological development, which will benefit all sections of the society.

REFERENCES

Abdalla, A. S., Rahman, T. F., & Marojevic, V. "UAVs with Reconfigurable Intelligent Surfaces: Applications, Challenges, and Opportunities", arXiv:2012.04775v1 [eess.SP] 8 Dec 2020.

Abdalla, A. S., Rahman, T. F., & Marojevic, V. "UAVs with Reconfigurable Intelligent Surfaces: Applications, Challenges, and Opportunities", arXiv:2012.04775v1 [eess.SP] 8 Dec 2020, doi: /arXiv.2012.04775.DOI: 10.48550

Bithas, P. S., Michailidis, E. T., Nomikos, N., Vouyioukas, D., & Kanatas, A. G. (2019). A Survey on Machine-Learning Techniques for UAV-Based Communications. *Sensors (Basel)*, 19(23), 5170. DOI: 10.3390/s19235170 PMID: 31779133

Bushnaq, O., Mishra, D., Natalizio, E., & Akyildiz, I. F. (2022). Unmanned aerial vehicles (UAVs) for disaster management. *Nanotechnology-Based Smart Remote Sensing Networks for Disaster Prevention*, (January), 159–188. DOI: 10.1016/B978-0-323-91166-5.00013-6

Chi, N. T. K., Phong, L. T., & Hanh, N. T. (2023, June). PanelNguyen Thi Khanh Chi, Le Thai Phong & Nguyen Thi Hanh, "The drone delivery services: An innovative application in an emerging economy". *The Asian Journal of Shipping and Logistics*, 39(2), 39–45. DOI: 10.1016/j.ajsl.2023.01.002

Dajer, M., Ma, Z., Piazzi, L., Prasad, N., Qi, X.-F., Sheen, B., Yang, J., & Yue, G. (2022, April). Reconfigurable intelligent surface: Design the channel – a new opportunity for future wireless networks. *Digital Communications and Networks*, 8(2), 87–104. DOI: 10.1016/j.dcan.2021.11.002

Goh, C. Y., Leow, C. Y., & Nordin, R. (2023). Energy Efficiency of Unmanned Aerial Vehicle with Reconfigurable Intelligent Surfaces: A Comparative Study. *Drones (Basel)*, 7(2), 98. DOI: 10.3390/drones7020098

Gong, Y., Yoon, I., & Noh, D. K. (2024). Optimizing the Charging Mobility of WPT-Enabled UAV to Enhance the Stability of Solar-Powered LoRaWAN Io. *Energies*, 17(7), 1617. DOI: 10.3390/en17071617

Hasan, S. R., Sabuj, S. R., Hamamura, M., & Hossain, M. A.SYED RAKIB HASAN. (2024, October 31). SAIFUR RAHMAN SABUJ, MASANORI HAMAMURA AND MD AKBAR HOSSAIN, "A Comprehensive Review on Reconfigurable Intelligent Surface for 6G Communications: Overview, Deployment, Control Mechanism, Application, Challenges, and Opportunities". *Wireless Personal Communications*, 139(1), 375–429. Advance online publication. DOI: 10.1007/s11277-024-11630-1

Hong, I.-P. (2023). Reviews Based on the Reconfigurable Intelligent Surface Technical Issues. *Electronics (Basel)*, 12(21), 4489. DOI: 10.3390/electronics12214489

Korakianitis, N., Vokas, G. A., & Ioannides, G. (2019). Review of Wireless Power Transfer (WPT) on Electric Vehicles (EVs) Charging. *AIP Conference Proceedings*, 2190, 020072. DOI: 10.1063/1.5138558

Li, J., Xue, H., Wu, M., Wang, F., Gao, T., & Zhou, F. (2023). Energy efficiency performance in RIS-based integrated satellite–aerial–terrestrial relay networks with deep reinforcement learning. *EURASIP Journal on Advances in Signal Processing*, 2023(1), 121. Advance online publication. DOI: 10.1186/s13634-023-01070-7

Liu, Z., Li, T., Li, S., & Mi, C. C. (2024, June 18). Advancements and challenges in wireless power transfer: A comprehensive review. *Nexus*, 1(2), 100014. DOI: 10.1016/j.ynexs.2024.100014

Nguyen, M. T., Amri, M. M., Park, J. H., Kim, D. I., & Choi, K. W. (2022, November). Security–Reliability Tradeoff Analysis for SWIPT- and AF-Based IoT Networks With Friendly Jammers. *IEEE Internet of Things Journal*, 9(21), 1–1. DOI: 10.1109/JIOT.2022.3182755

Nguyen, M. T., Nguyen, C. V., Truong, L. H., Le, A. M., Quyen, T. V., Masaracchia, A., & Teague, K. A. (2020). Electromagnetic Field Based WPT Technologies for UAVs: A Comprehensive Survey. *Electronics (Basel)*, 9(3), 461. DOI: 10.3390/electronics9030461

Nguyen, M. T., Nguyen, V.-C., Truong, L. H., Le, A. M., Van Quyen, T., & Masaracchia, A. (2020, March). Electromagnetic Field Based WPT Technologies for UAVs: A Comprehensive Survey. *Electronics (Basel)*, 9(3), 461. DOI: 10.3390/electronics9030461

Ojha, T., & Theofanis, P. (2023). Raptis· Andrea Passarella, Marco Conti, "Wireless power transfer with unmanned aerial vehicles: State of the art and open challenges". *Pervasive and Mobile Computing*, 93(June), 101820. DOI: 10.1016/j.pmcj.2023.101820

Pogaku, A. C., Do, D.-T., Lee, B. M., & Nguyen, N. D. (2022, January). UAV-Assisted RIS for Future Wireless Communications: A Survey on Optimization and Performance Analysis. *IEEE Access : Practical Innovations, Open Solutions*, 10(4), 1–1. DOI: 10.1109/ACCESS.2022.3149054

Rahmatov, N., & Baek, H. (2023, October). RIS-carried UAV communication: Current research, challenges, and future trends. *ICT Express*, 9(5), 961–973. DOI: 10.1016/j.icte.2023.03.004

Sharma, T., Chehri, A., & Fortier, P. (2021). Reconfigurable Intelligent Surfaces for 5G and beyond Wireless Communications: A Comprehensive Survey. *Energies*, 14(24), 8219. DOI: 10.3390/en14248219

Sziroczak, D., Rohacs, D., & Rohacs, J. (2022, October). Review of using small UAV based meteorological measurements for road weather management. *Progress in Aerospace Sciences*, 134(1), 100859. DOI: 10.1016/j.paerosci.2022.100859

Velusamy, P., Rajendran, S., Mahendran, R. K., Naseer, S., Shafiq, M., & Choi, J.-G. (2022). Unmanned Aerial Vehicles (UAV) in Precision Agriculture: Applications and Challenges. *Energies*, 15(1), 217. DOI: 10.3390/en15010217

Wang, C., & Ma, Z., "Design of wireless power transfer device for UAV", Institute of Electrical and Electronics Engineers Inc., ISBN: 978-150902394-3, DOI: . DOI: 10.3390/en15010217

Yan, Y., Shi, W., & Zhang, X. (2020). Design of UAV wireless power transmission system based on coupling coil structure optimization. *EURASIP Journal on Wireless Communications and Networking*, 2020(1), 67. Advance online publication. DOI: 10.1186/s13638-020-01679-4

Zhang, S., Li, M., Jian, M., Zhao, Y., & Gao, F. "AIRIS: Artificial Intelligence Enhanced Signal Processing in Reconfigurable Intelligent Surface Communications", arXiv:2106.00171v1 [eess.SP] 1 Jun 2021, DOI:/arXiv.2106.00171.DOI: 10.48550

Chapter 9

Reconfigurable Intelligence Surface-Aided Satellite Communication Design: Challenges and Applications

V. Bakyalakshmi
Dr. NGP Arts and Science College, India

V. Sowmiya
Dr. NGP Arts and Science College, India

ABSTRACT

This study looks into a new method for instantaneous Doppler positioning using Iridium signals, which are interpreted as satellite signals of opportunity (SOP). It begins by describing the structure and features of the Iridium SOP-based positioning system. The method for instantaneous Doppler positioning is then described, followed by a discussion of how measurement inaccuracies and satellite orbital errors affect positioning accuracy. Furthermore, the impact of constellation shape on accuracy, particularly in terms of positioning between receiver locations, is evaluated. Simulations and real-world trials with IRIDIUM NEXT signals demonstrate that static receivers can reach 22-meter (1σ) two-dimensional positioning accuracy with height assistance and unimpeded sky views.

DOI: 10.4018/979-8-3693-8099-4.ch009

1. INTRODUCTION

Global navigation satellite systems (GNSS) have evolved greatly in recent years, expanding their applications across a wide range of industries. However, academics have been working on alternative locating systems to serve as backups or to lessen reliance on GNSS, especially in cases where GNSS performance is impaired due to severe interference or complicated settings [1]. In other circumstances, GNSS is unable to give positional data because the navigation message cannot be decoded [2]. Inertial systems and other technologies are utilized to compensate for the lack of GNSS signals. More recently, innovative systems such as opportunistic navigation (OpNav) [3] and collaborative opportunistic navigation (COpNav) [4] have evolved, which make use of a wide variety of frequencies. These technologies aim to improve or replace existing systems such as GPS and may be very helpful. Similarly, the All Source Positioning and Navigation (ASPN) program, launched by the US Defense Advanced Research Projects Agency (DARPA), aims to provide dependable, continuous, and cost-effective navigation in any environment by utilizing all available signals, regardless of GPS availability. Non-GNSS satellite signals, known as signals of opportunity (SOPs), offer several advantages, including increased signal strength, lower prices, and more coverage. There are now several non-GNSS satellite constellations in operation, including IRIDIUM, GLOBALSTAR, and ORBCOMM. By 2019, IRIDIUM had finished deploying its second-generation constellation, which included 75 IRIDIUM NEXT satellites. In 2016, IRIDIUM unveiled its Satellite Time and position (STL) service, which is based on its satellite network and provides a reliable time and position solution that might supplement or possibly replace GPS, however the technical details are mainly unknown [7]. In 2015, One Web and Boeing announced plans to create additional constellations of low Earth orbit (LEO) satellites [8]. Similarly, China has initiated equivalent endeavors, including the HONYAN constellation and the HONYUN project. Satellite SOPs may have fully or partially known properties. In the case of non-cooperative satellite SOPs, receivers can measure Doppler changes even when range information is lacking [11], allowing Doppler positioning algorithms to deliver location services despite insufficient data. Hill [12] initially introduced the concept of GPS instantaneous Doppler location. This strategy can help speed up the time to first fix (TTFF) [14]. Van goes on to discuss how the instantaneous Doppler locating approach, when combined with the A-GNSS technology, can help with frequency and coding delay corrections [15]. Positioning problems in Iridium SOPs are caused mostly by satellite orbital inaccuracies and Doppler observation mistakes. Because Doppler location relies on satellite velocity information, assessing these mistakes is more difficult than in classic pseudorange positioning, as satellite velocity effects must be considered. The Doppler geometric dilution of precision (Doppler DOP) is a dimensional physical quantity that is determined by both the geometric configuration of the constellation and the satellites' dynamic paths. Reducing the Doppler DOP by carefully selecting GNSS satellites is critical for enhancing position accuracy. However, there has been little investigation on this topic in the literature. This study seeks to fill these gaps by providing an in-depth investigation of the Iridium SOPs positioning technique.

2. LITERATURE REVIEW

Recent research has increasingly focused on incorporating Reconfigurable Intelligent Surfaces (RIS) into satellite communication systems, demonstrating their potential for improving coverage, communication efficiency, and energy performance. Numerous studies have looked into the design issues and real-world uses of this technology. Wu and Zhang (2020) demonstrated RIS' ability to intelligently manipulate electromagnetic waves, enhancing spectrum efficiency in terrestrial systems—a concept later used for satellite communications. Zhang et al. (2021) highlight beamforming and channel estimation as a major design difficulty, citing the dynamic nature of satellite-ground communication, which challenges precise management of reflected beams, particularly with moving satellites. Furthermore, hardware constraints, such as the need for strong materials that can resist the harsh temperatures of space, hinder development. Zhao et al. (2022) stressed the need of RIS components that can withstand high temperatures and radiation, a topic that has received little attention in recent studies. Energy efficiency is also a key consideration. While the passive design of RIS helps minimize power consumption, research like those by Li et al. (2021) show that balancing energy economy and performance in low-power satellite systems is still a substantial challenge. Furthermore, RIS-induced interference needs advanced mitigation approaches, as described by Tang et al. (2022). The need for real-time reconfiguration, which is critical in satellite systems, complicates matters due to the high processing demands. On the application side, various studies have shown that RIS has the ability to enhance satellite system coverage. Di Renzo et al. (2020) demonstrated that RIS could improve signal reach in difficult-to-reach areas, improving service in places like urban canyons and rural areas. RIS has also been shown to increase data rates and capacity, with Jiang et al. (2022) demonstrating that focusing signal energy improves spectral efficiency. RIS is also projected to play a key role in the development of 6G networks and IoT connectivity, particularly in underserved areas, according to Zhang et al. (2021) and Tang et al. (2021). Despite these theoretical advances, practical deployment remains challenging, notably in hardware design and real-time flexibility, as shown in recent studies. As RIS technology advances, future research should address these challenges while also investigating new uses for energy-efficient communication and increased security, as proposed by Shi et al. (2022).

3.METHODOLOGY

3.1 Positioning Technique Using Iridium Satellite Sops

Iridium has demonstrated time transfers with accuracies in the microsecond and nanosecond ranges. Low Earth Orbit (LEO) satellite signals of opportunity (SOPs) can operate well in harsh environments such as emergency response, firefighting in deep structures, and combat scenarios. These SOPs can also be used as a comprehensive GNSS backup, with DARPA's ASPN project serving as a prime example of GPS backup study. Figure 1 depicts the concept of placement with Iridium SOPs. An antenna with a 5 dB gain that covers the frequency range of 380 MHz to 20 GHz connects to a radio frequency (RF) front end. The software module typically consists of three major processes: Doppler acquisition, satellite orbit reconstruction, and location determination. The receiver's location is derived using instantaneous Doppler positioning, with height assisting for added accuracy. The dotted part of Figure 1 depicts novel analytical methodologies, such as error analysis and research into how satellite geometry affects positioning errors. Measurements are taken during the acquisition phase because Iridium satellite burst signals are discontinuous and often substantially greater than background noise. Doppler measurements, both coarse and fine, are obtained using fast Fourier transform (FFT) and maximum likelihood estimator (MLE) techniques, with the MLE enhancing accuracy in the face of FFT resolution limitations. Orbital data for Iridium satellites is published in NORAD's two-linc element (TLE) set format, which includes orbital forecast parameters and times. . The SGP4 model is used to rebuild satellite orbits. Iridium TLE sets are updated on a regular basis, usually once or twice a day, and can be accessed via the NORAD website or celestrak.com. Position inaccuracies based on TLE data for Iridium satellites can range from 100 meters to several kilometres, with satellite velocity errors of roughly 3 meters per second, which are often smaller in the radial direction than in the cross-track and along-track directions.

FIGURE 1. Basic Iridium SOPs positioning principle.

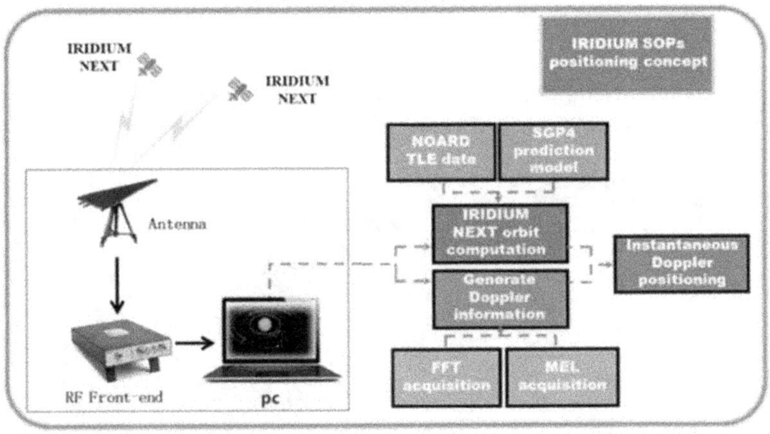

3.2 Instantaneous Doppler Positioning and Doppler Geometric Dilution of Precision (DOP)

This section describes an innovative way to analyzing placement errors using geometric analysis. It begins with an overview of the instantaneous Doppler locating method and the related Doppler dilution of precision (DOP). The novel technique is then used to assess the impact of measurement and orbital errors on positioning accuracy, together with the introduction of mathematical models for position faults.

3.3 Instantaneous Doppler Positioning and Error Analysis

When the Doppler shift from a satellite is detected by a stationary sensor on the Earth's surface at a certain time, all sites with the same Doppler measurement create a circular conical surface, assuming no mistakes. Point N represents the satellite's projection onto the base of this cone. The cone is called the iso-Doppler circular conical surface (IDCCS), and the angle θ between the satellite's velocity vector and the line of sight is known as the IDCCS field angle. The Doppler shift is expressed by the equation below...

Figure 2. Doppler effect in satellite positioning

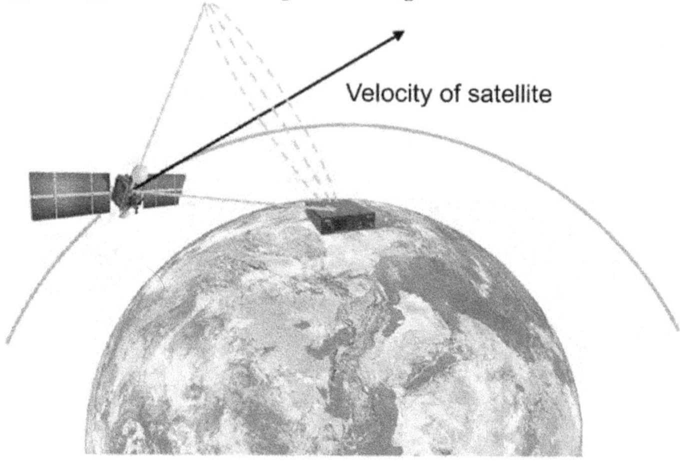

Here, rSatr_{Sat}rSat and vSatv_{Sat}vSat represent the satellite's position and velocity vectors, respectively. rUserr_{User}rUser signifies the user's position vector, while λ\lambdaλ is the signal wavelength. Equation (1) approximates the Doppler shift, assuming that the satellite's line-of-sight velocity is much slower than the speed of light [21]. Using the relation, vSat=||vSat||v_{Sat} = \|v_{Sat}\|vSat=||vSat||.

The user's location must be on the matching Iso-Doppler Contour Coordinate System (IDCCS) for a specific satellite, as determined by a static receiver's Doppler shift measurement. When the receiver simultaneously measures Doppler shifts from numerous satellites, the user's position is determined by the intersection of these IDCCSes. As long as there are enough satellites, the Doppler measurement equation can be used to calculate the receiver's position. There are several techniques to deriving the navigation equations for immediate Doppler locating. Van, for example, differentiated the linearized pseudorange positioning equations, and the least squares method is used to calculate the user's location. While different approaches present navigation solutions in various vector formats, their scalar formulations are constant. In Doppler positioning, the receiver's clock frequency drift is handled as an unknown variable, rather as the receiver clock error. If the exact signal transmission time is not available, the time inaccuracy is handled as unknown. This is known as the “snapshot technique” in pseudorange placement. Fernández-Hernández and Borre investigated this strategy, and their studies revealed substantial advances. In general, at least four satellites are required for immediate Doppler locating. Doppler measurements from the same satellite at different times or other satellites can be used with a stationary receiver because the receiver's coordinates do not vary with time. However, due to clock instability, the receiver's clock frequency drift may differ between measurements. To simulate this drift, a third-order polynomial is used, yielding six unknowns to solve. In this instance, at least six Doppler readings from the same or other satellites taken at varying times are required. Height helping reduces the number of unknowns to five, allowing for correct location with only five Doppler readings. Assuming measurement errors have a zero mean, are uncorrelated, and have equal variance (denoted as σv2\sigma_v^2σv2), the variance of positioning errors (σu2\sigma_u^2σu2) can be represented as...

DPDOP (Doppler Position Dilution of Precision) is an important component that affects positioning precision. The IDCCS in two dimensions is illustrated in the top section of Fig. 3, with points A and B representing the receiver's actual and estimated positions, respectively. The angle difference between genuine line of sight and measurement is indicated by Δθ\Delta \thetaΔθ. The middle graphic shows the difference as a position vector, with the vector AB→\overrightarrow{AB}AB representing the location mistake. In the bottom plot, the vector vAB=vB−vA\mathbf{v}_{AB} = \mathbf{v}_B - \mathbf{v}_AvAB=vB−vA illustrates the disparity between the true and observed Doppler-shift vector. As a result, measurement errors create a new velocity vector vAB\mathbf{v}_{AB}vAB, shifting the receiver's estimated position from the true position (A) to the estimated position (B) by a magnitude of |vAB|\|\mathbf{v}_{AB}\||vAB| over time. This conclusion is true even if the estimated position happens at different points along segment SB. The standard deviations σv\sigma_vσv and σu\sigma_uσu in equation (3) correspond to |vAB|\|\mathbf{v}_{AB}\||vAB| and AB→\overrightarrow{AB}AB, respectively. DPDOP reflects the time it takes the Doppler positioning solution to move from the real to the estimated position as a result of measurement mistakes. These findings are also applicable in three dimensions and are still relevant to Doppler Horizontal Dilution of Precision (DHDOP) and Doppler Vertical Dilution of Precision (DVDOP).

3.4 Geometric Analysis of the Impact of Satellite Orbital Errors

When performing a geometric analysis of how satellite orbital errors affect positioning accuracy, three major effects must be considered. The first effect is an inaccuracy in the satellite's tangential velocity. If the real satellite tangent velocity is designated as vSatRealv_{SatReal}vSatReal and the error is represented by ΔvSat\Delta v_{Sat}ΔvSat, then...

Equation (4) shows that mistakes in satellite tangent velocity affect the IDCCS in the same way that measurement errors do, because both types of errors change the field angle of the related IDCCS. The second effect to consider is the directional error in the satellite's tangent velocity. Let Δβ\Delta \betaΔβ signify the angular deviation of the satellite's tangent velocity. In this context, IDCCS1 is displaced by Δβ\Delta betaΔβ from the genuine IDCCS. As a result, the positioning solution will align with IDCCS1 rather than the actual IDCCS.

Given that these errors are stochastic, their magnitude and direction are unknown. IDCCS1 rotates around the line SO of the actual IDCCS at a constant angle θ\thetaθ. This transformation causes the IDCCS to morph into a Doppler Annulus Cone Geometry (DACG), demonstrating that the directional inaccuracy of the satellite's tangent velocity has the same effect as the tangent velocity error itself. Furthermore, it is critical to evaluate the consequences of satellite position mistakes. Let ΔrSat\Delta \mathbf{r}_{Sat}ΔrSat be the location error vector. Equation (4) shows that the field angle of the IDCCS is unaffected by the satellite position vector rSat\mathbf{r}_{Sat}rSat, indicating that the IDCCS does not experience distortion (Δθ=0\Delta \theta = 0Δθ=0). However, the placement of the IDCCS has changed. The erroneous satellite position may be within or on the edge of a sphere with a radius of LmaxL_{max}Lmax, representing the error limits. As a result, satellite position errors transform the typical IDCCS into an irregular geometry.

3.5 Impact of Constellation Geometric Configuration on Positioning Accuracy

This section looks at how the geometric distribution of the satellite constellation affects location accuracy. The geometric arrangement includes not only the relative positions of the satellites and the receiver, but also the directions of their respective velocities. Following the presentation of a key geometric notion, we shall provide a related analysis.

3.6 Experimental Results

We show experimental results based on both simulated and actual data. The simulations are intended to validate the previous theoretical analysis of Doppler DOP, while the real data collected from Iridium NEXT is studied to assess the positioning capabilities of the Iridium SOPs, as described below.

3.7 Simulation Evaluation of Doppler Dop Analysis

The DHDOP is shown to vary with S4 location. The dotted line depicts the variation in DHDOP when S4 is placed above the receiver, although the real anomalies of S1, S2, and S3 vary concurrently. Next, we look at the link between DHDOP, DVDOP, and satellite velocities. The velocity directions of S3 and S1 are changed at the same time by keeping S1, S2, and S3 in their original positions and altering their orbital inclinations. This analysis shows the relationship between Doppler DOP and satellite velocities, with the x-axis showing the angle between the orbital inclinations of S1 or S3 and that of S2. Furthermore, we investigate the link between Doppler

DOP and satellite positions. To simplify things, the genuine anomalies of S1, S2, and S3 are all adjusted to 3 degrees, and the shifting Doppler DOP curves as S4's elevation increases are shown. The movement of the true anomaly from left to right suggests that satellite elevation is increasing and will eventually plateau. The solid line indicates that S4's elevation has no effect on DHDOP when the other satellites are above the receiver. Higher elevation satellites effect the width of the horizontal intersection area between distinct DACGs, limiting the horizontal positioning error to a given range. The dotted line shows that DHDOP values drop dramatically as the heights of S1, S2, and S3 rise, with S4 positioned above the receiver. The DHDOP trend is linked to the positions and velocities of the other three satellites, with satellites at higher elevations having lower DHDOP values. This is due to a reduction in the width of the intersection region between all DACGs and the Earth's surface, which prevents a single high-elevation satellite from determining horizontal intersections. As a result, using more satellites at the top of the receiver can help reduce 2-D positioning errors. As satellite velocities scatter, Doppler DOP decreases. The DVDOP curve remains practically linear since the vertical intersection area does not vary appreciably. The results show that both DPDOP and DVDOP are extremely sensitive to the elevation of satellite S4; when S4's elevation increases, the vertical intersection broadens, demonstrating that elevation variations mostly effect upward positioning mistakes. Meanwhile, DHDOP is almost constant since the other satellites limit horizontal positional precision. These studies confirm the importance of constellation geometric distribution on positioning accuracy, and using the above approach improves Doppler DOP results.

4.CONCLUSION

This study investigated the use of Iridium signals as noncooperative signals of opportunity (SOPs) for locating purposes. A fundamental innovation provided was a new geometric analysis approach for analyzing positioning faults in instantaneous Doppler positioning and understanding how the constellation's geometric distribution influences positioning accuracy. The positioning approach based on Iridium SOPs, as well as the impact of constellation geometry on positioning errors, were validated using real data from IRIDIUM NEXT and simulations performed with STK software.

The key findings of this study are summarized as follows: The relevance of DPDOP, DHDOP, and DVDOP stems from the temporal delay that occurs when Doppler positioning solutions depart from their true position to their estimated position due to measurement error.

REFERENCES

Abosekeen, A., Noureldin, A., & Korenberg, M. J. (2019). Improving the RISS/GNSS land-vehicles integrated navigation system using magnetic azimuth updates. *IEEE Transactions on Intelligent Transportation Systems*. Advance online publication. DOI: 10.1109/TITS.2019.2905871

Hill, J. (2001). The principle of a snapshot navigation solution based on Doppler shift. In *Proceedings of the ION GPS* (pp. 3044–3051). Salt Lake City, UT, USA.

Kaplan, D. E., & Hegarty, C. J. (2006). *Understanding GPS: Principles and Applications* (2nd ed.). Artech House.

Kirkko-Jaakkola, M., Parviainen, J., Collin, J., & Takala, J. (2012). Improving TTFF by two-satellite GNSS positioning. *IEEE Transactions on Aerospace and Electronic Systems*, 48(4), 3660–3670. DOI: 10.1109/TAES.2012.6324754

Lawrence, D., Cobb, H. S., Gutt, G., Tremblay, F., Laplante, P., & O'Connor, M. (2016). Test results from a LEO-Satellite-Based assured time and location solution. In *Proceedings of the ION GNSS* (pp. 125–129). Monterey, CA, USA. DOI: 10.33012/2016.13416

Pesyna, K. M.Jr, Kassas, Z. M., Bhatti, J. A., & Humphreys, T. E. (2011). Tightly-coupled opportunistic navigation for deep urban and indoor positioning. In *Proceedings of the ION GPS* (pp. 3065–3616). Portland, OR, USA. DOI: 10.1109/GLOCOMW.2011.6162371

Peterson, B., Hartnett, R., & Ottman, G. (1995). GPS receiver structures for the urban canyon. In *Proceedings of the ION GPS* (pp. 1323–1332). Palm Springs, CA, USA.

Reid, T. G., Neish, A. M., Walter, T. F., & Enge, P. K. (2016). Leveraging commercial broadband LEO constellations for navigating. In *Proceedings of the ION GNSS* (pp. 2300–2314). Portland, OR, USA. DOI: 10.33012/2016.14729

Seco-Granados, G., López-Salcedo, J. A., Jiménez-Baños, D., & López-Risueño, G. (2012). Challenges in indoor global navigation satellite systems: Unveiling its core features in signal processing. *IEEE Signal Processing Magazine*, 29(2), 108–131. DOI: 10.1109/MSP.2011.943410

Van Diggelen, F. (2009). *A-GPS: Assisted GPS, GNSS, and SBAS*. Artech House.

Yang, C., & Nguyen, T. (2009). Self-calibrating position location using signals of opportunities. In *Proceedings of the ION GNSS* (pp. 1055–1063). Savannah, GA, USA. DOI: 10.1109/NAECON.2009.5426658

Chapter 10
Reconfigurable Intelligent Surface-Aided Physical Layer Security Techniques: Applications and Future Trends

Rajesh Kanna Rajendran
https://orcid.org/0000-0001-7228-5031
Christ University, India

Immanuel V. Ashok
https://orcid.org/0000-0002-2422-6261
Christ University, India

Priya T. Mohana
https://orcid.org/0000-0001-5637-7921
Christ University, India

V. B. Kirubanand
Christ University, India

V. Rohini
Christ University, India

B. Murugesakumar
https://orcid.org/0000-0003-4042-4265
Dr.SNS Rajalakshmi College of Arts and Science, India

ABSTRACT

Reconfigurable Intelligent Surfaces (RIS) are emerging as a groundbreaking technology in the realm of wireless communications, with significant implications for enhancing physical layer security. This chapter delves into the integration of RIS with advanced security techniques, exploring how this innovative technology can be harnessed to address the growing challenges of securing wireless networks. this chapter delves into the future trends and advancements in RIS technology, including next-generation RIS architectures and their potential integration with emerging technologies like 6G. It explores how RIS could pave the way for innovative security protocols and play a pivotal role in advancing secure wireless network

DOI: 10.4018/979-8-3693-8099-4.ch010

infrastructures. RIS technology enables the dynamic and intelligent modification of radio environments through programmable surfaces, which can adjust and optimize signal paths to improve both communication efficiency and security. This chapter provides valuable insights into the current applications and future prospects of RIS in enhancing wireless network security.

INTRODUCTION

Reconfigurable Intelligent Surfaces (RIS) are one of the revolutionary technologies in wireless communication that bring in a new dimension of control over the propagation of electromagnetic waves. RIS can be defined as a flat surface that has an array of programmable, passive reflective elements, also called meta-elements, that can dynamically change the phase, amplitude, or polarization of incident signals. Unlike conventional active devices, for example, antennas and relays, RIS is a passive system, meaning it reflects or scatters signals without any significant energy consumption. This gives RIS an exciting role as a promising candidate for improving the performance of a wireless system in line with energy-efficient and cost-effective design principles.

RIS operation is based on wave manipulation principles, allowing it to precisely control how signals propagate in the wireless environment. By intelligently adjusting the configuration of its meta-elements, RIS can steer signals toward desired locations, avoid obstacles, or cancel out interference. This functionality makes RIS particularly valuable in addressing NLOS communication challenges commonly encountered in urban and indoor environments. Moreover, RIS supports advanced features like beamforming and wavefront shaping that enhance signal strength and improve communication reliability for legitimate users while reducing the likelihood of interception by eavesdroppers. RIS, therefore, stands on a very important advantage of energy-efficient and low-cost deployment compared to conventional active devices. While technologies like massive MIMO and relays require significant power and complex hardware, simple passive RIS components are easily embedded in existing infrastructures such as walls, ceilings, or building facades. This scalability makes RIS ideal for widespread usage in both urban and rural scenarios, where improving wireless coverage and security are critical. In addition, it aligns with the objectives of green communications to minimize the carbon footprint of wireless networks.

RIS has wide practical applications across various domains, such as smart environments, vehicular networks, and industrial IoT systems. For example, in smart buildings, RIS optimizes indoor signal coverage, which improves the quality of service for users. In vehicular networks, RIS enhances communication reliability in V2X applications, which are critical for autonomous driving systems. Similarly, in industrial IoT, RIS provides secure and robust communication between interconnected devices, ensuring smooth operations in highly automated environments. Reconfigurable Intelligent Surfaces are the most transformative technology in wireless communication-novel, energy-efficient, and cost-effective solutions for present-day challenges. Their dynamical manipulation of wireless signals and integration into existing infrastructure make them a key enabler of 6G and beyond. The potential of RIS for reshaping the wireless landscape, enhancing physical layer security, and enabling sustainable communication will continue to grow as this research and development advances.

RELATED WORK

Reconfigurable Intelligent Surfaces (RIS) has recently gained significant attention in integrating its capabilities in improving Physical Layer Security (PLS). Wyner first proposed the fundamental concept of PLS by developing the wiretap channel model, which theoretically framed secure communication over channels by taking advantage of the channel quality disparity between a legitimate receiver and an eavesdropper in 1975. Following this, researchers worked on techniques such as artificial noise (AN) generation and cooperative jamming to improve secrecy rates (Goel & Negi, 2008). These methods, though efficient, often need considerable power resources and perfect CSI to effectively counter the eavesdropper.

With the RIS, new avenues for the improvement of PLS have opened. RIS technology allows for the dynamic control of the wireless environment by reconfiguring the phase, amplitude, and polarization of reflected signals. Studies such as Huang et al. (2020) have shown that RIS-assisted beamforming can substantially improve secrecy capacity by guiding signals towards legitimate users and away from eavesdroppers. Similarly, Wu and Zhang (2019) showed that integrating RIS with MIMO systems enhances spatial selectivity, which further reduces the opportunity for eavesdropping. RIS-enabled artificial noise has also been suggested as a power-efficient approach to degrade eavesdropper channels without interfering with the legitimate communications (Li et al., 2021).

There are also recent studies on RIS-assisted PLS in complex environments, including multi-user and multi-eavesdropper settings. For example, Pei et al. (2021) investigated RIS-assisted PLS in UAV communication networks, covering both LOS and NLOS conditions. The studies also explored the security of millimeter-wave and terahertz communications, as both these high-frequency bands are sensitive to eavesdropping in nature (Zhang et al., 2022). Realtime optimization frameworks of machine learning methods have been proposed to find optimal RIS configurations to provide maximum secrecy rates in the real environment according to Zheng et al.

(2021). Other issues include the hardware constraint of RIS, complexity in achieving high-precision CSI at the eavesdropper's side, and scalability of RIS in very large networks. Liu et al. (2023) stressed the energy-efficient designs and integration of RIS with 6G networks, which should ensure secure and sustainable communication. As the field advances, the increasing recognition of RIS-aided PLS as a transformative solution for future wireless communication systems in the context of IoT, autonomous vehicles, and industrial automation can be seen.

Recent breakthroughs in RIS-aided Physical Layer Security (PLS) have provided innovative solutions to evolving wireless security challenges. Chen et al. (2020) studied secure beamforming in RIS-assisted MISO systems with the joint optimization of active and passive beamforming. The authors have shown significant enhancements in secrecy rates by using RIS to increase signal directionality and decrease the eavesdropping capability of eavesdroppers. In a similar manner, Luo et al. (2021) studied RIS-aided PLS in multi-eavesdropper scenarios, proposing optimal phase shift and power allocation strategies to effectively counter the complexity of dense eavesdropper networks.

Machine learning techniques have also been applied to advance RIS configurations for PLS. Shao et al. (2022) developed a reinforcement learning-based framework for adaptive phase control in RIS, allowing real-time optimization of secrecy rates under dynamic channel conditions. This achieved nearly optimal performance with less overhead, hence suitable for practical implementation. In addition, Zhang et al. in 2022 explored the integration of RIS to secure millimeter-wave (mmWave) and terahertz (THz) communication systems. Their study presented the effectiveness of RIS in overcoming interception threats at high frequency bands by smartly shaping the propagation environment.

Energy efficiency in RIS-assisted PLS is another topic of the recent studies. Guo et al. (2023) presented an energy-efficient design framework for RIS-assisted systems that balance power consumption and secrecy performance. Their results demonstrate that optimized RIS configurations can enhance security while minimizing energy expenditure, a critical consideration for energy-constrained environments such as IoT and 6G networks. These studies collectively underline the transformative potential of RIS in enhancing physical layer security while overcoming practical challenges such as scalability, energy efficiency, and real-time adaptability Jiang et al. (2020).

The chapter highlight the revolution in the physical layer security by RIS that helps to overcome the weaknesses of traditional approaches and pave the way to more secure and efficient wireless networks.

PHYSICAL LAYER SECURITY: CONCEPTS AND CHALLENGES

PLS is a paradigm in wireless communication that leverages the inherent characteristics of the physical medium to secure data transmission. While traditional cryptographic methods function at higher layers of the communication stack, the focus here is on exploiting properties like channel fading, noise, and interference for secure communication. The secrecy rate is a key metric in PLS, which measures the maximum information rate that can be transmitted secretly without the eavesdropper intercepting it. Due to its ability to benefit from dynamic and unpredictable behavior of the wireless channel, the PLS offers a viable solution as a complement to or substitute for traditional security. The core idea of PLS is the wiretap channel model by Shannon and Wyner. It considers a legitimate receiver Bob and an eavesdropper Eve with different channel conditions. PLS techniques aim at maximizing the SNR at Bob and minimizing the SNR at Eve. Techniques include artificial noise generation, cooperative jamming, and beamforming. For example, it is possible to transmit artificial noise in directions that degrade Eve's reception without impacting Bob. Thus, it enhances the secrecy capacity. In a similar vein, beamforming in MIMO systems concentrates signal energy at the intended receivers and suppresses it elsewhere.

Even with such benefits, PLS remains quite challenging to be practical in real-world wireless environments. The first issue is that of channel state information, requiring that CSI of the channels between the transmitter, legitimate receiver, and eavesdropper be known accurately. Practically, CSI of the eavesdropper cannot be acquired in such accuracy and thus severely limits the implementation of PLS techniques. Other complications in implementing PLS strategies arise from multiple eavesdroppers, mobility of users, and different interference levels. Another challenge arises from the lack of power and spectrum resources in wireless networks. In many cases, secure communication requires strong power for artificial noise generation or cooperative jamming, which may not be possible to implement in energy-constrained devices such as those in IoT. The spectral efficiency of communication systems may also deteriorate when resources are allocated in enhancing security. These trade-offs in security, power consumption, and spectral efficiency have to be maintained carefully while designing PLS mechanisms.

Physical Layer Security is an encouraging approach to secure wireless communications and is resilient against eavesdropping using physical channel characteristics. However, it poses significant challenges in practical implementation: the acquisition of CSI, the constraints of resources, and environmental dynamics. Recent breakthroughs in technologies such as RIS, machine learning-based channel estimation, and multi-antenna systems will enable overcoming these challenges to offer more efficient and robust PLS solutions for future generations of wireless networks.

RECONFIGURABLE INTELLIGENT SURFACES (RIS) FOR ENHANCING PHYSICAL LAYER SECURITY

Reconfigurable Intelligent Surfaces is a transformative technology that can alter the wireless communication environment by redefining it. RIS is composed of passive or semi-passive elements that may dynamically control the phase, amplitude, and polarization of the reflected signals to customize the propagation environment. When applied to Physical Layer Security (PLS), RIS offers the possibility of radical rethinking of how wireless networks can potentially avoid eavesdropping by adapting signal paths to favor desirable users while disrupting adversarial ones. This functionality makes RIS an appealing addition to modern communication systems that target a robust and efficient security implementation.

Figure 1. Reconfigurable intelligent surfaces

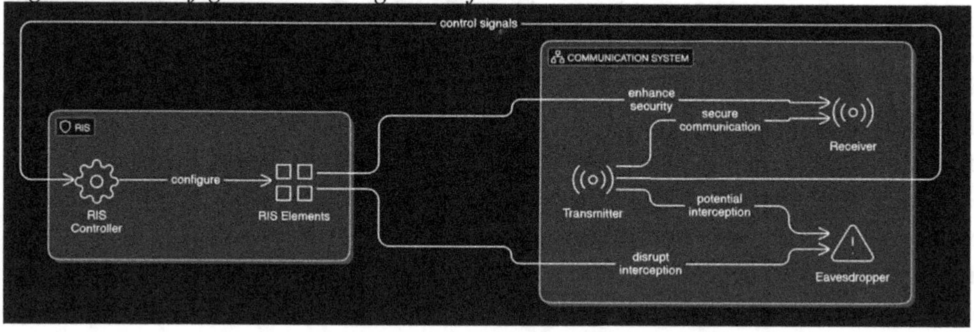

Beamforming and Signal Directionality

One of the major contributions of RIS to PLS is enhancement of beamforming. Since the traditional beamforming was dependent on the capability of a transmitter to steer signals, additional control is introduced by adding a layer of shaping the signal path after transmission. This is achieved when the reflective elements of an RIS adjust their phase shifts to produce highly concentrated beams toward legitimate receivers, which in turn create destructive interference toward eavesdroppers Basar et al. (2020). This exact control ensures that the legitimate receiver benefits from an enhanced SNR while the eavesdropper's interception probability is significantly reduced; hence, boosting the overall secrecy capacity of the system.

Artificial Noise and Jamming Techniques

RIS can provide additional enhancements to PLS in terms of artificial noise generation and jamming. Although traditional AN schemes are power-consuming to disturb the eavesdropper channels, thereby degrading the quality of legitimate communication, RIS can be low-power with smart reflection of noise signals toward the eavesdropper with minimal disruption to legitimate users. It maximizes the secrecy rate and reduces energy consumption, making this AN generation with RIS particularly very attractive for energy-constrained environments such as IoT networks.

Figure 2. Artificial noise and jamming techniques

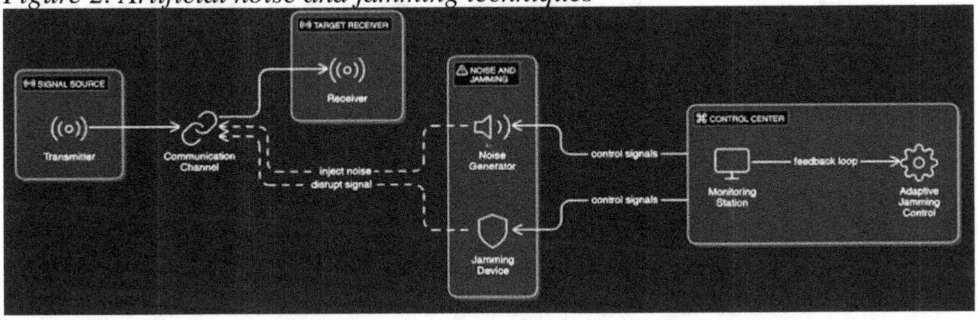

The integration of RIS into advanced communication technologies, such as MIMO systems and millimeter-wave communications, further enhances PLS. The RIS can complement the spatial diversity of antennas by creating virtual paths to make them reinforce legitimate communication and weaken eavesdropping channels within MIMO systems. In the same way, in mmWave and terahertz systems, signals are of extremely high directivity and subject to possible interceptions, and, thanks to RIS, it can dynamically shape the surrounding propagation environment toward enhanced security. These integrations demonstrate how versatile RIS is in facing challenges of security across a vast range of communication scenarios.

RIS presents significant benefits toward PLS while implementation is problematic. Achieving accurate channel state information is critical for optimal phase adjustment, but obtaining precise CSI in dynamic environments remains a very challenging task. Moreover, RIS hardware, although energy efficient, has limitations in terms of real-time adaptability and scalability for large networks. Advanced algorithms for real-time optimization and the integration of RIS with machine learning techniques are the future research to be carried out to overcome these challenges. Despite these challenges, RIS is a promising solution to improve PLS in next-generation communication networks and will provide unmatched control over the wireless environment.

APPLICATIONS OF RIS-AIDED PHYSICAL LAYER SECURITY

1.Securing IoT Networks

IoT is extremely prone to attacks as it consists of distributed nature with resource-constrained devices. RIS-aided PLS offers an energy-efficient and cost-effective solution as it dynamically shapes the propagation environment so that the likelihood of eavesdropping is at a minimum. RIS can enhance SNR for legitimate IoT devices and create destructive interference for the potential attacker. Furthermore, RIS can be implemented in passive modes, which makes it very suitable for low-power IoT applications, such as smart homes, industrial automation, and healthcare systems, where traditional cryptographic methods may not be possible.

Figure 3. Process flow of RIS-Aided physical security

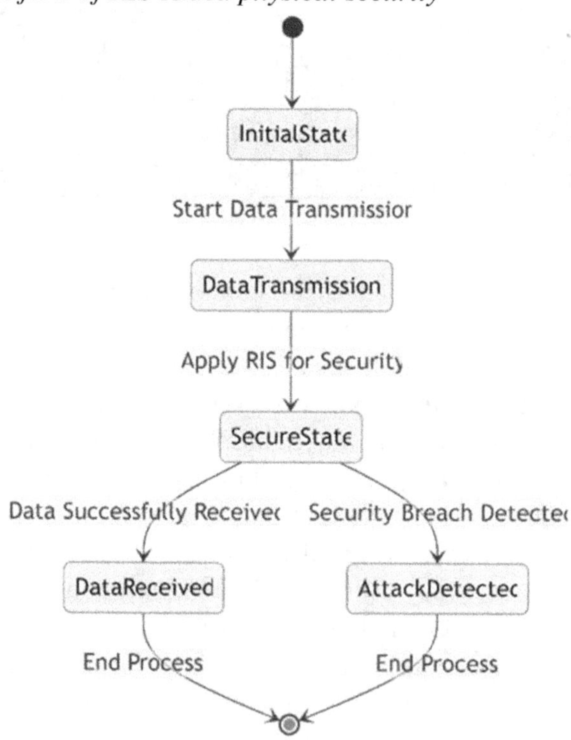

2. Security Enhancement in 6G Networks

With the advent of wireless communication into 6G, the requirement for ultra-reliable and secure communication has increased exponentially. RIS will be an important component of 6G networks as it will handle security issues in ultra-low latency, massive connectivity, and high-frequency bands such as terahertz (THz). By smartly manipulating the propagation paths, RIS can prevent unauthorized access and improve secrecy in dense urban environments, autonomous vehicle communication, and mission-critical applications. Its ability to support beamforming and artificial noise generation will ensure robust security in such high-demand scenarios.

3. Wireless Power Transfer Systems Security

WPT systems, which are used to charge devices wirelessly, are vulnerable to energy interception and data theft. RIS-aided PLS can counter these threats by focusing the energy transfer beams only towards legitimate devices and away from unauthorized users. This selective focusing reduces power wastage and ensures that only intended recipients benefit from the transfer. WPT systems with secure applications range across IoT sensors, medical implants, and electric vehicles, such as both security and power efficiency.

4. Consolidating UAV and Satellite Communications

Increased use of UAVs as well as satellite systems exists in disaster management, surveillance, and remote sensing domains but have open-air communication channels that are highly susceptible to interception. RIS-assisted PLS can adapt the communication environment in real time to improve secrecy by establishing safe communication channels for UAVs and satellites. In addition, RIS can reduce the vulnerability of LOS in UAV communications by deflecting signals through NLOS paths, which makes it difficult for eavesdroppers to intercept the communication.

5. Securing Edge Computing and Fog Networks

Edge and fog computing architectures bring the processing of data closer to the users. In these decentralized systems, new security challenges emerge. The RIS can enhance the security of such networks by allowing access to sensitive data only at authorized edge nodes and devices. It dynamically adjusts the signal propagation environment to prevent access and leakage of data. This capability is especially valuable in applications such as smart grids, autonomous vehicles, and augmented reality, where real-time data processing and security are critical.

RIS DESIGN FOR PHYSICAL LAYER SECURITY

The design of RIS for PLS is very careful to enhance the secrecy of wireless communications while overcoming practical implementation challenges. The main goal is to exploit the programmable nature of RIS to control the propagation environment and optimize security metrics such as secrecy capacity and secrecy outage probability. The key design elements include optimization of phase shifts, location of RIS, integration with other existing systems, and energy efficiency considerations. A properly adjusted phase in reflective elements enables RIS to bring about constructive interference at authorized receivers while causing destructive interference to eavesdroppers. For this purpose, high CSI accuracy is required for successful and accurate signal path forecasting and manipulation. Often the more advanced optimization algorithms of convex optimization and machine learning are implemented to improve the phase-shift design efficiency.

Another important design component relates to the placement of RIS. In environments wherein various physical obstructions characterize large areas, such as built cities or indoor spaces, careful placement maximizes the impairment created on the communication link. This requires considerations relating to geometric relationships between transceiver, receiver, RIS and possible eavesdroppers. This integration involves designing RIS controllers capable of interacting with base stations to align with the broader network objectives. For example, in MIMO systems, RIS can act as an additional layer of beamforming to complement the capabilities of the antenna arrays, while in mmWave systems, RIS helps mitigate line-of-sight dependencies and secures directional transmissions. Energy efficiency is another important design concern in the RIS for PLS. Unlike active jamming schemes, RIS uses passive operation, which considerably decreases its power consumption. Nonetheless, its performance in a massive scale network calls for finding appropriate trade-offs between the size of the reflecting elements and computational complexity. Often energy-efficient designs use AI in a configuration that adjusts RIS with respect to real-time changes in the environment. Meta-materials or software-defined surfaces in modern RIS prototypes are still under development. These designs need to be robust, cost-effective, and scalable for large-scale deployment in real-world applications. With advancements in research, RIS design is expected to adopt adaptive mechanisms that combine enhancements in the physical layer with upper-layer security protocols for a holistic approach to securing next-generation wireless networks.

Energy-Efficient Designs for RIS-Aided Physical Layer Security

Energy efficiency is a critical consideration in designing Reconfigurable Intelligent Surfaces (RIS) for Physical Layer Security (PLS), especially in applications like IoT networks and next-generation wireless systems where power resources are limited. Unlike traditional active jamming systems, RIS operates passively, reflecting and shaping signals without requiring power-intensive signal generation. However, achieving optimal performance while maintaining low energy consumption requires innovative approaches to RIS design. This includes optimization of the number of reflecting elements, minimization of power consumption through control circuitry, and the adoption of intelligent algorithms for real-time operation.

Optimization of RIS Configurations

Phase shift optimization: it is an efficient technique toward energy-efficient RIS. Methods like convex optimization and reinforcement learning are utilized for real-time phase-shift adaptation of elements in RIS in the pursuit of enhanced security alongside energy efficiency. In addition, the methods allow one to maximize the secrecy rate at the same time because more focused signal power will reach intended users, and other parts of the path accessible by eavesdroppers is deteriorated. But energy-efficient designs also take into account the trade-off between the number of reflective elements and computational complexity because a greater number of elements offers greater security at the cost of higher energy consumption as well as latency in optimization processes Zhang et al. (2022).

Low-Power Control and Hardware Design

Hardware advancements are included in energy-efficient RIS designs as well. Current RIS prototypes utilize passive or even semi-passive elements developed based on meta-materials, consuming minimal amounts of energy to reflect. Controlling circuits that adjust phase shifts consume power, but reducing these consumptions must be pursued. Ultra-low-power electronics and energy-harvesting technologies allow standalone operation for RIS units by exploiting ambient energy resources such as solar or RF signals. These developments make RIS particularly suitable for deployment in resource-constrained environments, such as remote IoT networks or UAV communications.

INTEGRATION OF RIS WITH ADVANCED WIRELESS TECHNOLOGIES

RIS and Massive MIMO Systems

In the integration of Reconfigurable Intelligent Surfaces (RIS) with massive Multiple-Input Multiple-Output (MIMO) systems, many challenges of modern wireless communication are addressed. In massive MIMO systems, the performance of the systems relies on deploying numerous antennas at the transmitter to exploit spatial diversity. However, their performance degrades in environments with significant physical obstructions or weak channel conditions. RIS improves these architectures with an extra layer of passive beamforming that produces virtual paths reinforcing the SNR between legitimate receivers. This integration optimizes spatial degrees of freedom, making massive MIMO systems able to achieve higher secrecy rates with less influence by eavesdropping effects. The use of RIS reduces power consumption, complementing the energy-hungry nature of massive MIMO systems.

RIS in mmWave and THz Communication Systems

The millimeter-wave (mmWave) and terahertz (THz) frequency bands are considered crucial parts of future wireless technologies, such as 6G, with significant bandwidth for ultra-high-speed data rates. However, the high-frequency signals have a highly directional nature, and their LOS dependency and susceptibility to interception make them unreliable. RIS tackles these weaknesses by offering smart propagation control, such that the mmWave and THz signals can maintain high strength even in NLOS scenarios. RIS provides additional benefits to the physical layer security of high-frequency communication by redirecting signals around obstacles while simultaneously cutting paths available to eavesdroppers. This is useful for applications such as URLLC and real-time industrial automation, where the security and robustness of the links are critical.

RIS with Cooperative Communication and Relay Networks

Cooperative communication and relay networks introduce a new dimension to security and efficiency through RIS integration. Conventional relay systems amplify and forward signals, increasing the chances of consuming power and adding noise. The RIS operates passively, reflecting the signals towards the desired destination without further amplification, which helps in interference and energy use reduction. In a cooperative network, RIS can be used as a virtual relay, changing its phase shifts dynamically to facilitate the legitimate user communication and weaken eavesdropper links. This integration can benefit various scenarios, including multi-hop networks, vehicular communication, and disaster recovery operations, where communication across extensive areas is secure and efficient.

RIS for UAV and Satellite Communications

Integration with Unmanned Aerial Vehicles (UAVs) and satellite systems is inherent for advanced wireless technologies, however the open-air communication link for UAVs and satellites has a weakness in their system that makes them particularly susceptible to interception. Smart RIS can be seen to enhance the security when combined with UAV and satellites systems. For UAV systems, it can counter LOS-based vulnerability by providing secure paths as NLOS, guaranteeing robust communication even during high-mobility, dynamic scenarios. Similarly, in satellite communications, RIS can be used to compensate for the atmospheric losses and improve the beamforming capability of ground terminals. In such scenarios, the overall reliability, security, and efficiency of aerial and space-based communication networks are improved considerably when these technologies are integrated together with RIS.

MACHINE LEARNING FOR RIS-AIDED PHYSICAL LAYER SECURITY (PLS)

Improvement in RIS Configuration with Machine Learning

The main function of machine learning (ML) is optimization in the configuration of Reconfigurable Intelligent Surfaces (RIS) for Physical Layer Security (PLS). This paper involves supervised and unsupervised learning models in ML algorithms that can predict channel state information and dynamically change RIS phase shifts to achieve a maximization of secrecy rates. Conventional optimization techniques used in traditional ways for RIS configurations can be challenged by complexity and dynamism in various environments; these include delays in computing and scalability problems. The above limitations are overcome by ML-based approaches that learn the propagation environment and adapt the RIS configuration in real-time. This helps RIS enhance the quality of the signal received by the legitimate user while disrupting the eavesdropper channels efficiently.

Reinforcement Learning for Adaptive Security

The adaptive decision-making in RL elevates the performance of RIS-aided PLS further, given the uncertain and dynamic wireless environments. In RL, the RIS controllers play the role of agents who continually learn optimal strategies from interaction with the wireless environment. For example, algorithms based on RL can learn which phase-shift configurations would be optimal to counteract the effects of variations in eavesdropper positions or changes in channel conditions. This adaptability becomes essential to secure next-generation networks like 6G, wherein real-time reconfiguration will be required in order to counter dynamic security threats. Techniques such as deep Q-learning and actor-critic methods are useful in managing large state and action spaces in RIS-aided PLS scenarios.

FUTURE PROSPECTS OF ML IN RIS-AIDED PLS

The ML-based integration into RIS-aided PLS is at a point when network evolution is going to lead to very promising development. This includes federated learning on distributed ML models for training of the RIS controllers without transmission of sensitive information to enhance not only security but also privacy in implementations. The hybrid application of ML using supervised learning for predictions of CSI and RL, in real-time decision, could unlock levels of efficiency as well as security. As RIS technology matures, ML will be the foundation for managing complex environments, reducing computational overhead, and ensuring robust physical layer security in applications ranging from IoT to satellite communications.

Challenges and Practical Implementation of RIS-Aided Physical Layer Security

1. Channel State Information (CSI) Acquisition and Accuracy

One of the most severe challenges in implementing RIS-aided Physical Layer Security (PLS) is the acquisition of precise and real-time Channel State Information (CSI). Since the RIS works on manipulating the environment of wireless propagation by adjusting the phase precisely, CSI plays a pivotal role in optimizing the reflection coefficients of the RIS elements. CSI is quite tough to obtain for RIS, and particularly in dynamic environments; because it is a passive setup with a large number of elements. Conventional ways to acquire CSI incur computational complexity and overheads with most of them intractable for real-time realization. Innovative techniques of CS estimation or machine learning modeling must be used in the techniques so that accuracy and delay trade-offs are optimally made considering complexity.

2. RIS Placement and Deployment Strategy

RIS placement and deployment significantly determine the performance and the level of enhancement the system would achieve on top of PLS. Indeed, properly deploying the RIS in very complicated environments like urban settings or dense propagation scenarios with heavy obstructed physical environments ensures legitimate channels have strength, whereas the eavesdropping ones remain weak. However, determining the optimal placement involves solving complex geometric optimization problems that account for multiple factors like user mobility, eavesdropper location, and environmental conditions. Additionally, the deployment of RIS in practical scenarios must consider physical constraints, such as wall compatibility, power supply, and integration with existing infrastructure, which can limit its flexibility.

3. Energy Efficiency and Scalability

Although RIS is energy-efficient by nature, as it does not rely on any active components, the scale of RIS-assisted PLS brings up new concerns. With more reflecting elements, the amount of configuration for the RIS also increases significantly. This implies that the growing computational and signaling overhead to configure RIS may neutralize its energy-saving features. Further, the power consumed by controllers of RIS, adjusting the phase shifts, has to be minimized to ensure the whole system energy-efficient. Energy-harvesting techniques and lightweight optimization algorithms are needed to overcome these challenges, especially in large-scale deployments.

4. Hardware Limitations and Robustness

Practical implementation of RIS faces hardware-related limitations, including the design and manufacturing of cost-effective, robust meta-materials capable of precise and reliable signal manipulation. Current RIS prototypes are typically plagued by non-idealities, including low phase resolution, mutual coupling between elements, and sensitivity to environmental factors like temperature and humidity. These imperfections can compromise the system's performance and prevent effective PLS enhancement. Improvements in materials science and manufacturing processes will be necessary to create high-performance, low-cost RIS units suitable for large-scale deployment.

5. Security in Dynamic and Adverse Environments

RIS-assisted PLS is designed to offer robust security in dynamic environments with changing user positions, channel conditions, and eavesdropper strategies. However, it is challenging to ensure consistent security performance in such scenarios. For instance, highly mobile users or rapidly changing eavesdropper locations can make pre-configured RIS settings ineffective. Moreover, adversarial attacks on the RIS itself, such as spoofing or jamming its control signals, pose new security threats. To overcome these challenges, RIS systems must have a mechanism for real-time adaptation, such as reinforcement learning, and robust anti-jamming strategies to support secure communication.

Practical Implementation Strategies

To overcome these challenges, practical implementation of RIS-aided PLS must be integrated with advanced algorithms, robust hardware designs, and intelligent deployment strategies. Hybrid optimization techniques that combine deterministic and machine learning approaches may be applied to improve the adaptability and efficiency of RIS. Cooperative RIS networks, where multiple RIS units are used in collaboration, can enhance scalability and coverage while mitigating individual hardware limitations. The use of cloud-based or edge computing platforms can distribute the computational burden of complex optimizations, allowing for real-time and large-scale deployments.

CONCLUSION

Reconfigurable Intelligent Surfaces (RIS) is a transformative technology with great potential to improve Physical Layer Security (PLS) in modern wireless communication systems. It is capable of intelligent manipulation of the wireless propagation environment, optimizing signal transmission paths, achieving higher secrecy rates, and reducing the possibility of an interception, thus ensuring a cost-effective and energy-efficient method of security in communication systems. The integration of RIS with advanced wireless technologies, such as massive MIMO, mmWave, and THz communication, better supports performance in high-frequency and high-speed and complex environments. Machine learning techniques also perform a crucial role in dynamically optimizing configurations of RIS, offering the adaptability required for real-time and secure communication in dynamic scenarios.

However, the practical implementation of RIS-aided PLS is faced with various challenges, such as accurate Channel State Information (CSI), optimal placement and deployment strategies, and overcoming hardware limitations. Scalability and energy efficiency in RIS systems are very important to ensure that its performance does not degrade in a large-scale or resource-constrained environment. These challenges can be overcome only through further research and development in advanced CSI estimation techniques, energy-efficient hardware designs, and intelligent algorithms that will be integrated for real-time adaptation. Robust security measures to protect RIS systems from adversarial attacks are also important for their resilience in practical deployments.

As RIS technology matures, its applications in next-generation networks, including 6G, IoT, and satellite communications, will be expanded, ensuring greater security, reliability, and efficiency. The combination of RIS with machine learning and advanced wireless technologies has the potential to address the growing security demands of future communication systems. With continuous innovation in hardware, optimization algorithms, and deployment strategies, RIS-aided PLS will find its place at the forefront of secure, sustainable, and high-performance wireless communication networks.

REFERENCES

Alamzadeh, S., & Rahmati, N. (2021). Reconfigurable intelligent surfaces in physical layer security: Eavesdropper analysis. *IEEE Access : Practical Innovations, Open Solutions*, 9, 120321–120332.

Basar, E., & Di Renzo, M. (2020). Reconfigurable intelligent surfaces for PLS: Current research and future directions. *Proceedings of the IEEE*, 108(9), 1654–1685.

Chen, J., Zhou, G., & Yuan, X. (2020). Secure beamforming design for MISO systems assisted by reconfigurable intelligent surfaces. *IEEE Wireless Communications Letters*, 9(1), 108–112.

Cui, Y., Zhang, X., & Yuan, X. (2021). Secure transmission in RIS-aided MIMO systems: Joint beamforming design. *IEEE Transactions on Signal Processing*, 69, 1759–1773.

Goel, S., & Negi, R. (2008). Guaranteeing secrecy using artificial noise. *IEEE Transactions on Wireless Communications*, 7(6), 2180–2189. DOI: 10.1109/TWC.2008.060848

Guo, J., Tang, J., & Xu, L. (2023). Energy-efficient design of RIS-aided physical layer security systems. *IEEE Transactions on Communications*, 71(4), 2234–2245.

Huang, C., Mo, R., & Zhang, C. (2020). Reconfigurable intelligent surface-assisted multi-user MISO systems with secrecy rate optimization. *IEEE Transactions on Wireless Communications*, 19(8), 4939–4952.

Jiang, W., & Yuan, J. (2020). Secure communication via RIS: Outage probability analysis. *IEEE Transactions on Wireless Communications*, 19(12), 8125–8136.

Li, X., Zhang, Y., & Cui, S. (2021). Secure transmission using reconfigurable intelligent surfaces: Artificial noise approach. *IEEE Transactions on Communications*, 69(3), 1837–1850.

Lin, S., Shen, X., & Cao, Y. (2023). Reconfigurable intelligent surfaces in 6G networks: Physical layer security perspectives. *IEEE Transactions on Green Communications and Networking*, 7(2), 215–229.

Liu, Y., Zhang, Q., & Wu, H. (2023). Energy-efficient RIS-aided physical layer security in 6G networks: Challenges and solutions. *IEEE Communications Surveys and Tutorials*, 25(1), 34–56.

Luo, H., Xu, J., & Tang, W. (2021). Secure communication with RIS in multi-eavesdropper scenarios. *IEEE Transactions on Vehicular Technology*, 70(7), 6912–6924.

Pei, X., Zeng, J., & Yu, Y. (2021). RIS-aided secure UAV communication: Secrecy rate maximization with flight path optimization. *IEEE Transactions on Vehicular Technology*, 70(2), 1256–1267.

Shao, Y., Liu, Z., & Wang, F. (2022). Reinforcement learning-based design of RIS for physical layer security. *IEEE Internet of Things Journal*, 9(5), 3897–3909.

Tang, W., Chen, M., & Guo, Y. (2021). RIS-aided physical layer security for massive IoT: System design and performance analysis. *IEEE Internet of Things Journal*, 8(10), 8672–8686.

Wu, Q., & Zhang, R. (2019). Towards smart and reconfigurable environment: Intelligent reflecting surface aided wireless networks. *IEEE Communications Magazine*, 58(1), 106–112. DOI: 10.1109/MCOM.001.1900107

Wyner, A. D. (1975). The wire-tap channel. *The Bell System Technical Journal*, 54(8), 1355–1387. DOI: 10.1002/j.1538-7305.1975.tb02040.x

Zhang, H., Liu, Y., & Wu, Q. (2022). RIS-assisted physical layer security for mmWave and THz communication. *IEEE Communications Letters*, 26(3), 401–405.

Zhang, Z., Li, M., & Yu, W. (2022). RIS-assisted secure millimeter-wave communications: Challenges and opportunities. *IEEE Wireless Communications*, 29(4), 72–79.

Zheng, B., Zhang, Q., & Zhu, H. (2021). Intelligent reflecting surface-assisted secure communication: A machine learning approach. *IEEE Transactions on Communications*, 69(9), 5987–6000.

Chapter 11
Reconfigurable Intelligent Surface for mmWaves and Multi-Antenna Systems: Design Challenges and Applications

R. Sivaprasad
Sri Sairam Engineering College, India

ABSTRACT

Reconfigurable Intelligent Surfaces (RIS) are innovative, engineered structures composed of a large array of passive or semi-passive reflecting elements, known as meta-atoms, capable of controlling the behaviour of electromagnetic waves. These elements can be dynamically reconfigured to manipulate the amplitude, phase, polarization, and direction of the reflected signals. RIS systems are typically controlled via software or electronic components that adjust how each meta-atom interacts with incoming signals, thus enabling real-time control of the radio propagation environment. Unlike conventional communication elements like active antennas, which directly transmit or receive signals, RIS modifies the characteristics of an impinging signal without active power amplification.

DOI: 10.4018/979-8-3693-8099-4.ch011

1. INTRODUCTION TO RECONFIGURABLE INTELLIGENT SURFACES (RIS)

1.1 Definition and Concept

Reconfigurable Intelligent Surfaces (RIS) are innovative, engineered structures composed of a large array of passive or semi-passive reflecting elements, known as meta-atoms, capable of controlling the behaviour of electromagnetic waves. These elements can be dynamically reconfigured to manipulate the amplitude, phase, polarization, and direction of the reflected signals. RIS systems are typically controlled via software or electronic components that adjust how each meta-atom interacts with incoming signals, thus enabling real-time control of the radio propagation environment.

Unlike conventional communication elements like active antennas, which directly transmit or receive signals, RIS modifies the characteristics of an impinging signal without active power amplification. This results in more energy-efficient designs that reshape the radio wave environment, improving coverage, quality, and data transmission rates (Huang, Zappone, Alexandropoulos, Debbah, Yuen, 2020). The surfaces are intelligent because they can adaptively respond to the wireless environment, optimizing signal paths and reducing signal losses.

Figure 1. Concept of RIS

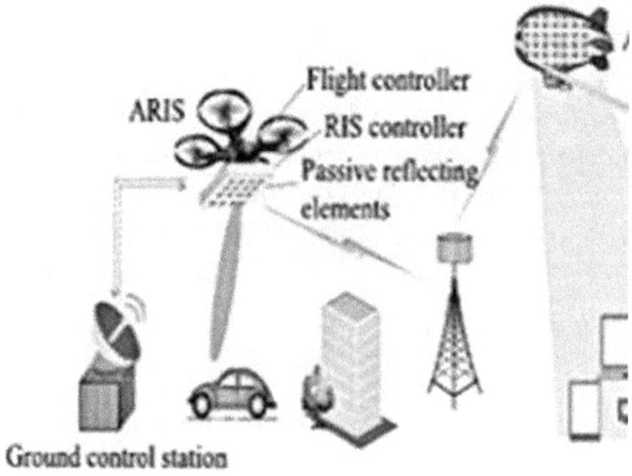

RIS primarily functions by redirecting electromagnetic waves toward targeted areas to enhance or diminish the strength of wireless signals in specific locations. It allows precise control of beam direction and reflection angles, enabling communication systems to circumvent obstacles, reduce interference, and extend the range of wireless transmissions. This dynamic reconfiguration can also mitigate issues such as path loss, multipath fading, and poor signal quality, which are common in high-frequency bands like millimeter waves (mm Wave).

1.2 Importance in Wireless Communications

The growing demand for high-speed wireless communications, driven by applications like video streaming, virtual reality, autonomous vehicles, and the Internet of Things (IoT), necessitates the adoption of new technologies to maximize spectrum utilization and network capacity (Basar et al., 2021). In this context, RIS is gaining significant attention due to its potential to address several challenges in modern wireless communication systems.

Key aspects where RIS can make a substantial impact include:

1. **Enhancing Coverage**: RIS can improve the coverage of wireless networks by intelligently reflecting signals into areas with weak or blocked signals, such as urban canyons or indoor environments. This is particularly useful in high-density cities where buildings and other structures obstruct line-of-sight communication.
2. **Improving Signal Quality**: The reflective elements of RIS can focus signals toward specific users or devices, boosting the strength of the received signal. This ensures better communication reliability and reduces the likelihood of dropped connections in environments with heavy interference.
3. **Energy Efficiency**: Since RIS operates passively, it consumes significantly less power compared to active relays and base stations. This makes it an attractive solution for improving communication efficiency in energy-constrained scenarios, such as rural areas or IoT networks where energy conservation is a critical requirement.
4. **Mitigating Interference**: In dense networks, where multiple devices are communicating simultaneously, interference can degrade the quality of signals. RIS can be used to steer signals in directions that minimize interference, improving overall network performance.
5. **Cost-Effectiveness**: Deploying traditional base stations, repeaters, and antennas to cover large areas can be prohibitively expensive. RIS offers a low-cost alternative, as it can be integrated onto existing infrastructure (e.g., walls, ceilings, or even outdoor environments), extending the reach of wireless networks without the need for additional power or complex hardware.
6. **Adaptability in Dynamic Environments**: One of the defining features of RIS is its reconfigurability. The wireless environment is often dynamic, with moving objects, changing weather conditions, and variable user density. RIS

systems can adapt to these changes in real time, providing optimal signal paths and maximizing the efficiency of the communication network.

1.3 Relevance to Millimeter Waves (mm Wave) and Multi-Antenna Systems

Millimeter-wave (mm Wave) communication, which operates in the frequency range of 30-300 GHz, is a cornerstone of 5G and future 6G networks. It offers enormous bandwidth potential, supporting data rates of several gigabits per second, making it ideal for high-speed applications. However, mm Wave technology faces several challenges that limit its wide adoption, such as short transmission ranges, high propagation losses, susceptibility to obstacles, and the need for line-of-sight (LOS) communication.

Figure 2. RIS communication

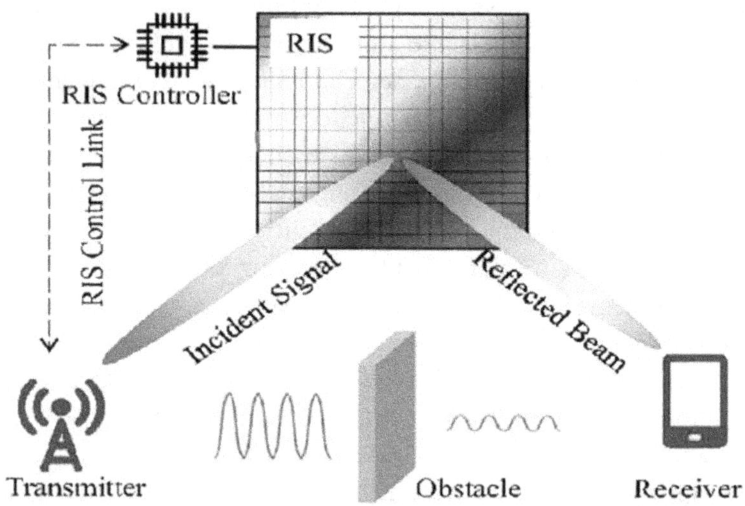

RIS presents a viable solution to many of these challenges, making it highly relevant to mm Wave systems in the following ways:

1. **Overcoming Path Loss**: mm Wave signals suffer from significant path loss due to their high frequencies. RIS can effectively redirect mm Wave signals around obstacles or toward areas with weak reception, mitigating the path loss issue and extending the coverage area.

2. **Enabling Non-Line-of-Sight (NLOS) Communication**: One of the primary limitations of mm Wave communication is its reliance on LOS paths, as obstacles like buildings, trees, or even human bodies can easily block these high-frequency signals. RIS allows for NLOS communication by reflecting mm Wave signals off its surface, creating alternative paths that bypass obstacles and maintain connectivity.
3. **Supporting Beamforming**: Beamforming is a technique used in mm Wave and multi-antenna systems to direct signals toward specific users or devices, maximizing signal strength and minimizing interference. RIS can be used as a reflective beamformer, further enhancing the directionality and precision of mm Wave signals, improving spectral efficiency, and supporting high data rates for multiple users simultaneously.
4. **Integration with Multi-Antenna Systems**: Multi-antenna systems, particularly MIMO (Multiple Input Multiple Output) and massive MIMO, are key technologies for improving network capacity and performance. In MIMO systems, multiple antennas transmit and receive signals, enhancing spatial diversity and spectral efficiency. RIS can complement these systems by providing additional degrees of freedom in controlling signal paths, especially in environments where conventional MIMO arrays might struggle with interference or coverage gaps.
5. **Reducing Hardware Complexity**: In mm Wave communication systems, the use of large arrays of antennas is necessary to achieve high beamforming gains. However, deploying such arrays can be costly and complex. RIS offers a simpler and more scalable solution, as it can emulate the effect of multi-antenna systems without the need for additional active hardware, providing similar benefits at a fraction of the cost and energy consumption.

2. FUNDAMENTALS OF MILLIMETER WAVE (MM WAVE) TECHNOLOGY

2.1 Overview of mm Wave Technology

Millimeter-wave (mm Wave) technology refers to the use of electromagnetic spectrum in the frequency range of 30 GHz to 300 GHz, corresponding to wavelengths between 1 mm and 10 mm. This spectrum offers significantly wider bandwidth compared to traditional microwave frequencies (below 6 GHz), enabling much higher data transmission rates, which is essential for supporting next-generation applications such as ultra-high-definition video streaming, virtual reality, autonomous vehicles, and massive machine-type communications (mMTC).

In the context of wireless communication, mm Wave technology is a key enabler of 5G networks and beyond due to its ability to deliver multi-gigabit data rates and reduce network congestion in densely populated areas. However, mm Wave signals behave differently compared to lower frequency signals. They are highly directional, propagate over shorter distances, and are more susceptible to environmental blockages and interference. These unique characteristics necessitate advanced techniques such as beamforming, the use of multi-antenna systems, and the integration of Reconfigurable Intelligent Surfaces (RIS) to overcome inherent challenges.

Figure 3. RIS in mm wave communication

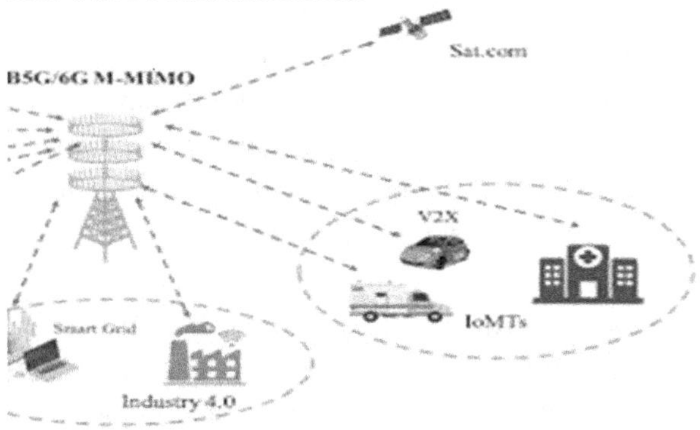

2.2 Advantages and Challenges in mm Wave Communication

2.2.1 Advantages of mm Wave Communication

1. **High Data Rates and Capacity**: The wider bandwidth available in the mm Wave spectrum enables significantly higher data transmission rates, supporting speeds of up to 10 Gbps and beyond. This makes mm Wave ideal for high-bandwidth applications such as ultra-high-definition streaming, augmented reality, and large-scale IoT deployments.
2. **Reduced Network Congestion**: Traditional frequency bands below 6 GHz are increasingly crowded due to the growing number of wireless devices. mm Wave communication alleviates this congestion by utilizing underutilized frequency bands, providing more spectrum for wireless communications.

3. **Directional Communication and Beamforming**: mm Wave signals are highly directional, which allows for precise beamforming techniques. This leads to better signal quality and reduced interference in environments with many users or devices, particularly in dense urban areas.
4. **Support for Dense Deployments**: mm Wave is ideal for supporting dense networks, such as in smart cities or urban environments, where a high concentration of users and devices are connected simultaneously. The ability to reuse frequencies across closely packed small cells enhances the overall network capacity.

2.2.2 Challenges in mm Wave Communication

1. **High Propagation Loss**: One of the main challenges of mm Wave communication is its higher free-space path loss, which increases as the frequency rises (Rappaport, 2013). The short wavelength of mm Wave signals leads to more rapid attenuation as the signal propagates through space, limiting the effective transmission range compared to lower frequencies.
2. **Sensitivity to Blockages**: mm Wave signals are easily blocked by obstacles such as buildings, trees, walls, and even human bodies. This requires line-of-sight (LOS) conditions for optimal performance, making it difficult to maintain a reliable connection in environments where obstacles frequently obstruct the signal path.
3. **Limited Coverage Range**: Due to high propagation losses and the susceptibility to blockages, mm Wave communication has a limited effective range, often covering only a few hundred meters. This necessitates the use of dense small cell deployments, which increases infrastructure costs and complexity.
4. **Penetration Issues**: Unlike lower-frequency signals, mm Wave signals have poor penetration capabilities. They struggle to pass through walls, windows, and other physical barriers, making it challenging to provide indoor coverage from outdoor base stations or through multiple partitions within buildings.
5. **Atmospheric Absorption**: mm Wave signals are more susceptible to absorption by atmospheric gases, particularly oxygen and water vapor, which further diminishes their range and reliability, especially in humid or rainy conditions.
6. **High Energy Consumption**: To overcome path loss and ensure reliable connections, mm Wave systems often require high-power transmitters and densely deployed small cells, leading to increased energy consumption. This challenge has spurred research into more energy-efficient transmission methods, including the use of RIS to reduce energy demands.

2.3 Role of mm Wave in 5G and Beyond

Millimeter-wave technology is fundamental to the design of 5G networks and will play an increasingly critical role in 6G and future wireless communication systems. Its ability to deliver high data rates, support ultra-low latency, and accommodate massive numbers of connected devices makes it indispensable for next-generation applications.

2.3.1 mm Wave in 5G Networks

In the deployment of 5G, mm Wave frequencies complement traditional sub-6 GHz bands to meet the growing demand for faster data speeds and higher capacity. While sub-6 GHz frequencies provide broader coverage and better penetration, mm Wave offers the extreme data rates and capacity needed for hotspot areas such as stadiums, transportation hubs, and city centers.

Key roles of mm Wave in 5G networks include:

1. **Enhanced Mobile Broadband (eMBB)**: mm Wave enables ultra-high-speed data transmission for mobile devices, enhancing user experiences for bandwidth-intensive applications such as 4K/8K video streaming, cloud gaming, and virtual reality.
2. **Massive Machine-Type Communication (mMTC)**: The high capacity of mm Wave technology makes it well-suited for supporting massive numbers of IoT devices in smart cities, industrial IoT environments, and large-scale sensor networks, where traditional communication methods might fail due to spectrum congestion.
3. **Ultra-Reliable Low-Latency Communication (URLLC)**: mm Wave technology supports mission-critical applications requiring ultra-reliable, low-latency communication, such as autonomous driving, remote surgery, and industrial automation. The combination of high data rates and low latency ensures that real-time data can be transmitted and processed almost instantaneously.
4. **Fixed Wireless Access (FWA)**: mm Wave frequencies provide an alternative to fiber-optic connections in areas where laying physical cables is too expensive or impractical. FWA delivers broadband-level connectivity to homes and businesses using mm Wave technology, providing gigabit-speed internet to areas with insufficient fiber infrastructure.

2.3.2 mm Wave in 6G and Beyond

Looking beyond 5G, mm Wave will continue to be a foundational technology for 6G networks, which are expected to operate in even higher frequency bands, including the terahertz (THz) range. These future networks aim to deliver data rates up to 100 Gbps and support advanced applications such as holographic telepresence, immersive extended reality (XR), and space-air-ground integrated networks.

Potential roles for mm Wave in 6G and beyond include:

1. **High-Speed, Low-Latency Communication for Holographic Applications**: mm Wave frequencies, with their immense bandwidth and low-latency capabilities, will be essential for transmitting the massive amounts of data required for real-time holographic communication, enabling applications like virtual meetings, remote education, and entertainment.
2. **Smart Cities and Autonomous Systems**: mm Wave will play a pivotal role in enabling real-time communication between smart city infrastructure, autonomous vehicles, and drones, where ultra-low latency and reliable connectivity are paramount for safety and efficiency.
3. **Integration with Satellite Communication**: In 6G, mm Wave frequencies may be used for satellite communication, creating global connectivity systems that integrate space, air, and ground networks. These systems could enable seamless, high-speed internet access in remote and underserved areas.
4. **Advanced Industry 4.0 Applications**: mm Wave technology will support the next wave of industrial automation, providing ultra-reliable, low-latency communication for advanced manufacturing, robotics, and supply chain management.

3. OVERVIEW OF MULTI-ANTENNA SYSTEMS

3.1 Introduction to MIMO (Multiple Input Multiple Output)

Multiple Input Multiple Output (MIMO) is a wireless communication technology that uses multiple transmitting and receiving antennas to improve the reliability and performance of data transmission. In traditional communication systems, signals are transmitted from a single antenna (Single Input Single Output or SISO), which limits capacity and performance. MIMO, on the other hand, leverages spatial diversity by transmitting multiple data streams simultaneously across multiple antennas, increasing the system's spectral efficiency and data rates without requiring additional bandwidth.

The key principles of MIMO include:

- **Spatial Multiplexing**: MIMO enables the transmission of multiple independent data streams through different antennas, thereby increasing the data throughput without requiring additional frequency resources.
- **Diversity Gain**: MIMO enhances signal reliability by using multiple antennas to receive signals along different paths, reducing the impact of fading or interference on any one path.
- **Beamforming**: By intelligently directing transmission power toward specific users or devices, MIMO systems can improve signal strength and reduce interference, leading to better quality communication.

MIMO technology has become fundamental in modern wireless standards like 4G LTE, Wi-Fi, and 5G, playing a critical role in improving network capacity, coverage, and data throughput. The technology continues to evolve, with higher-order MIMO systems (using more antennas) being introduced to meet the increasing demand for high-speed wireless communication.

3.2 Evolution of Multi-Antenna Technologies: From SISO to Massive MIMO

The evolution of multi-antenna systems has been driven by the need to increase wireless network capacity and efficiency, especially as the demand for mobile data has exploded. This evolution can be summarized in several key stages:

3.2.1 SISO (Single Input Single Output)

In traditional wireless systems, such as early cellular networks (1G, 2G), communication occurred between a single transmitter antenna and a single receiver antenna (SISO). This system design was simple but inherently limited in capacity and performance, as it was susceptible to signal fading, interference, and lower data rates due to the lack of spatial diversity.

3.2.2 SIMO (Single Input Multiple Output) and MISO (Multiple Input Single Output)

The next evolution in antenna systems introduced multiple antennas at either the receiver or the transmitter:

- **SIMO (Single Input Multiple Output)** systems use multiple antennas at the receiver to improve signal reception by taking advantage of the different propagation paths that a transmitted signal follows (Larsson, Edfors, Tufvesson,

Marzetta, 2001). This increases reliability, as the receiver can combine signals from multiple antennas to mitigate fading and interference.

- **MISO (Multiple Input Single Output)** systems use multiple antennas at the transmitter to send the same signal across multiple paths, improving the probability that at least one signal will be received clearly, thus enhancing reliability and signal quality.

While SIMO and MISO improve link reliability through diversity gains, they do not increase data throughput as effectively as MIMO.

3.2.3 MIMO (Multiple Input Multiple Output)

MIMO technology emerged as a breakthrough by employing multiple antennas at both the transmitter and the receiver. This enabled spatial multiplexing, where multiple data streams are transmitted simultaneously, significantly increasing the capacity of the communication system. MIMO also introduced diversity and beamforming gains, which improved reliability and reduced interference.

With the advent of 4G LTE, MIMO became a central feature for enhancing wireless networks. For example, 4x4 MIMO (using four antennas at both ends) became common in many modern devices, allowing higher data throughput and better signal reception in various environments.

3.2.4 Massive MIMO

Massive MIMO is the next evolution of MIMO technology, introduced with 5G networks. It refers to the use of a very large number of antennas (often in the hundreds) at base stations to serve multiple users simultaneously. Massive MIMO further extends the benefits of MIMO by providing even greater spectral efficiency, capacity, and signal robustness.

Key features of massive MIMO include:

- **Higher Capacity**: By serving multiple users with spatially separated beams, massive MIMO allows for higher user density and greater overall network capacity.
- **Beamforming Accuracy**: With more antennas, massive MIMO systems can form highly directional beams, reducing interference between users and improving signal quality.
- **Spectral Efficiency**: Massive MIMO can improve spectral efficiency by allowing the same frequency resources to be reused across multiple users in the same geographic area.

Massive MIMO is a cornerstone of 5G and beyond, allowing these networks to meet the growing demand for faster, more reliable, and higher-capacity wireless communication.

3.3 Multi-Antenna System Performance in mm Wave Networks

The integration of multi-antenna systems, especially MIMO and massive MIMO, with millimeter-wave (mm Wave) technology is crucial for overcoming the unique challenges associated with high-frequency communication (Marzetta, 2019). mm Wave communication, while offering much higher data rates due to the larger available bandwidth, suffers from several limitations, such as higher path loss, limited coverage range, and increased sensitivity to blockages. Multi-antenna systems help mitigate these challenges in several ways:

3.3.1 Overcoming Path Loss with Beamforming

In mm Wave networks, high propagation losses due to the shorter wavelength necessitate highly directional communication to maintain signal strength over longer distances. Multi-antenna systems, especially those using massive MIMO, enable precise beamforming, focusing the transmission power in narrow beams directly toward the receiver. This focused energy compensates for the path loss and extends the effective communication range of mm Wave signals.

Beamforming in massive MIMO systems can dynamically adjust beam directions to follow users as they move, ensuring continuous high-quality communication even in challenging environments like urban areas with frequent blockages.

3.3.2 Supporting High User Densities in mm Wave Networks

mm Wave frequencies are well-suited for dense urban environments, where a large number of users require high-speed connectivity. Multi-antenna systems provide spatial multiplexing capabilities, allowing multiple data streams to be transmitted simultaneously over the same frequency band. This improves the overall network capacity and supports more users without increasing spectrum usage, a critical feature in densely populated areas like city centres and public venues.

Massive MIMO further enhances this capability by leveraging a larger number of antennas to serve many users at the same time, with each user benefiting from its own spatially separated beam.

3.3.3 Addressing Signal Blockages with NLOS Communication

One of the biggest challenges of mm Wave communication is its reliance on line-of-sight (LOS) paths, as obstacles can easily block or reflect the high-frequency signals. Multi-antenna systems can use techniques such as beam steering to redirect signals around obstacles, creating non-line-of-sight (NLOS) communication paths. In addition, MIMO's ability to leverage multiple propagation paths (multipath diversity) helps improve signal resilience in environments where direct LOS communication is difficult to maintain.

By combining beamforming and multi-path transmission, multi-antenna systems enable mm Wave networks to maintain robust connectivity even in complex environments with frequent obstructions.

3.3.4 Enhancing Spectral Efficiency and Throughput

mm Wave networks are expected to support high data throughput for demanding applications such as augmented reality, ultra-high-definition streaming, and real-time gaming. Multi-antenna systems enhance the spectral efficiency of mm Wave communication by transmitting multiple data streams simultaneously (spatial multiplexing), effectively multiplying the data rate within the same bandwidth.

This capability, combined with the large bandwidth available in mm Wave frequencies, enables multi-antenna systems to deliver the ultra-fast data rates required for next-generation applications.

4. RECONFIGURABLE INTELLIGENT SURFACES (RIS): ARCHITECTURE AND WORKING PRINCIPLE

Reconfigurable Intelligent Surfaces (RIS) are emerging as a revolutionary technology in wireless communications, enabling the dynamic control of the electromagnetic environment to improve signal quality, enhance coverage, and increase spectral efficiency (Basar et al., 2021). By strategically reflecting, refracting, or absorbing electromagnetic waves, RIS can shape the propagation characteristics of wireless signals, providing more flexibility and control in wireless network design.

This section delves into the architecture of RIS, its working principle, control mechanisms, and its role in electromagnetic beamforming, along with a comparison to traditional beamforming techniques.

4.1 Basic RIS Structure: Meta-Surfaces and Reflective Elements

At the core of a Reconfigurable Intelligent Surface is a meta-surface, which is an artificial, ultra-thin surface consisting of numerous sub-wavelength reflective elements (referred to as "meta-atoms"). These reflective elements can dynamically manipulate the phase, amplitude, and polarization of incoming electromagnetic waves. The meta-surface is typically composed of a two-dimensional grid of passive or active elements, such as low-cost antennas or reflectors, arranged in a lattice.

Key components of RIS architecture include:

- **Meta-Surfaces**: Meta-surfaces are engineered materials with unique electromagnetic properties that allow for precise control over the behavior of incident signals. By tuning the response of each element on the surface, RIS can alter the reflection, refraction, or absorption of radio waves in a manner that enhances signal strength and quality.
- **Reflective Elements**: Each meta-surface element can act as a reflector that adjusts the phase and amplitude of incoming waves. The reflective elements are electronically tunable, enabling them to dynamically adjust how they reflect or refract signals based on environmental conditions or network requirements.
- **Substrate**: The reflective elements are usually mounted on a substrate, which can be flexible or rigid, depending on the deployment. The substrate provides structural support and can influence the overall electromagnetic behavior of the RIS.
- **Control Layer**: Beneath the meta-surface, there is typically an embedded control layer that interfaces with the wireless network. This layer adjusts the properties of the reflective elements in real-time, based on external commands from network controllers, enabling intelligent adaptation to changing network conditions.

4.2 RIS Control Mechanisms: Passive vs. Active RIS

RIS can be broadly categorized into **passive** and **active** systems, depending on how the reflective elements are controlled and powered.

4.2.1 Passive RIS

In a **passive RIS**, the meta-surface elements do not generate or amplify signals. Instead, they rely on external radio signals, passively reflecting or refracting them in a controlled manner (Di Renzo, Debbah, Phan-Huy, Zappone, Alouini, Yuen, 2019). Each element adjusts its phase and reflection characteristics using tunable components, such as varactor diodes, liquid crystals, or MEMS (Micro-Electro-Mechanical Systems). Passive RIS is characterized by its energy efficiency and simplicity, as it does not require power-hungry amplification circuits.

Advantages of Passive RIS:

- **Low Power Consumption**: Since passive RIS elements do not require active signal generation, they consume very little power, often operating without external power sources or relying on minimal energy harvesting.
- **Low Complexity and Cost**: Passive RIS is simpler and cheaper to manufacture, making it highly scalable for large deployments.

Limitations of Passive RIS:

- **Limited Signal Enhancement**: Passive RIS can only redirect and modify existing signals; it cannot amplify them, which may limit its performance in scenarios where the signal strength is low or attenuation is severe.

4.2.2 Active RIS

In contrast, **active RIS** includes active components, such as amplifiers, that can not only reflect but also enhance the strength of incoming signals. Active RIS is more powerful than passive RIS, as it can compensate for signal loss due to propagation or reflection, improving the overall signal-to-noise ratio (SNR).

Advantages of Active RIS:

- **Signal Amplification**: Active RIS can boost signal power, overcoming propagation losses, which is especially useful in non-line-of-sight (NLOS) environments or when signals are weak.
- **Better Performance in High-Loss Environments**: By amplifying the reflected signals, active RIS can improve connectivity in areas with significant signal degradation, such as dense urban environments or indoor settings.

Limitations of Active RIS:

- **Higher Energy Consumption**: The use of active components increases the power requirements, potentially reducing the energy efficiency of the system.
- **Increased Complexity and Cost**: Active RIS systems are more complex to design and manufacture, and their components are more expensive, leading to higher deployment and operational costs.

4.3 Electromagnetic Beamforming through RIS

One of the most powerful capabilities of RIS is **electromagnetic beamforming**, which refers to the control of signal propagation through the intelligent manipulation of the phase and direction of electromagnetic waves. Unlike traditional beamforming, which is achieved at the transmitter or receiver using antenna arrays, RIS can perform beamforming at the surface level, effectively shaping how signals reflect off the environment.

RIS beamforming works by altering the phase shift applied to each meta-surface element, causing the reflected waves to constructively interfere at desired locations (to enhance signal strength) or destructively interfere at other locations (to reduce interference). This allows RIS to create a highly directional beam of radio waves, guiding the signal toward intended users or devices.

The key processes involved in RIS-enabled beamforming include:

- **Phase Control**: Each reflective element on the RIS can be independently controlled to introduce specific phase shifts to the reflected signal. By coordinating the phase shifts across the surface, RIS can direct the reflected signal toward a target user.
- **Adaptive Beam Steering**: RIS can dynamically adjust its reflective properties in response to changing user locations or environmental conditions, ensuring that the beam is always directed where it is needed most, even as users move.
- **Interference Management**: By carefully controlling the phase and direction of reflected waves, RIS can reduce interference between users or networks, improving overall spectral efficiency.

This ability to manipulate the propagation environment gives RIS a distinct advantage in managing multipath effects and improving coverage in difficult-to-reach areas, such as indoor environments, urban canyons, or areas with heavy obstacles.

4.4 Comparison with Traditional Beamforming Techniques

Beamforming has traditionally been performed using **antenna arrays** at the transmitter or receiver, which adjust the phase and amplitude of the transmitted signals to form narrow beams directed toward specific users. However, RIS introduces a new paradigm for beamforming by enabling environmental control of signal propagation. The key differences between RIS-enabled beamforming and traditional beamforming techniques are outlined below:

4.4.1 Traditional Beamforming

- **Location of Beamforming**: Traditional beamforming is typically performed at the base station or user device using antenna arrays. Signals are transmitted or received in specific directions by adjusting the phase and amplitude of the signals at each antenna element.
- **Complexity**: Beamforming in traditional systems requires complex hardware and algorithms to process signals in real-time, often involving digital or hybrid beamforming techniques to optimize the signal paths.
- **Power Requirements**: Traditional beamforming requires significant power, as it is performed at the transmitter or receiver, where amplification and signal processing occur.
- **Performance in NLOS**: Traditional beamforming can struggle in non-line-of-sight (NLOS) environments, as it relies on direct line-of-sight communication paths between the transmitter and receiver.

4.4.2 RIS-Enabled Beamforming

- **Environmental Control**: RIS shifts beamforming from the transmitter or receiver to the environment itself. By using passive or active meta-surfaces, RIS can redirect signals from arbitrary locations, enabling more flexibility in controlling signal propagation paths.
- **Lower Complexity**: RIS does not require complex signal processing or amplification at the transmitter or receiver, making it a more energy-efficient solution. The meta-surfaces are also relatively low-cost and easy to deploy in large numbers.
- **Enhanced NLOS Capabilities**: RIS excels in NLOS scenarios by intelligently reflecting signals around obstacles, ensuring that communication can be maintained even when direct line-of-sight paths are unavailable.

- **Power Efficiency**: Especially in passive RIS systems, power consumption is minimal compared to traditional beamforming techniques, as RIS relies on controlling existing signals without generating or amplifying them.

5. DESIGN CHALLENGES IN RECONFIGURABLE INTELLIGENT SURFACES FOR MM WAVE AND MULTI-ANTENNA SYSTEMS

The deployment of Reconfigurable Intelligent Surfaces (RIS) in millimeter-wave (mm Wave) and multi-antenna systems offers significant potential for enhancing wireless communication, but it also presents several design challenges. These challenges span multiple layers of the system architecture, from the physical hardware to real-time control, optimization, and integration into existing networks. This section explores the key design issues that need to be addressed to fully realize the potential of RIS in mm Wave and multi-antenna systems.

5.1 Physical Layer Challenges

The physical layer of RIS-assisted mm Wave systems is particularly sensitive to several factors, including the inherent propagation characteristics of high-frequency waves, path loss, and the complexity of RIS hardware design (Huang, Zappone, Alexandropoulos, Debbah, Yuen, 2020; Zappone, Di Renzo, Shams, Yu, Poor, 2019). The following sub-sections outline the primary physical layer challenges.

5.1.1 High-Frequency Propagation Losses in mm Wave

Millimeter-wave communication operates at frequencies between 30 GHz and 300 GHz, which offer wider bandwidths for higher data rates. However, these frequencies suffer from significant propagation losses due to the shorter wavelength. This leads to:

- **High free-space path loss**, which limits the coverage range of mm Wave signals.
- **Sensitivity to blockages**, as objects like buildings, foliage, and even humans can significantly attenuate or block the signals.
- **Atmospheric absorption**, particularly due to oxygen and water vapor, further attenuates mm Wave signals over long distances.

RIS can mitigate some of these issues by redirecting signals and enabling non-line-of-sight (NLOS) communication, but the effectiveness of RIS in overcoming high-frequency losses depends on careful system design.

5.1.2 Channel Modeling and Estimation for RIS-assisted Systems

Accurate **channel modeling and estimation** is crucial for the optimal operation of RIS-assisted mm Wave systems. The complexity arises from the fact that RIS introduces additional paths for signal reflection, which significantly complicates the channel model. Some key challenges include:

- **Modeling multi-path propagation** in the presence of RIS, as each RIS element introduces a new reflection point.
- **Dynamic channel conditions**: As users and obstacles move, the channels between the transmitter, RIS, and receiver change dynamically, requiring real-time channel estimation and adaptation.
- **High-dimensional channel estimation**: In massive MIMO systems, the number of RIS elements can be very large, leading to a substantial increase in the number of channel parameters that need to be estimated, which becomes computationally expensive and time-consuming.

5.1.3 Path Loss and Penetration Issues in RIS-aided mm Wave Networks

RIS can help alleviate some of the path loss and penetration issues in mm Wave networks by intelligently reflecting signals around obstacles. However, the design still faces challenges:

- **Limited reflection efficiency**: Each RIS element has a finite reflection efficiency, and a significant amount of energy may be lost during reflection, particularly when using passive RIS.
- **Coverage constraints**: RIS placement must be optimized to ensure that reflected signals reach users efficiently, without incurring excessive path loss. Poorly placed RIS elements may not provide adequate coverage or signal enhancement.
- **Penetration issues**: mm Wave signals have difficulty penetrating walls and other obstacles, which can limit indoor and outdoor connectivity. RIS must be strategically placed to mitigate these penetration losses, requiring careful planning in heterogeneous environments.

5.1.4 RIS Hardware Design: Scalability, Energy Efficiency, and Cost

Designing scalable and energy-efficient RIS hardware is a significant challenge. The performance of RIS depends on the number of reflective elements on the surface, but increasing the number of elements introduces several issues:

- **Scalability**: As the number of RIS elements increases, so does the complexity of control and reconfiguration, as well as the computational and communication overhead between the RIS and the network.
- **Energy efficiency**: While passive RIS is highly energy-efficient, active RIS consumes more power due to amplification circuits. Designing energy-efficient RIS hardware that can scale to support large numbers of elements is crucial, especially in massive MIMO systems.
- **Cost**: Manufacturing RIS at scale, with hundreds or thousands of controllable elements, is expensive. Reducing the cost of materials and manufacturing processes, while maintaining performance, is essential for widespread deployment.

5.2 Control and Optimization Challenges

Effective control and optimization of RIS are key to unlocking its potential in mm Wave and multi-antenna systems. These challenges involve the real-time reconfiguration of RIS elements, managing algorithmic complexity, coordinating with base stations, and optimizing resource allocation.

5.2.1 Real-Time RIS Reconfiguration

One of the fundamental features of RIS is its ability to dynamically adjust the reflective properties of each element in response to changing network conditions. However, real-time reconfiguration presents several challenges:

- **Latency**: The time required to adjust the reflection coefficients of RIS elements must be minimized to ensure that the system can adapt to fast-changing environments, such as when users are moving rapidly.
- **Coordination**: RIS must coordinate with base stations and user devices to ensure that the reconfiguration aligns with the current communication needs. This requires efficient communication protocols between RIS and the network.

- **Computational complexity**: Real-time optimization of the phase shifts and reflection coefficients for thousands of RIS elements is computationally intensive, particularly in large-scale networks.

5.2.2 Algorithmic Complexity in RIS Beamforming

Beamforming through RIS involves adjusting the phase shifts of individual elements to direct the signal toward a specific target. While this is theoretically straightforward, the algorithmic complexity grows exponentially with the number of RIS elements and users. Challenges include:

- **Optimal beamforming algorithms**: Developing efficient algorithms that can compute optimal phase shifts in real-time is critical, especially in environments with multiple users and dynamic channels.
- **Trade-offs between performance and complexity**: Many beamforming algorithms that achieve optimal performance are computationally prohibitive for real-time implementation. Balancing performance and complexity is a key challenge.

5.2.3 Coordination of RIS with Base Stations and Mobile Devices

RIS must be tightly integrated with base stations (BS) and mobile devices to maximize its effectiveness. This requires coordinated control of multiple system components, including:

- **RIS-BS communication**: Base stations must control RIS configurations, requiring efficient communication protocols to exchange channel state information and configuration commands.
- **Multi-user scenarios**: In multi-user environments, RIS must simultaneously optimize reflections for multiple devices, which increases the complexity of coordination and requires sophisticated multi-user beamforming strategies.

5.2.4 Resource Allocation and Spectrum Efficiency

Optimizing resource allocation in RIS-aided networks is critical for improving spectrum efficiency. Key challenges include:

- **Dynamic spectrum allocation**: RIS can improve spectrum efficiency by enabling better control over interference and improving signal-to-noise ratios, but this requires dynamic resource allocation strategies that adapt to real-time traffic conditions.

- **Balancing RIS placement and power allocation**: The placement of RIS in the network and the allocation of power resources between the base station, RIS, and users need to be optimized for maximum spectral efficiency.

5.3 Interference Management

Interference is a major concern in multi-antenna systems, especially in dense urban environments or multi-cell networks. RIS can help mitigate interference, but also introduces new interference management challenges.

5.3.1 RIS and Inter-Cell Interference in Multi-Antenna Systems

In multi-cell systems, inter-cell interference is a significant issue, particularly when base stations are located close to each other. RIS can help reduce inter-cell interference by steering reflected signals away from interfering cells, but this requires precise control and coordination across the network. Challenges include:

- **Coordinated beamforming**: RIS-assisted systems need to coordinate beamforming across multiple cells to minimize interference while maximizing signal quality for users in each cell.
- **Cross-cell interference**: In multi-cell scenarios, reflections from RIS in one cell may interfere with signals in neighboring cells, requiring sophisticated interference management algorithms.

5.3.2 Self-Interference Cancellation in RIS-aided MIMO Systems

In RIS-aided MIMO systems, self-interference can occur when signals reflected by RIS interfere with the original transmitted signals. This self-interference is especially problematic in full-duplex communication systems. Addressing this issue requires:

- **Self-interference cancellation techniques**: Algorithms that can identify and cancel self-interference in real-time must be developed, especially in complex multi-path environments.
- **RIS configuration**: The phase shifts of RIS elements need to be carefully controlled to minimize self-interference while optimizing beamforming performance.

5.4 Network Integration and Deployment

The deployment and integration of RIS into existing wireless networks present several challenges, particularly in terms of network architecture, standardization, and regulatory concerns.

5.4.1 Integration of RIS into Existing Wireless Networks

Integrating RIS into current wireless networks (such as 5G) requires careful planning to ensure compatibility with existing infrastructure. Challenges include:

- **Hybrid architectures**: RIS needs to be integrated with base stations, user equipment, and network controllers, which may require hybrid architectures that combine traditional infrastructure with new RIS elements.
- **Backward compatibility**: Ensuring that RIS can work with existing standards and devices is crucial for gradual deployment without disrupting current services.

5.4.2 Standardization and Regulatory Issues

The widespread adoption of RIS technology will require clear **standardization** and **regulatory** frameworks, including:

- **Frequency allocation**: Regulatory bodies must determine how RIS can be used within existing frequency bands, especially in shared spectrum environments.
- **Standardization of RIS protocols**: New communication protocols for controlling and managing RIS need to be standardized to ensure interoperability between different manufacturers and network operators.

5.4.3 Trade-offs in Network Performance and RIS Placement

The performance of RIS-assisted networks is highly dependent on the placement of RIS in the environment. Trade-offs must be considered between network performance, coverage, and cost. Challenges include:

- **Optimal placement**: Determining the optimal locations for deploying RIS to maximize coverage and signal enhancement while minimizing costs is a complex problem.

- **Balancing cost and performance**: Deploying large numbers of RIS elements can improve network performance, but at a higher cost. Network operators must balance the cost of deploying RIS with the expected performance gains.

6. APPLICATIONS OF RECONFIGURABLE INTELLIGENT SURFACES IN WIRELESS NETWORKS

Reconfigurable Intelligent Surfaces (RIS) offer a range of innovative applications across modern wireless networks, enhancing various aspects of communication systems, from improving coverage to boosting energy efficiency. This section discusses the key applications of RIS in different contexts, particularly within 5G and beyond networks, massive MIMO systems, Internet of Things (IoT), smart city deployments, and security applications.

6.1 5G and Beyond Networks

RIS is poised to play a critical role in the enhancement of 5G networks and the development of future wireless communication systems like 6G. The ability of RIS to intelligently control the electromagnetic environment offers numerous benefits, particularly in overcoming the challenges associated with mm Wave propagation.

6.1.1 Enhancing Coverage in Urban and Dense Environments

Urban and dense environments pose significant challenges to wireless communication, such as severe signal attenuation due to buildings and obstacles. RIS can be strategically deployed in urban areas to:

- **Improve coverage** by reflecting signals around obstacles and providing enhanced non-line-of-sight (NLOS) communication.
- **Boost signal strength** in areas with poor line-of-sight (LOS), ensuring better connectivity for users in dense environments, such as city centers, office complexes, and shopping malls.

6.1.2 Reducing Shadowing Effects in mm Wave Networks

The high-frequency signals used in mm Wave communication are particularly susceptible to shadowing effects, where large objects block the signal path, causing significant signal loss. RIS can mitigate these shadowing effects by:

- **Redirecting signals** around obstacles, ensuring continuous connectivity even in environments where direct LOS is blocked.
- **Enhancing signal quality** in areas that experience shadowing from tall buildings or other large structures, thereby maintaining a reliable communication link.

6.1.3 RIS for Non-Line-of-Sight (NLOS) Communications

One of the most promising applications of RIS is in facilitating **NLOS communications**. By intelligently controlling signal reflections, RIS can:

- **Create alternative signal paths** that bypass obstacles and provide reliable communication in scenarios where direct LOS is not available.
- **Improve signal quality** for users in challenging environments, such as indoors or in areas with heavy urban infrastructure, by reflecting signals toward the receiver through indirect paths.

6.1.4 Energy-Efficient Communications with RIS in 5G

Energy efficiency is a critical concern in 5G networks, where the demand for higher data rates and increased capacity often leads to higher power consumption (Basar et al., 2021). RIS can help in achieving energy-efficient communications by:

- **Reducing the need for power-hungry base stations** by reflecting signals more efficiently over long distances, thus lowering the power requirements for signal transmission.
- **Optimizing beamforming** to focus energy on specific users or areas, minimizing waste and improving the overall energy efficiency of the network.

6.2 Massive MIMO Systems

Massive Multiple Input Multiple Output (MIMO) systems are essential for meeting the high capacity and performance demands of modern wireless networks. RIS can further enhance the capabilities of massive MIMO systems in several ways.

6.2.1 Improving Beamforming Capabilities in Massive MIMO Systems

RIS can significantly enhance the beamforming capabilities of massive MIMO systems by:

- **Providing additional degrees of freedom** for directing signals. By reflecting signals from multiple directions, RIS can assist in creating highly focused beams, improving signal quality and reducing interference.
- **Enhancing coverage** in environments where massive MIMO antennas face challenges in forming narrow beams, such as in highly cluttered or dynamic environments.

6.2.2 RIS-Assisted Spatial Multiplexing

Spatial multiplexing, where multiple data streams are transmitted simultaneously, is a key feature of massive MIMO systems. RIS can improve spatial multiplexing by:

- **Enhancing signal separation** between different data streams through intelligent reflection and phase adjustment, enabling higher data throughput.
- **Improving user capacity** by facilitating more efficient use of spatial resources in densely populated networks, ensuring that multiple users can be served simultaneously without significant interference.

6.2.3 Load Balancing in Dense Networks with RIS

In dense networks, load balancing between base stations is crucial to avoid network congestion. RIS can contribute to effective load balancing by:

- **Redirecting user traffic** to less congested base stations or access points, optimizing network resources and improving user experience.
- **Enhancing handovers** between different cells by ensuring smoother transitions for users moving between coverage areas, reducing dropped connections and improving overall network performance.

6.3 IoT and Smart City Applications

The Internet of Things (IoT) and smart city applications require robust and reliable connectivity, especially in environments with high device density and varied communication requirements. RIS offers several advantages for IoT and smart city deployments.

6.3.1 Smart Environments with RIS

RIS can be used to create **smart environments** where the wireless propagation is actively controlled to optimize communication between IoT devices. Applications include:

- **Smart homes and offices**, where RIS can improve indoor connectivity by reflecting signals around walls and obstacles, ensuring that IoT devices receive strong and reliable signals.
- **Smart factories**, where RIS can optimize connectivity for industrial IoT devices in complex environments with heavy machinery and metallic objects that cause signal degradation.

6.3.2 Enhancing Connectivity for IoT Devices in mm Wave

Many IoT devices operate at lower power levels, making them susceptible to signal degradation in mm Wave environments. RIS can enhance connectivity for IoT devices by:

- **Amplifying and redirecting signals**, ensuring that low-power IoT devices can communicate effectively even in environments with high signal attenuation.
- **Improving battery life** for IoT devices by reducing the need for high transmission power, as RIS can optimize the signal path and reduce energy consumption.

6.3.3 RIS in Smart Cities: Applications in Public Infrastructure

RIS can play a vital role in **smart city infrastructure**, enhancing communication in public spaces such as transportation hubs, stadiums, and urban roads. Applications include:

- **Optimizing public Wi-Fi networks** in crowded areas by improving signal coverage and reducing interference.
- **Enhancing vehicle-to-everything (V2X) communication** for smart transportation systems, ensuring reliable communication between vehicles, traffic signals, and other infrastructure elements.

6.4 Security and Privacy Applications

In addition to improving network performance, RIS can also enhance security and privacy in wireless communications by controlling signal propagation in a way that mitigates potential security risks.

6.4.1 Enhancing Physical Layer Security with RIS

RIS can be used to enhance **physical layer security** by controlling the direction of signal propagation and making it harder for malicious users to intercept signals (Di Renzo et al., 2020). Key applications include:

- **Directional beamforming**: RIS can focus signals toward legitimate users while minimizing signal leakage in other directions, reducing the risk of eavesdropping.
- **Secrecy rate optimization**: By adjusting the reflection patterns, RIS can increase the secrecy rate, ensuring that sensitive information is transmitted securely even in the presence of eavesdroppers.

6.4.2 Secure Communications in mm Wave and Multi-Antenna Systems

In mm Wave and multi-antenna systems, security risks are heightened due to the use of highly directional beams, which can be intercepted by adversaries. RIS can enhance secure communication by:

- **Shaping the propagation environment** to prevent unauthorized access to signals, creating a secure communication channel between legitimate users.
- **Intelligently redirecting signals** to avoid interception, dynamically adjusting the beam direction in response to potential security threats.

6.4.3 Mitigating Eavesdropping Risks through RIS-Controlled Channels

RIS can help mitigate **eavesdropping risks** by creating controlled communication channels that are difficult for eavesdroppers to access. By continuously reconfiguring the RIS to change the signal path, RIS can:

- **Confuse potential eavesdroppers** by dynamically altering the reflection patterns, making it difficult for unauthorized users to predict or intercept the signal.
- **Ensure secure data transmission** in environments where the risk of eavesdropping is high, such as public Wi-Fi networks or crowded urban areas.

7.CONCLUSION

The deployment and application of Reconfigurable Intelligent Surfaces (RIS) represent a significant breakthrough in the evolution of wireless communication systems. As the demand for faster, more reliable, and energy-efficient networks continues to grow, RIS emerges as a powerful tool to overcome many of the inherent limitations faced by current and future wireless technologies. This section summarizes the key takeaways, highlights the role of RIS as a transformative enabler for future wireless networks, and outlines a roadmap for future research and development in RIS technology.

7.1 Summary of Key Takeaways

- **Enhanced Wireless Performance:** RIS technology offers the ability to dynamically control the wireless environment, improving signal propagation, coverage, and quality, especially in challenging conditions such as urban areas or dense environments where line-of-sight (LOS) communication is difficult.
- **Key Role in mm Wave Communication:** With mm Wave frequencies becoming more prevalent in 5G and beyond, RIS plays a critical role in overcoming the limitations associated with high-frequency signal propagation, such as severe path loss, shadowing, and limited penetration.
- **Seamless Integration with Multi-Antenna Systems:** RIS complements and enhances the capabilities of massive MIMO systems by improving beamforming, spatial multiplexing, and load balancing, making it an ideal partner for multi-antenna technologies.
- **Wide Range of Applications:** RIS technology is highly versatile and can be applied in various domains, including improving energy efficiency in 5G networks, enabling smart city and IoT applications, and enhancing physical layer security in wireless communication systems.
- **Design and Deployment Challenges:** Despite its potential, RIS technology faces several challenges, including the need for scalable, energy-efficient

hardware, complex real-time control algorithms, efficient network integration, and standardized protocols for deployment.

7.2 RIS as a Key Enabler for Future Wireless Networks

RIS is set to become a **key enabler for next-generation wireless networks**, including 6G and beyond. Its unique ability to manipulate electromagnetic waves in real-time opens up new opportunities for:

- **Improved network coverage and capacity**: By intelligently reflecting signals, RIS can enhance network coverage, especially in areas where traditional infrastructure struggles to deliver reliable connectivity, such as dense urban environments, rural areas, or locations with heavy obstructions.
- **Energy-efficient communications**: RIS significantly reduces the energy consumption of base stations and devices by minimizing the need for high transmission power, enabling sustainable and green communication solutions.
- **Seamless user experiences**: The ability of RIS to adapt to dynamic environments in real-time, providing optimized signal paths for moving users, ensures a seamless communication experience in high-mobility scenarios such as vehicular communication and high-speed rail networks.

7.3 Roadmap for Future Research and Development in RIS Technology

To fully realize the potential of RIS, future research and development efforts must focus on overcoming the remaining technical and practical challenges. The following areas are critical for advancing RIS technology:

- **Scalable and Low-Cost RIS Hardware**: Developing cost-effective, scalable RIS hardware that can support thousands of controllable elements is essential for widespread adoption. Research should focus on improving materials, fabrication processes, and energy efficiency to make RIS economically viable for large-scale deployments.
- **Advanced Control Algorithms**: Real-time control of RIS elements requires sophisticated algorithms that can quickly and efficiently compute optimal phase shifts and reflection coefficients. Future research should aim to develop lightweight, scalable algorithms capable of operating in dynamic environments with minimal computational overhead.
- **Integration with Existing Networks**: One of the key challenges is integrating RIS with current and emerging wireless networks. Research should explore

hybrid architectures, seamless integration protocols, and backward compatibility to ensure that RIS can coexist with traditional network infrastructure.

- **Standardization and Regulatory Frameworks**: For RIS technology to achieve commercial success, clear standardization and regulatory guidelines are necessary. Industry collaboration and the establishment of global standards for RIS communication protocols and deployment strategies will be crucial.
- **Interference Management and Security**: As RIS becomes a vital part of future networks, ensuring robust interference management and enhanced security will be critical. Future research should focus on mitigating inter-cell interference, enhancing physical layer security, and developing techniques to safeguard against eavesdropping and cyber threats.

REFERENCES

Basar, E., Di Renzo, M., De Rosny, J., Debbah, M., Alouini, M. S., & Zhang, R. (2021). Smart radio environments enabled by reconfigurable intelligent surfaces: How it works, the challenges, and the opportunities. *IEEE Access : Practical Innovations, Open Solutions*, 9, 2234–2252.

Di Renzo, M., (2020). Reconfigurable intelligent surfaces for 6G and beyond: The smart radio environment, principles, and challenges. *IEEE Journal on Selected Areas in Communications*, 38(11), 2450–2525. DOI: 10.1109/JSAC.2020.3007211

Di Renzo, M., Debbah, M., Phan-Huy, D. T., Zappone, A., Alouini, M. S., & Yuen, C. (2019). Reconfigurable intelligent surfaces: An emerging paradigm for wireless communications. *EURASIP Journal on Wireless Communications and Networking*, 2019(1), 129.

Huang, C., Zappone, A., Alexandropoulos, G. C., Debbah, M., & Yuen, C. (2020). Reconfigurable intelligent surfaces for energy efficiency in wireless communication. *IEEE Journal on Selected Areas in Communications*, 38(11), 2282–2291.

Larsson, E. G., Edfors, O., Tufvesson, F., & Marzetta, T. L. (2001). Space-time processing for wireless communications: An overview. *IEEE Communications Magazine*, 39(1), 30–38.

Marzetta, T. L. (2019). Massive MIMO: The next wireless revolution. *IEEE Communications Magazine*, 57(1), 136–141.

Rappaport, T. S., Sun, S., Mayzus, R., Zhao, H., Azar, Y., Wang, K., Wong, G. N., Schulz, J. K., Samimi, M., & Gutierrez, F. (2013). Millimeter wave wireless communications: A review. *IEEE Transactions on Communications*, 61(9), 3619–3652.

Zappone, A., Di Renzo, M., Shams, F., Yu, X., & Poor, H. V. (2019). Reconfigurable intelligent surfaces for wireless communications: From phase shifters to baseband precoding and computing. *IEEE Transactions on Signal Processing*, 67(19), 4809–4826.

Chapter 12
Reconfigurable Intelligent Surface for mMIMO and NOMA Networks: Applications and Research Challenges

Rajesh Kanna Rajendran
https://orcid.org/0000-0001-7228-5031
Christ University, India

L. Vinay Kumar Reddy
https://orcid.org/0009-0001-6740-7791
Christ University, India

Justin Ziv Patil
Christ University, India

Immanuel V. Ashok
https://orcid.org/0000-0002-2422-6261
Christ University, India

Chitra S. Banu
NPR Arts and Science College, India

ABSTRACT

Reconfigurable Intelligent Surfaces are emerging as a transformative technology in wireless communication, particularly in the context of massive Multiple-Input Multiple-Output (mMIMO) systems and Non-Orthogonal Multiple Access (NOMA) networks. This chapter provides an in-depth exploration of how RIS can enhance

DOI: 10.4018/979-8-3693-8099-4.ch012

the performance of mMIMO and NOMA networks, focusing on both practical applications and research challenges. RIS technology enables dynamic control over the wireless environment by adjusting signal reflections and enhancing signal propagation, which can significantly improve the efficiency and effectiveness of mMIMO and NOMA systems. For mMIMO, RIS can optimize spatial beamforming and mitigate interference, leading to enhanced capacity and coverage. In NOMA networks, RIS can This chapter offers a comprehensive overview of the potential of RIS to revolutionize mMIMO and NOMA networks, while also addressing the critical research challenges that must be overcome to fully realize its benefits.

INTRODUCTION

The rapid proliferation of wireless communication technologies has been fueled by the increasing demand for higher data rates, improved spectral efficiency, and better coverage. As the next-generation networks, including 5G and beyond, continue to evolve, they must accommodate a growing number of users and increasingly diverse services. Two prominent technologies in this context are Massive Multiple-Input Multiple-Output (mMIMO) and Non-Orthogonal Multiple Access (NOMA). Both mMIMO and NOMA promise to enhance network performance significantly by improving spectral efficiency, reducing latency, and facilitating massive connectivity. However, the quest for further performance gains, particularly in complex, dynamic environments, has led to the development of Reconfigurable Intelligent Surfaces (RIS) as an innovative solution.

RIS is a transformative technology that represents a paradigm shift in wireless communications. It consists of passive, planar surfaces embedded with a large number of low-cost, programmable elements capable of dynamically adjusting the phase, amplitude, and polarization of incident electromagnetic waves. Unlike conventional network infrastructure components such as base stations or relays, RIS operates by intelligently controlling the radio environment, effectively turning it into a programmable entity. This capability enables the enhancement of wireless signals, reduction of interference, and improvement of overall communication performance without the need for active power consumption at the surface. In combination with mMIMO and NOMA, RIS can substantially improve the spectral efficiency, energy efficiency, and coverage in dense urban or indoor environments, where conventional techniques face significant challenges.

LITERATURE REVIEW

Di Renzo et al. (2019) foundational paper provides an overview of the RIS technology and its potential applications in wireless communication. It highlights the concept of reconfigurable metasurfaces, explaining how RIS can improve the spectral efficiency of mMIMO systems by controlling the propagation of electromagnetic waves. The authors propose several research directions, particularly focusing on how RIS can be integrated with mMIMO and NOMA to optimize wireless communication systems.

Huang et al. (2020) explores the integration of RIS with mMIMO systems to enhance signal coverage in urban environments. The paper presents a mathematical model for optimizing RIS phase shifts in the presence of multiple antennas, demonstrating that RIS can effectively mitigate interference and enhance signal strength. This work also discusses the scalability challenges in mMIMO systems when deploying large RIS arrays.

Wu and Zhang (2020) the authors investigate the joint design of active and passive beamforming in RIS-assisted NOMA systems. They provide a detailed analysis of the power allocation problem in RIS-NOMA networks, showing that RIS can improve spectral efficiency and energy savings when user fairness and channel conditions are considered. This work also highlights the challenge of optimizing passive beamforming in real-time. Yang et al. (2020) focuses on the use of RIS to improve the performance of NOMA networks in dense urban environments. The authors propose a model for optimizing RIS-assisted NOMA communication to minimize interference and enhance user capacity. They address challenges related to channel estimation and feedback overhead, providing a detailed framework for RIS-assisted NOMA systems that ensures improved performance with minimal resource consumption.

Zhang et al. (2021) the authors assess the potential of RIS for improving the coverage and efficiency of mMIMO networks. They compare RIS with traditional relaying techniques, showing that RIS offers superior performance in terms of energy efficiency and scalability. The paper also discusses the practical challenges of deploying RIS in real-world mMIMO systems, such as channel estimation complexity and hardware limitations. Basar et al. (2019) presents an overview of RIS in the context of wireless communication, with a particular focus on the use of metasurfaces for improving the energy efficiency of mMIMO and NOMA systems. The authors propose a novel framework for optimizing RIS-assisted wireless networks and highlight the need for efficient algorithms to handle the high computational complexity associated with large-scale deployments.

Liu et al. (2021) investigates the performance gains achievable with RIS-assisted mMIMO systems in high-mobility environments, such as vehicular networks. The authors develop a new model that accounts for the Doppler effect and rapid channel variations, demonstrating that RIS can significantly enhance link reliability and reduce latency. The paper also emphasizes the need for more advanced channel estimation techniques in such dynamic environments. Guo et al. (2021) authors explore the potential of RIS for improving the energy efficiency of NOMA networks. They present a detailed analysis of how RIS can dynamically adjust signal paths to improve power efficiency and reduce interference. The paper also introduces a resource allocation framework that accounts for user mobility and varying channel conditions, highlighting the practical challenges in real-time RIS configuration.

Pan et al. (2021) examines the combination of RIS with hybrid beamforming techniques in mMIMO systems. The authors propose a joint optimization algorithm that adjusts the RIS phase shifts and active beamforming vectors in mMIMO systems. They show that the integration of RIS can reduce the hardware complexity of traditional beamforming schemes while enhancing overall system performance.

Yu et al. (2020) presents a thorough analysis of the security implications of deploying RIS in mMIMO and NOMA systems. The authors address potential vulnerabilities introduced by RIS, such as signal eavesdropping and jamming. They propose several mitigation strategies, including secure beamforming techniques and encryption schemes, to safeguard the RIS configurations and ensure privacy in RIS-assisted networks. Renzo et al. (2020) propose an innovative RIS-based approach to improve the capacity and reliability of NOMA systems, particularly in crowded indoor environments. They introduce a multi-user resource allocation scheme that leverages RIS to reduce interference and improve the signal-to-noise ratio (SNR) for users with poor channel conditions. The study also addresses the computational challenges of real-time RIS configuration for large-scale NOMA deployments.

Tang et al. (2020) investigate the use of RIS to extend the coverage of mMIMO systems in challenging indoor environments, such as large factories or shopping malls. They propose a channel estimation technique that accounts for the passive nature of RIS elements and reduces the feedback overhead associated with real-time configuration. The results demonstrate that RIS can enhance both coverage and spectral efficiency in indoor mMIMO systems. Zhang et al. (2020) focuses on the integration of RIS into mMIMO networks for improving the reliability of wireless communication in high-density user environments. The authors introduce a dynamic RIS control algorithm that adjusts the reflective elements to optimize the signal strength for each user, minimizing interference and ensuring reliable communication in scenarios with large numbers of users.

Li et al. (2020) explores the use of RIS for enhancing the performance of downlink NOMA systems. The authors propose an optimization framework that jointly designs the RIS configuration and power allocation for multiple users. Their results show that RIS can significantly improve the spectral efficiency and fairness of NOMA systems, particularly when serving users with varying channel conditions. Yang et al. (2021) presents a comprehensive study on the deployment of RIS in heterogeneous networks (HetNets) that combine mMIMO and NOMA. They propose a hierarchical resource allocation scheme that optimizes the use of RIS to enhance the performance of both small-cell and macro-cell users in HetNets. The paper also discusses the challenges of deploying RIS in heterogeneous environments, particularly regarding interference management and user coordination.

RECONFIGURABLE INTELLIGENT SURFACES (RIS) HARDWARE ARCHITECTURES

The deployment of Reconfigurable Intelligent Surfaces (RIS) in modern wireless communication networks, especially in conjunction with mMIMO and NOMA, demands innovative hardware designs that can provide flexibility, scalability, and cost-efficiency. RIS hardware architectures play a critical role in determining the performance, energy efficiency, and feasibility of the overall system. In this context, the RIS hardware must not only be capable of intelligently reflecting and manipulating electromagnetic signals but also be cost-effective and energy-efficient to ensure large-scale deployment in 5G and beyond networks.

Figure 1. Reconfigurable Intelligent Surfaces (RIS) Hardware Architectures

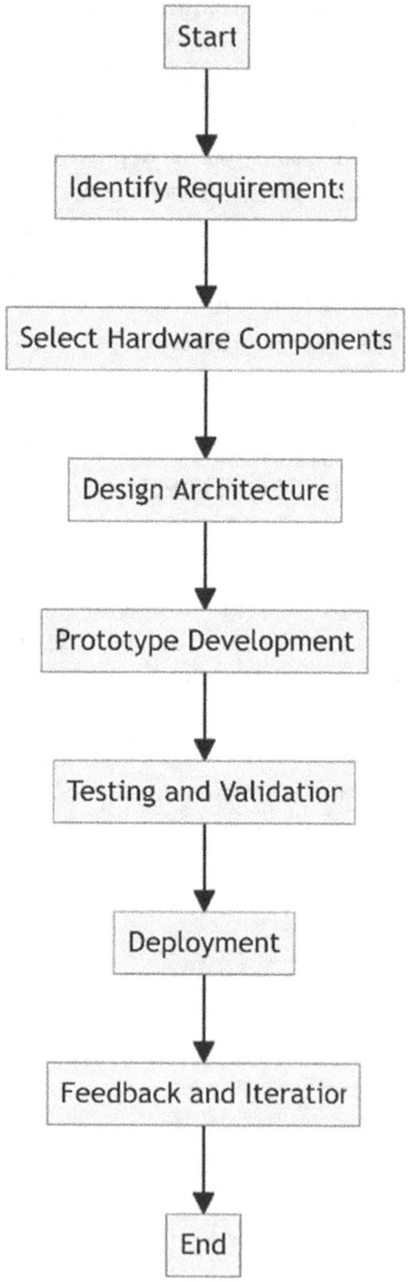

1. Metasurface-Based Architectures

A fundamental building block of RIS is the metasurface, a planar structure composed of sub-wavelength elements, or "meta-atoms," that interact with electromagnetic waves. These meta-atoms are designed to manipulate the phase, amplitude, and polarization of incident signals, offering unprecedented control over the wireless environment.

Passive Metasurfaces: Traditional metasurfaces are passive and lack the ability to dynamically adjust their response to changing signal conditions. They are typically composed of metallic or dielectric patches, designed to reflect signals with predetermined properties. While cost-effective and energy-efficient, passive metasurfaces are limited in their application as they cannot be dynamically reconfigured in real-time.
Tunable Metasurfaces: To enable dynamic control of wireless channels, tunable metasurfaces are integrated into RIS architectures. These metasurfaces use components such as varactor diodes, micro-electro-mechanical systems (MEMS), and liquid crystals to dynamically adjust their electromagnetic response. By changing the properties of these components, each element of the metasurface can be individually tuned, enabling real-time control over phase and amplitude.

2. Element Control Mechanisms

In RIS architectures, precise control of each individual element is essential for optimizing the reflection of signals and configuring the propagation environment. There are two primary categories of control mechanisms in RIS hardware:

Analog Control: In analog-controlled RIS, the reflective elements (such as varactor diodes) are continuously adjustable, offering smooth control over the phase and amplitude of reflected signals. Analog control is favored for its fine granularity in controlling signal reflection; however, it suffers from higher complexity in terms of signal processing and may introduce additional noise and imperfections.
Digital Control: Digital-controlled RIS architectures, on the other hand, employ quantized control over the reflective elements. Each element can take on a finite set of states, typically corresponding to discrete phase shifts (e.g., 0°, 90°, 180°, and 270°). Digital control offers the advantage of simplicity and robustness, making it easier to implement in practical systems. However, the

limited number of discrete states results in less precision in controlling signal reflections, which can impact overall system performance.

3. Hybrid Control Architectures

Some RIS architectures employ a hybrid control approach that combines the benefits of both analog and digital control mechanisms. In such architectures, a subset of the elements is controlled digitally, while the remaining elements are adjusted using analog mechanisms. This hybrid approach allows for a trade-off between system complexity and performance. The digital elements provide coarse control, while the analog elements fine-tune the signal reflection, optimizing both efficiency and performance.

4. RIS Controller

To operate the RIS and manage the configuration of its reflective elements, an RIS controller is integrated into the hardware architecture. The RIS controller plays a pivotal role in coordinating the real-time adjustments of the metasurface elements based on the channel state information (CSI) and network demands.

Centralized Control: In centralized RIS architectures, the controller collects CSI and configures the entire metasurface based on this information. Centralized control offers the benefit of simplicity but may suffer from higher latency, especially when dealing with large-scale metasurfaces.
Distributed Control: Distributed control architectures divide the metasurface into smaller, independently controlled regions, each with its own controller. This reduces the overall latency and enables the metasurface to react more quickly to local changes in the wireless environment. Distributed control also offers greater scalability, making it suitable for large-scale RIS deployments.

5. Power Management and Energy Efficiency

A critical aspect of RIS hardware architecture is power management, particularly given the emphasis on energy efficiency in next-generation wireless networks. While RIS typically requires low energy compared to active relays or base stations, the power consumption of tunable metasurfaces and control circuits must be optimized.

Passive RIS: In passive RIS architectures, the energy consumption is minimal since the metasurface elements do not actively amplify signals. The only power required is for the control circuitry that adjusts the elements' states.

Active RIS: Some RIS architectures may incorporate active elements to boost the reflected signal's power, effectively acting as low-power relays. However, active RIS architectures come with higher energy consumption, which can offset their benefits in terms of performance enhancement.

6. Interfacing with Communication Systems

RIS hardware architectures must be designed to interface seamlessly with existing wireless communication systems, including base stations, user equipment, and mMIMO/NOMA systems. This involves developing appropriate hardware interfaces and protocols that allow the RIS to receive configuration commands and CSI from the network.

Wireless Interfacing: Some RIS designs propose wireless interfacing between the base station and the RIS controller, enabling the RIS to receive CSI and control commands over the air. This approach simplifies installation and reduces deployment costs by eliminating the need for wired connections.
Wired Interfacing: Wired interfacing, on the other hand, offers greater reliability and reduces latency, which is critical for real-time signal manipulation. Wired connections are particularly beneficial for RIS deployments in fixed environments, such as building facades or indoor ceilings.

APPLICATIONS OF RIS IN MMIMO AND NOMA NETWORKS

Reconfigurable Intelligent Surfaces (RIS) have a wide range of promising applications in wireless communication systems, with researchers actively investigating their potential in various scenarios. Some of the most notable applications include wireless power transfer, beamforming, and electromagnetic cloaking, each of which benefits from the unique capabilities of RIS. One significant application is wireless power transfer, which involves the transmission of electrical energy from a power source to a device without the need for physical cables. Typically, this process relies on electromagnetic waves, traditionally generated and received using antennas. RIS can enhance this process by acting as an intelligent surface that efficiently directs and transfers energy, offering several advantages over conventional antennas. Unlike traditional antenna-based systems, RIS can be made smaller, more compact, and easily integrated into devices, making them ideal for use in portable and wearable technologies. RIS can dynamically adjust its properties to optimize energy transfer based on the position and orientation of the receiver, significantly improving efficiency in real-time wireless power transfer scenarios.

Figure 2. RIS in mMIMO and NOMA Networks

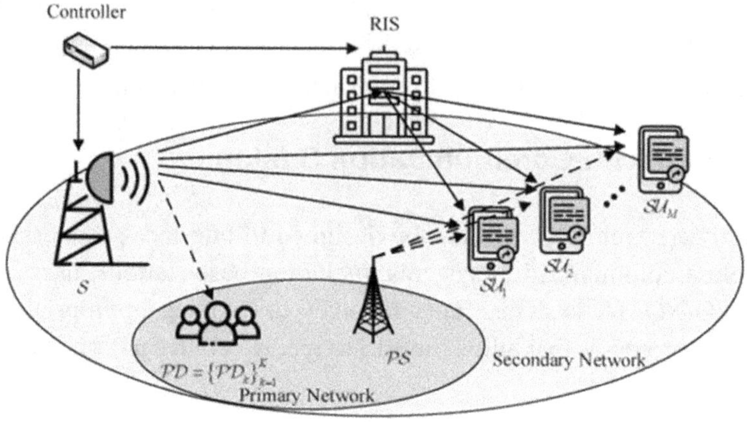

1. **Enhancing Spectral Efficiency**: In mMIMO systems, the capacity of a wireless network is significantly enhanced by employing multiple antennas at the transmitter and receiver ends. However, physical and environmental limitations such as signal blockages and interference can impede this efficiency. RIS can mitigate these issues by intelligently reflecting and directing signals to the desired receivers, thus optimizing the communication links. Similarly, in NOMA, where users share the same frequency spectrum, RIS can help improve user separation and reduce co-channel interference by dynamically adjusting the signal propagation environment.
2. **Energy Efficiency and Green Communications**: The integration of RIS into mMIMO and NOMA networks also holds the potential for substantial energy savings. By actively shaping the wireless environment, RIS can help steer signals toward their intended users, reducing the need for high transmission power and excessive signal processing. This is particularly important in NOMA systems, where users with different channel conditions are served simultaneously. RIS can optimize the signal strength for weaker users, thus reducing overall energy consumption. This energy-efficient approach aligns well with the growing emphasis on sustainable and green communication networks.
3. **Coverage Extension in Urban and Indoor Environments**: Urban areas and indoor environments are notoriously challenging for wireless communication due to dense buildings, walls, and other physical obstructions that cause severe signal degradation. RIS can be deployed on building facades or walls to dynamically reconfigure signal paths, allowing the extension of coverage into areas that were previously hard to reach. This application is critical for both mMIMO

and NOMA networks, where maintaining a stable and strong signal for a large number of users in such environments is challenging.

4. **Improvement in Latency and Reliability**: The ability of RIS to control the electromagnetic environment in real-time can contribute to reducing the end-to-end latency in communication systems. By optimizing the propagation paths and minimizing interference, RIS can ensure more reliable and low-latency communication, which is crucial for mission-critical applications like autonomous vehicles, industrial automation, and real-time gaming. NOMA networks, which often deal with varied user demands and channel conditions, can particularly benefit from RIS in ensuring reliable communication for users with poor channel conditions without significantly increasing latency.

MULTIUSER STAR-RIS-ASSISTED NOMA SYSTEM

A Multiuser STAR-RIS-Assisted NOMA System refers to a communication setup that combines several advanced technologies to enhance wireless network performance.

Multiuser System

A multiuser system in wireless communications allows multiple users to access and share the same network resources simultaneously. Efficient multiuser communication is crucial for improving network capacity and ensuring seamless service in crowded environments.

STAR-RIS (Simultaneously Transmitting and Reflecting Reconfigurable Intelligent Surface)

STAR-RIS is an advanced technology used to enhance signal propagation by intelligently controlling the reflection and transmission of electromagnetic waves. A reconfigurable intelligent surface (RIS) can adapt its behavior dynamically to optimize the wireless environment, allowing it to direct signals in a more efficient manner, whether through reflection or transmission. In the STAR variant, the surface can simultaneously transmit and reflect signals, enhancing both the coverage and capacity of the system.

Figure 3. STAR-RIS-Assisted NOMA System

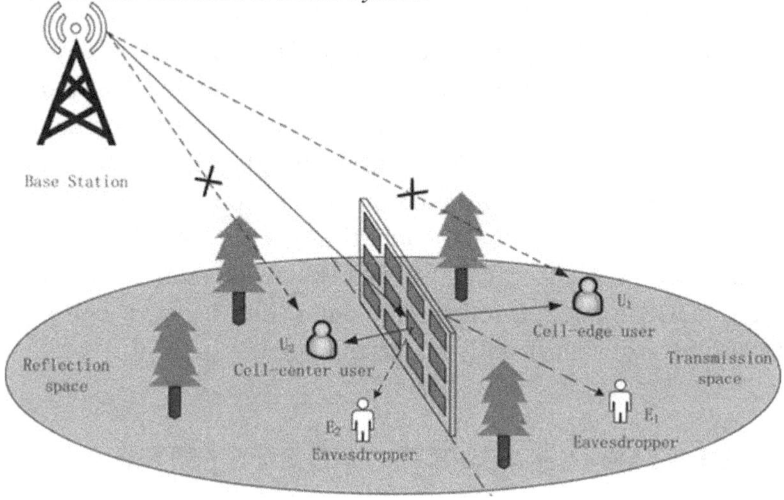

NOMA (Non-Orthogonal Multiple Access)

NOMA is a multiple access technique where multiple users share the same time, frequency, and code resources but are separated based on their power levels. This allows for more efficient use of the available spectrum compared to traditional orthogonal multiple access (OMA) systems, improving throughput and accommodating more users in the same bandwidth.

Integration in the System

In a **STAR-RIS-assisted NOMA system**, the RIS is deployed to intelligently control the wireless environment and improve signal quality for users by both reflecting and transmitting signals. This enhances the performance of NOMA, which already improves network efficiency by allocating power differently to various users. By integrating these two technologies, the system can increase spectral efficiency, improve signal strength in challenging environments, and provide better coverage for multiple users simultaneously.

This type of system is particularly useful in next-generation wireless networks (like 6G) to address the growing demand for high data rates, energy efficiency, and enhanced connectivity in densely populated areas.

NOMA-MIMO TECHNOLOGIES

To advance the development of intelligent wireless systems, designing an energy-efficient Massive multiple-input-multiple-output (MIMO)-non-orthogonal multiple access (NOMA)-aided Internet of Things (IoT) network is crucial. This network should accommodate numerous distributed users and IoT devices to ensure seamless data transfer and connectivity. Massive MIMO, with its use of a substantial number of antennas, is a fitting technology for an energy-efficient IoT network in beyond 5G (B5G) communications. However, the challenge lies in achieving swift data transfer and maintaining hyper-connectivity among IoT devices in B5G communications, posing an energy challenge. Numerous studies have shown that Non-Orthogonal Multiple Access (NOMA) outperforms Orthogonal Multiple Access (OMA) in various performance aspects. NOMA excels in meeting high-demand requirements, including extremely low latency, high spectral efficiency, increased network capacity, and elevated connectivity demand. Additionally, MIMO is recognized as a highly adaptable technology capable of enhancing capacity by increasing potential data rates. When compared to the combination of MIMO and OMA, MIMO and NOMA (MIMO-NOMA) are anticipated to achieve a higher capacity than MIMO and OMA (MIMO-OMA).

The requirements for next-generation technology include high data rates, substantial spectrum efficiency, successive interference cancellation (SIC), and ultra-reliable low latency (URLL). In the realm of next-generation technologies, non-orthogonal multiple access (NOMA) is preferred over orthogonal multiple access (OMA), offering advantages such as multi-user scaling, optimal spectral efficiency (SE), significant user-pairing improvement, and the capability for multiple users to share a single resource block. Researchers conducted comparative analyses of various power allocation algorithms to identify the optimal one for multiple-input multiple-output-NOMA (MIMO-NOMA) technology [28]. A multiple-input multiple-output (MIMO) structure enhances the overall capacity of modern communication networks without excessive power or bandwidth requirements. To meet the demands for higher user data rates and improved spectrum efficiency, the non-orthogonal multiple access (NOMA) configuration is a suitable candidate for integration with the MIMO structure. The dynamic uniform channel gain difference (DUCGD) user pairing technique plays a significant role in maximizing the capacity of all paired and served users.

RESEARCH CHALLENGES

Despite the numerous advantages and applications of RIS in mMIMO and NOMA networks, there are several technical and research challenges that need to be addressed to fully realize the potential of this technology.

1. **Joint Optimization of RIS and Network Resources**: One of the most significant challenges lies in the joint optimization of RIS configuration with existing network resources such as power allocation, beamforming in mMIMO, and user clustering in NOMA. The passive nature of RIS means that traditional active beamforming techniques may not apply directly, and novel optimization algorithms need to be developed to simultaneously optimize both the RIS parameters and network performance metrics.
2. **Channel Estimation and Feedback Overhead**: Effective utilization of RIS in mMIMO and NOMA networks requires accurate channel state information (CSI) between the base station, RIS, and users. However, channel estimation in RIS-assisted systems is more complex due to the large number of reflective elements and their passive nature. The sheer amount of feedback required to optimize RIS can also lead to excessive overhead, which can negate the benefits of the technology if not carefully managed.
3. **Scalability and Computational Complexity**: As the size of RIS increases, and the number of users in mMIMO and NOMA networks grows, the complexity of configuring the surface elements in real-time becomes a significant issue. Developing scalable solutions that can dynamically adjust the RIS configuration without overwhelming the network's computational resources remains a critical research focus.
4. **Hardware Implementation and Cost**: The practical deployment of RIS on a large scale requires cost-effective hardware that can operate in diverse environments and under varying conditions. Ensuring that the RIS elements are robust, reliable, and capable of functioning with minimal energy consumption is a key engineering challenge. Additionally, integrating RIS into existing network architectures without requiring significant changes to infrastructure is another practical consideration.
5. **Security and Privacy Concerns**: While RIS can enhance network performance, it also introduces new security challenges. The programmable nature of the surface can potentially be exploited by malicious actors to manipulate signal propagation or intercept communications. Ensuring that RIS configurations are secure and resistant to tampering is an important area of ongoing research.

WORKING PRINCIPLES OF RIS

Reconfigurable Intelligent Surfaces operate on the principle of wavefront manipulation, which involves modifying the phase, amplitude, and polarization of electromagnetic waves as they interact with the surface. Here's how it works:

Wave Reflection, Refraction, and Absorption: The metasurface elements on an RIS are capable of independently controlling the reflection, refraction, or absorption of incoming electromagnetic waves. By carefully adjusting these properties, the surface can redirect or concentrate the waves in specific directions or alter their properties to suit the communication or sensing objectives.
Phase Control: By changing the phase of electromagnetic waves, RIS can adjust the direction in which the waves propagate. This enables precise beamforming and beam steering, allowing the surface to focus signals toward the intended receiver or suppress interference from unwanted directions.
Real-Time Adaptation: The key advantage of RIS technology lies in its adaptability. It can respond to changes in the wireless environment and communication requirements in real-time. Algorithms continuously analyze the incoming signals and adjust the metasurface elements accordingly, optimizing signal strength, quality, and coverage.

EXPLORING RECONFIGURABLE INTELLIGENT SURFACES FOR 6G

Exploring Reconfigurable Intelligent Surfaces for 6G refers to the research, development, and investigation of the potential use of Reconfigurable Intelligent Surfaces (RISs) in the context of sixth-generation (6G) wireless communication systems. It highlights the idea that RISs could play a pivotal role in shaping the future of 6G wireless networks by creating what is often referred to as a "smart radio environment" (SRE).

Model-Based Constrained ML for NGMA

The design of Next-Generation Multiple Access (NGMA) communication often deals with complex non-convex, coupled, and mixed-integer constraints. Learning algorithms typically handle constraint violations by converting them into loss functions or utilizing projection operations to find feasible solutions, although their ability to strictly enforce constraints is limited. Recently, the integration of the Lagrangian dual method and the interior point method into model-based machine learning has shown promise in guiding machine learning with constrained optimization theory. This has generated interest among researchers in exploring model-based constrained machine learning for NGMA communication design.

ML Empowered Dynamic Multi-Objective Optimization for NGMA

In the context of next-generation wireless systems with time-variant and heterogeneous characteristics, communication design faces multiple competing optimizations, including system rate objectives, energy consumption, traffic latency, outage probability, and more. The dynamic nature of wireless environments introduces challenges in predicting the evolving Pareto optimal front for these competing objectives and constraints over time. To address this, efficient multitask machine learning methods are crucial for facilitating dynamic multi-objective optimization.

Accelerating AutoML for NGMA

While machine learning efficiently predicts favorable solutions through low-complexity forward propagation, the training phase, usually conducted through back-propagation, requires substantial datasets and imposes significant computational demands. Integrating AutoML techniques like meta-learning and NAS can further increase the time and computational intensity of the training process. The challenge is to address the research problem of constructing high-performance lightweight models and improving AutoML to reduce training costs for Next-Generation Multiple Access (NGMA).

Multiple-Input Multiple-Output (MIMO)

Deploying multiple antennas in both the transmitter and receiver, known as multiple-input multiple-output (MIMO) technology, significantly enhances the capacity of a radio communication channel, allowing the management of numerous independent channels within the same bandwidth. However, this is effective only in a sufficiently rich propagation environment and with specific antennas [58,59]. Recent research has shown significant interest in integrating MIMO and non-orthogonal multiple access (NOMA) to improve efficiency. The downlink (DL) NOMA network's bit error rate (BER) in both perfect and deficient successive interference cancellation (SIC) states, the study utilized additive white Gaussian noise (AWGN) and Rayleigh fading channels, with a focus on BPSK modulation. Notably, the exploration did not include BER-influencing variables such as distance and power location coefficients.

Bandwidth, Average Capacity Rate, Outage Probability, Bit Error Rate and Spectrum Efficiency

Researchers delved into the impact of varying bandwidth and the number of antennas in a 5G network on key performance metrics. The study explored the uplink's average capacity rate, outage probability (OP), and the downlink's bit error rate (BER) and spectrum efficiency (SE), considering the influence of Rayleigh fading. Analytical processes generated integral expressions for BER, SE, capacity rate, and OP, while modeling was used to explore diverse system configurations. The study focused on the NOMA approach, a fundamental element of 5G technology, and proposed a novel NOMA-MIMO power domain architecture. By integrating NOMA and MIMO to support users, system enhancements were observed. The central idea behind the research was to redesign NOMA-MIMO in the power domain to improve data rate, capacity, and throughput in a 5G network.

Combining MIMO and NOMA The swift progress in wireless communication, driven bythe widespread use of smart mobile devices and multimedia applications, has led to the exploration of non-orthogonal multiple access (NOMA) with successive interference cancellation (SIC) as a promising strategy for next-generation communication systems. Ongoing evaluations of various access technologies aim to determine the most efficient spectrum utilization, especially as new features emerge. The combination of NOMA with multi-input multi-output (MIMO) technology, known as MIMO-NOMA, enhances performance by grouping users into clusters. Despite the challenge of finding optimal user pairings, random and greedy pairing methods have been proposed to reduce computational costs. Users within the same cluster share a common precoding vector, effectively converting the MIMO channel into parallel single-input single-output (SISO) channels and maintaining NOMA's superiority over orthogonal multiple access (OMA). A comprehensive MIMO-NOMA framework utilizing zero-forcing precoding and signal alignment-based detection has been explored to eliminate inter-cluster interference, albeit assuming perfect instantaneous channel state information (CSI). Sun and his colleagues. investigate power distribution strategies for enhanced power efficiency, and the ergodic capacity of a two-user MIMO-NOMA system is analyzed with considerations for total transmit power and the minimum rate constraint of the weaker user.

ANOMA strategy for large MIMO systems and its performance in multi-cell networks, addressing inter-cell interference (ICI) through synchronized beamforming strategies, have been explored. Researchers are extending NOMA applications to communication using visible light and millimeter-wave (mmWave) technologies, showcasing the versatility of NOMA in diverse communication scenarios.

RIS FOR 6G NETWORKS

Reconfigurable Intelligent Surfaces (RIS) have garnered significant attention in both academia and the wireless industry in recent years, owing to the wide array of benefits they offer for both modern millimeter wave (mmWave) and traditional sub-6 GHz wireless communication systems. RIS provide several advantages, including substantial improvements in radio coverage and reductions in energy consumption for the radio subsystem. One of the key advantages of RIS is their ability to address the challenging coverage issues inherent in mmWave communications by enabling configurable reflection, refraction, and scattering of electromagnetic waves. This allows RIS to support radio signals, such as those from base stations, as they propagate toward mobile devices, thereby improving communication reliability.

RIS can enhance the multipath propagation characteristics of wireless signals, which is critical for applications like spatial multiplexing, where multiple data streams are transmitted simultaneously over the same frequency. By optimizing the propagation environment, RIS contribute to increased network throughput and efficiency. In addition to these benefits, RIS can be strategically employed to create dead zones, minimizing interference or preventing eavesdropping by unauthorized devices. These capabilities open up numerous potential applications for the deployment of RIS in wireless networks, making them a highly promising solution for improving overall network performance and security.

How to Optimize the Deployment of Reconfigurable Intelligent Surfaces (RIS)

While RIS technology holds immense promise, harnessing its potential requires meticulous planning and optimization. This is where network planning software comes into play. Simulation and Modelling: Network planning software utilizes sophisticated electromagnetic field simulation techniques to simulate the interaction between RIS elements and incoming electromagnetic waves. This powerful tool provides network planners with precise data, enabling them to meticulously adjust the placement and configuration of RIS elements. This fine-tuning process is indispensable for optimizing signal quality and expanding coverage to its fullest potential.

Coverage and Capacity Optimization: Utilising detailed propagation models, including ray tracing and path loss models, network planning software analyses signal strength, interference levels, and data rates across the coverage area. These models take into account various environmental factors, such as building materials, obstacles, and reflections. With this technical analysis, network planners can identify areas with poor signal quality or high interference and strategically place RIS elements to enhance coverage and capacity. This optimization process ensures that RIS elements are positioned where they can have the most significant impact on improving communication.

Based on the ray tracing model and path loss model, the RISs are precisely deployed to optimize the signal coverage by analysing various environmental factors.

- **Resource Allocation:** Efficient resource allocation in RIS involves determining which elements should modify the incoming signals and to what extent. Network planning software uses algorithms based on signal-to-noise-plus-interference ratios (SNIR) and channel quality indicators to dynamically allocate RIS resources. This means RIS can adapt to changing network conditions and communication demands, optimizing overall network performance.
- **Comparing Different Manufacturers:** As the adoption of RIS grows, an influx of manufacturers entering the market will lead to a diverse range of offerings. Network planning software will be crucial in evaluating and comparing factors such as hardware specifications, software compatibility, scalability and pricing structures to make informed decisions. These considerations extend beyond technical requirements and encompass cost-effectiveness and alignment with specific objectives, to guide engineers in deploying the most suitable RIS solution.
- **Interference Management:** In crowded wireless environments, interference management is critical to maintaining high-quality communication. Network planning software leverages interference prediction models that consider the behaviour of nearby wireless networks and devices. It also fine-tunes RIS settings based on real-time interference measurements, allowing RIS elements to work harmoniously with existing wireless infrastructure, ensuring the reliability of wireless communication.
- **Scalability and Future-Proofing:** To ensure that RIS deployments remain adaptable and scalable as networks evolve, network planning software incorporates scalability algorithms and design flexibility. It also considers factors such as spectrum allocation and evolving standards like 5G and beyond. These scalability features enable network planners to easily plan the expansion of RIS deployments as network demands grow, ensuring that RIS remains relevant and effective in the face of evolving technologies and communication standards.

FUTURE DIRECTIONS

- **Energy Efficiency:** Research into energy-efficient RIS designs and control mechanisms can further enhance the sustainability of wireless networks.

Low-power consumption RIS systems will be critical for battery-powered devices and IoT applications.

- **Miniaturization:** Advancements in nanotechnology and materials science may lead to the miniaturization of metasurface elements, allowing for more compact and integrated RIS solutions that are easier to deploy.
- **Machine Learning Integration:** Integrating machine learning and AI algorithms into RIS control systems can enhance their ability to adapt and optimize wireless environments. This could lead to even more efficient and autonomous RIS operation.
- **5G and Beyond:** RIS has the potential to play a pivotal role in the evolution of 5G and future wireless communication standards. Research and development efforts should focus on aligning RIS technology with the requirements of next-generation networks.
- **Cross-Disciplinary Collaboration:** Collaboration between researchers, engineers, and regulatory bodies is essential to drive innovation and address the multifaceted challenges associated with RIS technology.

CONCLUSION

Reconfigurable Intelligent Surface (RIS) technology has the potential to revolutionize wireless communication, particularly in the context of massive MIMO (mMIMO) and Non-Orthogonal Multiple Access (NOMA) networks. By intelligently manipulating electromagnetic waves, RIS can enhance connectivity, improve spectral efficiency, and increase energy efficiency in these advanced wireless systems. In mMIMO, RIS can help mitigate interference, improve beamforming, and provide better coverage, while in NOMA, it can enhance user fairness and optimize power allocation, enabling more efficient use of the available spectrum.

RIS integration into mMIMO and NOMA networks faces challenges related to cost, system complexity, deployment strategies, and regulatory concerns. Despite these obstacles, the potential of RIS to transform wireless communication in these contexts is undeniable. With continued research, development, and collaboration, RIS is expected to play a critical role in the evolution of mMIMO and NOMA systems, addressing some of the most pressing challenges in modern telecommunications. Its advancement promises to bring more efficient, secure, and scalable wireless communication solutions for the future.

REFERENCES

Basar, E., Di Renzo, M., De Rosny, J., Debbah, M., Alouini, M.-S., & Zhang, R. (2019). Wireless communications through reconfigurable intelligent surfaces. *IEEE Access : Practical Innovations, Open Solutions*, 7, 116753–116773. DOI: 10.1109/ACCESS.2019.2935192

Di Renzo, M., Debbah, M., Phan-Huy, D.-T., Zappone, A., Alouini, M.-S., Yuen, C., & Tretyakov, S. A. (2019). Smart radio environments empowered by reconfigurable intelligent surfaces: How it works, state of research, and road ahead. *IEEE Journal on Selected Areas in Communications*, 38(11), 2450–2525. DOI: 10.1109/JSAC.2020.3007211

Ding, Z., Lei, X., Karagiannidis, G. K., Schober, R., Yuan, J., & Bhargava, V. K. (2017). A survey on non-orthogonal multiple access for 5G networks: Research challenges and future trends. *IEEE Journal on Selected Areas in Communications*, 35(10), 2181–2195. DOI: 10.1109/JSAC.2017.2725519

Ding, Z., Zhao, Z., Peng, M., & Poor, H. V. (2017). On the spectral efficiency and security enhancements of NOMA assisted multicast-unicast streaming. *IEEE Transactions on Communications*, 65(7), 3151–3163. DOI: 10.1109/TCOMM.2017.2696527

Guo, X., Zhu, X., Yang, H., & Jin, S. (2021). RIS-enhanced NOMA networks: Joint passive beamforming and resource allocation design. *IEEE Transactions on Communications*, 69(6), 3662–3676.

Hassan, M., Singh, M., & Hamid, K. (2021). Survey on NOMA and spectrum sharing techniques in 5G. In *Proceedings of the 2021 IEEE International Conference on Smart Information Systems and Technologies (SIST)* (pp. 1-4). Nur-Sultan, Kazakhstan. DOI: 10.1109/SIST50301.2021.9465962

Huang, C., Alexandropoulos, G. C., Yuen, C., & Debbah, M. (2020). Indoor signal focusing with deep learning designed reconfigurable intelligent surfaces. *IEEE 21st International Workshop on Signal Processing Advances in Wireless Communications (SPAWC)*, 1-5.

Li, X., Liu, C., & Zhao, Y. (2020). RIS-aided downlink non-orthogonal multiple access for massive connectivity: A practical design. *IEEE Internet of Things Journal*, 8(1), 571–583.

Liu, Y., Liu, X., Mu, X., Hou, T., Li, J., Al-Dhahir, N., & Hanzo, L. (2021). Reconfigurable intelligent surface (RIS)-aided vehicular networks: A progressive survey of recent advancements and road ahead. *IEEE Communications Surveys and Tutorials*, 23(3), 1576–1608.

Makki, B., Chitti, K., Behravan, A., & Alouini, M.-S. (2020). A survey of NOMA: Current status and open research challenges. *IEEE Open Journal of the Communications Society*, 1, 179–189. DOI: 10.1109/OJCOMS.2020.2969899

Pan, C., Ren, H., Wang, K., Magedanz, T., Di Renzo, M., & Zhao, M. (2021). Reconfigurable intelligent surfaces for 6G systems: Principles, applications, and research directions. *IEEE Communications Magazine*, 59(6), 16–22. DOI: 10.1109/MCOM.001.2001076

Rehman, B. U., Babar, M. I., Ahmad, A. W., Alhumyani, H., Abdel Azim, G., Saeed, R. A., & Abdel Khalek, S. (2021). Joint power control and user grouping for uplink power domain non-orthogonal multiple access. *International Journal of Distributed Sensor Networks*, 17(12), 15501477211057443. DOI: 10.1177/15501477211057443

Renzo, M. D., Zappone, A., & Debbah, M. (2020). RIS-assisted communications with NOMA: Benefits and challenges. *IEEE Wireless Communications Letters*, 9(12), 2041–2045.

Shahab, M. B., Johnson, S. J., Shirvanimoghaddam, M., Chafii, M., Basar, E., & Dohler, M. (2020). Index modulation aided uplink NOMA for massive machine type communications. *IEEE Wireless Communications Letters*, 9(12), 2159–2162. DOI: 10.1109/LWC.2020.3015920

Tang, W., Dai, J. Y., Xu, W., Yuan, X., Zhao, L., & Zuo, C. (2020). Smart and reconfigurable environment: Indoor THz wireless communication enabled by intelligent reflecting surfaces. *IEEE Transactions on Communications*, 68(8), 5195–5209.

Wu, Q., & Zhang, R. (2020). Towards smart and reconfigurable environment: Intelligent reflecting surface aided wireless network. *IEEE Communications Magazine*, 58(1), 106–112. DOI: 10.1109/MCOM.001.1900107

Yang, P., Zhang, C., Zhou, X., & Pan, C. (2021). Hierarchical resource allocation for intelligent reflecting surface-assisted heterogeneous networks. *IEEE Transactions on Communications*, 69(8), 5222–5237.

Yang, Y., Wu, Q., Wang, S., Zhang, R., & Zheng, X. (2020). Intelligent reflecting surface-aided non-orthogonal multiple access. *IEEE Transactions on Wireless Communications*, 20(7), 4585–4600.

Yu, X., Shen, J., Jin, S., & Zhu, G. (2020). An overview of security issues in intelligent reflecting surface (IRS) aided wireless systems. *IEEE 91st Vehicular Technology Conference (VTC2020-Spring)*, 1-5.

Zhang, S., Zhang, R., & Cui, S. (2021). Cooperative interference cancellation and relaying for intelligent reflecting surface enhanced wireless networks. *IEEE Wireless Communications Letters*, 9(10), 1703–1707.

Zhang, Z., Wang, L., Xu, J., & Liu, H. (2020). Intelligent reflecting surface assisted massive MIMO systems: Capacity maximization through joint beamforming. *IEEE Transactions on Wireless Communications*, 19(8), 5357–5371.

Chapter 13
Reconfigurable Intelligent Surfaces in Smart Cities: Challenges and Future Directions

Swapnil Waghela
https://orcid.org/0000-0003-0917-8932
Shivajirao Kadam Institute of Technology and Management, India

Nitisha Waghela
https://orcid.org/0009-0003-7002-5377
Sage University, India

ABSTRACT

Rapidly growing wireless communication technologies have promoted the use of reconfigurable intelligent surfaces (RIS) in smart cities. This chapter explains how RIS technology can help transform urban environments, improving connectivity, efficiency, and sustainability. Reconfigurable intelligent surface technology has the ability to control electromagnetic waves in a flexible and dynamic way. This ability can solve many problems of modern cities, such as network congestion, energy shortages, and the integration of various IoT devices. This chapter thoroughly analyzes the main challenges associated with the implementation of RIS, such as the design and optimization of RIS systems, real-time reconfiguration algorithms, and compatibility with existing communication infrastructure. It also discusses future research and development directions of RIS technology and new advances that can further enhance its capabilities.

DOI: 10.4018/979-8-3693-8099-4.ch013

13.1 INTRODUCTION TO RECONFIGURABLE INTELLIGENT SURFACES (RIS) IN SMART CITIES

Reconfigurable Intelligent Surface (RIS) is a new technology that aims to improve wireless communications in cities. This means that radio waves (signals) can be controlled through this technology in such a way that they travel in the best way and reach where they are most needed. In smart cities, where a large number of people and devices (such as smart phones, cameras, sensors, etc.) are connected, RIS technology can improve the quality of connectivity and data transfer.

1. How Does RIS Work?

RIS is a special type of surface called a "metasurface". This surface is made of special types of materials that can control electromagnetic waves. For example, this surface can bend the signal, change its direction or increase its strength so that it can reach farther. This technology can be used in areas where the signal is weak or obstructed such as areas surrounded by tall buildings.

Figure 1. Reconfigurable Intelligent Surfaces (RIS) deployment

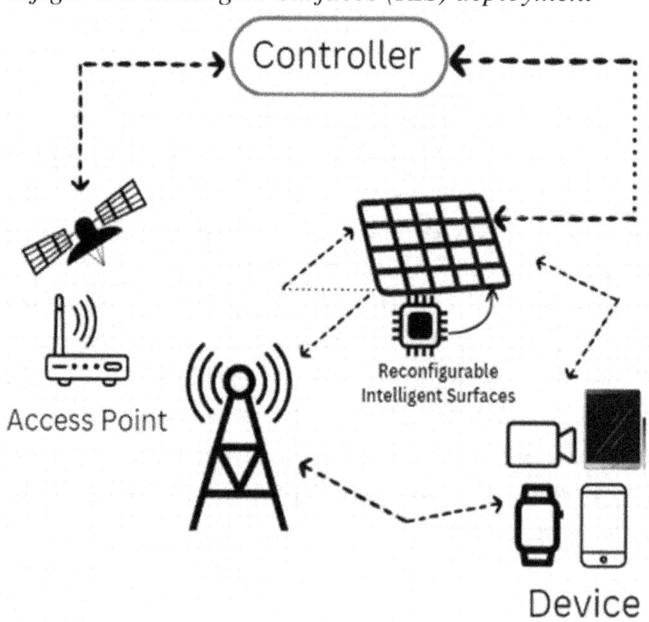

RIS consists of intelligent reflection surfaces that alter base station signal properties to achieve better network capacity and coverage. By smartly reconfiguring the wireless propagation environment and tuning the incident signals, RIS is transforming the wireless environment from a passive to an intelligent actor. (ETSI Industry Specification Group [ISG], 2023) Figure 1 illustrates the working of a Reconfigurable Intelligent Surface (RIS) system in wireless communication. The RIS interacts with incoming signals from an access point, satellite, or base station, redirecting them toward various connected devices. A central controller manages the RIS, enabling it to optimize signal direction, strength, and coverage in real-time, enhancing connectivity for devices in urban environments.

2. Why do Smart Cities Need RIS?

In today's smart cities, a lot of devices are connected to the Internet, which we call the "Internet of Things" (IoT). As the number of these devices increases, the load on the network increases, which can slow down the network and cause connectivity problems. RIS can solve these problems, as it helps send the signal in the right direction and reduce congestion in the network. In addition, RIS can also be used to reduce energy consumption. When the signal reaches its destination easily and efficiently, less energy is spent. Thus, RIS technology can not only improve connectivity in smart cities, but it can also be beneficial for the environment. (Wu & Zhang, 2019)

3. Historical Perspective of RIS and its Evolution in Wireless Communication

There have already been many technological advancements in wireless communication, such as 4G, 5G and now 6G. But, as the demand for communication devices and data in cities is increasing, the limitations of traditional communication systems are being exposed. In this context, RIS has been developed, in which metamaterials (special types of substances) are used and its surfaces are designed in such a way that they can bend, change the direction of the signal and also strengthen it. This technology is gradually becoming an important part of the future of wireless communication.

4. Potential of RIS: Solution to the Existing Challenges of Urban Communication

Problems like connectivity, network congestion and high energy consumption are common in today's modern cities. In smart cities where efforts are being made to connect every object to the Internet (such as IoT devices), RIS can provide a new direction. With the help of RIS technology, the signal can be distributed effectively, thereby reducing network congestion and increasing the quality of data transfer. Also, RIS technology also promotes efficient use of energy, which can reduce energy consumption in smart cities. (Huang et al., 2020)

Thus, RIS technology can prove to be a powerful means of improving the communication system in smart cities and making it more efficient. This chapter introduces readers to the fundamentals of this emerging technology and explains how RIS can address various challenges of smart cities.

13.2 BASIC PRINCIPLES OF RIS TECHNOLOGY AND ITS STRUCTURE

Reconfigurable Intelligent Surface (RIS) is an advanced technology to improve connectivity and communication in smart cities. In this chapter, we will understand how RIS technology works, what is the physics principle behind it, and what are the key components involved in its design.

- Basic Principles of RIS and its Physics

The main principle of RIS technology is that these surfaces can control electromagnetic waves (such as radio waves). When we send a wireless signal, it travels from one place to another in the form of waves. RIS technology is able to control the direction, intensity, and focus of these waves so that they reach the destination in the best way. (Liu et al., 2020)

For this, special types of materials are used on the RIS surface, called "metamaterials". These materials are designed in such a way that they can interact with the waves and bend or reflect them. In this way, RIS technology can provide good signals even in places where signals are usually weak, such as between buildings or in densely populated areas.

1. Main Components and Structures in RIS Design

The design of reconfigurable intelligent surfaces (RIS) is based on several main components and structures that enable it to effectively control electromagnetic waves. These components make the RIS flexible and adaptable so that it can improve the quality of communication. Below is a simple description of the main components of the RIS design and their structure. (Selvaraj et al., 2023)

1. Metasurface

The metasurface is the most important part of the RIS. It is a special type of surface consisting of many small elements (unit cells or meta-atoms). These unit cells together form a surface that can interact with incoming electromagnetic waves and control their direction, intensity, and phase.

Metasurfaces are usually made of metamaterials (a type of artificial material), which can control waves differently than normal materials. This feature of metasurfaces makes them highly effective for communication.

2. Unit Cells (Meta-atoms)

Each tiny element built on the metasurface is called a unit cell or meta-atom. Each unit cell acts like a tiny reflector or phase-changing device that can be programmed.

By controlling these unit cells, the signal can be bent or focused in any direction. This allows the RIS surface to send waves in the direction where they are most needed, thereby achieving better connectivity.

3. Control Circuitry

The RIS has a control circuit that controls each unit cell by sending electrical signals. These circuits are connected to an external controller that can change the settings of the unit cells so that they optimize the signal.

This control circuit allows the system to change the signal in real time. This ability allows the RIS to adapt to changes in the environment and adjust the direction of the waves according to the needs of the network.

4. Controller or processing unit

The controller or processing unit is an important part of the RIS, which manages the entire system. It processes external signals or network requirements and sends instructions to the control circuits, so that the RIS surfaces can be adjusted according to the needs of users and the network.

The controller also has the task of coordinating the RIS with other components of the network, so that the RIS is fully compatible with the network infrastructure and can provide high performance.

5. Power source

The control circuits and processing units of the RIS require energy to work. A power source is necessary for this. Some RIS designs feature energy storage (such as solar cells) to make the system more sustainable.

In places with low energy consumption, the RIS can obtain energy from surrounding radio signals, reducing the need for power supply.

6. Support structure and encasing

The support structure or framework of the RIS holds all its components together. It is made of lightweight and robust materials so that RIS can be easily mounted on buildings or pillars. The structure is designed in such a way that it can easily fit in the urban environment.

Additionally, an encasing or protection layer can also be used to protect the RIS so that it can be protected from rain, dust or extreme temperatures. This is essential for the safety of RIS, especially in outdoor areas.

RIS design includes key components such as metasurface, unit cell, control circuitry, controller, power source and support structure. The coordination of these components leads to a flexible and adaptable RIS system, which helps in improving communication in urban areas. These capabilities of RIS have made it an impressive technology for smart cities, capable of taking connectivity to new heights.

2. RIS's Ability to Modulate Electromagnetic Waves

The most important feature of RIS is that it can modify electromagnetic waves in several ways:

Directionalization - RIS can bend the direction of the waves so that they reach the right place. For example, if a building is obstructing the signal, RIS can bend the signal in another direction so that it can pass around the building.
Increasing or decreasing signal strength - RIS technology can also modulate the signal intensity. This means that the signal strength can be increased or decreased according to the need.
Focusing - RIS can also focus the signal to a point. This is useful when we need strong connectivity at a particular location, such as in a crowded area.

These features of RIS technology make it a unique and powerful device. Its design consists of a combination of metasurfaces and control circuits that together make communication more efficient and flexible. Due to these capabilities of RIS, it can play an important role in improving the quality of communication in smart cities. (Basar et al., 2019)

13.3 RIS IN URBAN CONNECTIVITY: INCREASING NETWORK COVERAGE AND CAPACITY

Reconfigurable Intelligent Surface (RIS) technology is becoming an effective tool for improving communication networks in urban areas. Especially in high-density urban areas, where there is a lot of load on the network, the use of RIS can help increase network coverage and data transmission rates. In this chapter, we will learn how RIS can reduce network congestion, what techniques are useful for optimizing connectivity, and what practical considerations should be made when implementing RIS in large cities.

1. The Role of RIS in Reducing Network Congestion in Urban Areas

A large number of people and devices use the Internet in large urban areas, which puts a lot of load on the network. This leads to problems such as call drops, slow Internet speeds, and interruptions in connectivity.

The main advantage of RIS technology is that it provides the ability to direct electromagnetic waves. Through RIS, network signals can be directed to areas where more traffic is needed.

Figure 2. RIS in urban connectivity

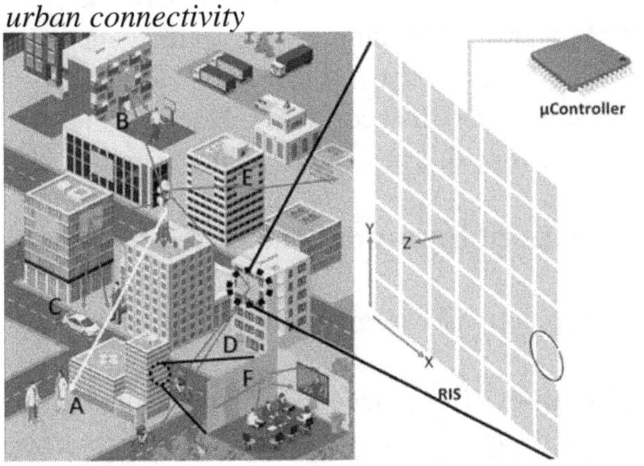

For example, in crowded places (such as railway stations, shopping malls or major intersections), RIS surfaces can focus the network signal to that area, thereby maintaining network connectivity in the crowded area and reducing the load on the network. Various indoor and outdoor application scenarios of RISassisted wireless communications are illustrated in Figure 2. The typical structure of a unit cell, a PIN diode with ON/OFF control, the lumped-element models of the PIN diode, an RIS comprising of multiple unit cells, and a micro-controller to turn ON/OFF the PIN diodes for the generation of specified radiation patterns are also shown in this figure. Non-Line of Sight (NLoS) scenarios dominate urban environments where buildings often block the signals. In such situations, single beam steering and multi-beam forming with equal/unequal power levels at different angles may be achieved with the help of an RIS to reduce outages and improve the spectral efficiency (Rafique et al., 2023).

2. Techniques to Optimize RIS to Improve Connectivity and Data Rates

The purpose of RIS is not only to change the direction of the signal, but also to improve the speed and quality of the network. Several techniques can be used for this:

Beamfocusing: RIS surfaces can focus the signal in a particular direction, increasing the signal strength. This gives users faster data rates and a more reliable connection.
Multiple beamforming: RIS can send signals in multiple directions at the same time. This technique allows multiple users to receive high quality service at the same time, thereby efficiently using the network.
Real-time optimization: RIS surfaces can optimize in real-time. This means that as the user's location or network demand changes, the RIS can adjust the signal direction and strength accordingly. This helps to balance the network distribution among users in urban areas. (Singh et al., 2022)

3. Practical Aspects of Implementing RIS in Metros

Several important points must be kept in mind while installing RIS in urban areas:

Selection of location: RIS surfaces should be installed at such locations from where it can effectively transmit the signal in the right direction. Such as on the walls of tall buildings or on pillars, from where it can transmit the signal in open and busy areas.

Weather and environmental factors: In cities, it is important to provide protection to the RIS from the external environment such as rain, dust and extreme temperatures. For this, strong and durable casing is necessary so that the RIS can work for a long time.

Energy source: RIS requires energy to operate. In urban areas, this energy can be obtained from solar energy or nearby radio signals apart from conventional sources. Choosing a good source of energy storage improves the sustainability and cost of RIS.

Compatibility with network infrastructure: RIS should be designed to work effectively with the existing network. It also requires making RIS compatible with 5G or future technologies so that the use of RIS remains profitable in the coming years.

RIS technology can play a vital role in improving connectivity in urban areas, increasing data rates and maximizing network capacity. This chapter highlights various RIS technologies in urban areas, points to be considered in its deployment and its compatibility with the existing infrastructure of the network. The right use of RIS can help improve network quality in cities and build smart cities.

13.4 ENERGY EFFICIENCY WITH RIS IN SMART CITIES

Energy efficiency is an important aspect in smart cities as these cities use a large number of appliances and connected devices. Reconfigurable Intelligent Surfaces (RIS) provide an effective solution to reduce energy consumption in smart cities. In this chapter, we will understand how RIS can make urban environments sustainable, what technologies can be adopted to reduce energy consumption in IoT and wireless networks, and how RIS can contribute to building green cities.

1. Contribution of RIS in Sustainable Urban Environment

Improving Urban Environment: One of the major benefits of RIS is that it can make communication in urban areas more efficient, thereby reducing energy consumption. Using RIS technology, communication signals can be sent to areas where it is needed the most, allowing devices to operate with less power.

Reduce Energy Consumption: RIS surfaces adjust the direction and quality of the signal required to provide connectivity to the network. Thus, it reduces the energy consumption for signal transmission, and there is no need to increase the signal power too much, especially in small or remote areas. This saves energy in the network. (Rajaan et al., 2024)

2. Ways to Reduce Energy Consumption in IoT and Wireless Networks

Low Power Mode: IoT devices usually have the ability to operate on very low power. With RIS, these devices can be optimized as per the need so that they work more efficiently while using less power. For example, the direction of the signal can be corrected by RIS, so that the device in the wireless network does not need to spend much energy.

Smart Resource Allocation: RIS manages the communication resources smartly, thereby optimizing network resources. It can be designed in such a way that it is active only when it needs to be used and turned off when it is not needed, which can reduce energy consumption.

Multi-beamforming and data transfer optimization: With the help of RIS, network traffic can be divided in a better way. By using multi-beamforming technology, signals can be sent as per the requirement, which will reduce energy consumption and increase the speed of data transfer.

3. Green City Applications for RIS

Use of Renewable Energy Sources: RIS can be designed in such a way that it is powered by solar panels or other renewable energy sources. This reduces both energy consumption and pollution. For example, RIS devices installed in urban areas can be charged with solar energy, reducing the need for energy from external sources (Kisseleff et al., 2020).

Eco-Friendly Networks: The use of RIS technology can make urban networks more eco-friendly. By reducing energy consumption in the network with RIS, smart cities can get lower carbon emissions and more environmental benefits.

Green Transportation and Infrastructure: RIS can also be used in smart transportation systems. For example, RIS in roads and public transport can help optimize traffic conditions and vehicle speeds, thereby reducing fuel consumption and greenhouse gas emissions.

Smart devices with low energy consumption: RIS can be used in smart homes and buildings where it can reduce the energy consumption of devices by optimizing signals between them. This reduces power consumption in smart buildings and improves lifestyles.

RIS technology not only enhances the connectivity and speed of urban networks, but it also promotes energy efficiency and sustainability in smart cities. In this chapter, we discussed ways to reduce energy consumption using RIS and take important steps towards a green city. By using RIS technology, we can create a more environmentally friendly and stable urban environment, which will not only be energy efficient but also ideal for a smart and sustainable future.

13.5 INTEGRATING RIS WITH IOT ECOSYSTEM IN SMART CITIES

Nowadays, the number of Internet of Things (IoT) devices in smart cities is growing rapidly. These devices are connected to make various functions of the city smart, efficient and automated. Reconfigurable Intelligent Surfaces (RIS) can play a vital role in improving connectivity, increasing network efficiency and speeding up communication among IoT devices. In this chapter, we will understand the challenges in integrating RIS in IoT ecosystem, how RIS can make smart city systems more efficient and interconnected, and see some examples of RIS-enhanced IoT networks in the real world. (Hameed & Koo, 2024)

1. Challenges in integrating RIS with Various IoT Devices

Diversity and communication protocols: There are a large number of IoT devices in smart cities and these devices are of various types such as smart lights, security cameras, traffic sensors and smart meters. Different types of protocols and technologies are used for communication between these devices. Integrating RIS with all these devices without any technical barriers can be a big challenge.

Structural barriers: Physical and structural problems can arise when integrating RIS with various IoT devices. For example, RIS surfaces can be installed on city buildings or other infrastructure, but ensuring that they remain suitable and connected to all devices can be a complex task.

Security and privacy: Security and privacy issues also arise when growing a network of IoT devices in smart cities. Since RIS technology manages the signals of the network, ensuring that communications remain secure and data privacy is maintained is a major challenge.

Performance and scalability: The number of IoT devices in smart cities can reach millions. Deploying RIS at a large scale and ensuring that it can effectively connect all devices can be a technical challenge. With the increasing size and complexity of the network, it can be difficult to maintain the performance of RIS.

2. Benefits of RIS Making Smart City Systems Efficient and Interconnected

Improved connectivity and data transmission: RIS technology can improve connectivity between IoT devices. By controlling the direction of the signal by RIS, the speed and quality of data transfer can be improved. This ensures effective communication between various IoT devices in smart cities.

Low energy consumption: With the help of RIS, the signal can be controlled in such a way that the network device does not require much energy. RIS can send the signal in the direction where it is needed more so that the energy consumption is reduced, which also reduces the energy consumption of the IoT device.

Resilient network: RIS makes the network more flexible and adaptable. For example, if an IoT device has connectivity issues, RIS can reroute the signal near that device and provide it with better communication.

Improved network efficiency: The use of RIS helps in increasing network efficiency. RIS technology balances the traffic of the network and provides high-quality service for users, thereby accelerating various functions of smart cities and, as a result, making the system more efficient.

3. Real-World Examples of RIS-Enhanced IoT Networks

Smart traffic management: RIS is being used for traffic management in many smart cities. RIS optimizes traffic signals to control traffic in a better way. RIS works with IoT networks connected to smart traffic lights and sensors to optimize the speed of vehicles on the road and reduce traffic jams.

Smart environmental monitoring: RIS can also be used in environmental monitoring. For example, RIS can be integrated with a network of IoT devices tracking pollution levels and weather conditions in cities. RIS receives data from these devices and with its help transmits the signal in a better way, so that more accurate and faster data can be obtained.

Smart energy management: RIS can optimize data transfer between IoT networks of smart meters and energy devices. The signals required by RIS to monitor energy usage can be sent in such a way that it reduces energy consumption and increases energy efficiency.

Smart Parking System: RIS can also be implemented in smart parking systems. In this, information about parking spaces is obtained through a network of IoT devices and the data is sent in the right direction by controlling the RIS signals. This gives drivers information about vacant parking spaces and saves time spent in parking.

Integrating RIS technology with the IoT ecosystem provides many benefits in smart cities. It helps in improving connectivity, reducing energy consumption and interconnecting various IoT devices. Although there are some challenges in integrating RIS with IoT, it plays an important role in the development of smart cities due to its benefits. Real-world examples show that RIS technology can make various functions of smart cities more efficient and effective.

13.6 DESIGN AND OPTIMIZATION OF RIS SYSTEM

The design and optimization of reconfigurable intelligent surfaces (RIS) in smart cities is a complex process that requires consideration of many factors. To effectively deploy RIS in urban environments, it is important to understand its design and functioning. In this chapter, we will understand the key points to be considered during RIS design and implementation, ways to optimize RIS for different applications, and ways to address limitations in hardware, software, and scalability. (Qian et al., 2023)

1. Key Points of RIS Design and Implementation in Urban Environment

Selection of location and surface orientation: The RIS system should be installed at a location where it can provide maximum signal coverage. The orientation of the surface is determined by taking into account the height of buildings, roads, and other obstructions in urban areas so that the RIS can cover as much area as possible and promote connectivity.

Design of Metasurface: The metasurface is the most important part of the RIS, which works to control the signal. It is designed in such a way that it can adapt to the direction, intensity and phase of the signal. The metasurface is made up of small unit cells whose structure is such that they can effectively control electromagnetic waves.

Hardware design: The hardware of the RIS should be designed in such a way that it can withstand the changing weather and external influences in the urban environment. Also, it should be lightweight and robust so that it can be easily installed in buildings and other places. Also, the hardware should have low energy consumption so that it can work smoothly for a long time. **Energy efficiency:** Since the energy demand is high in urban areas, the RIS should be designed in such a way that it can reduce energy consumption. For this, low energy consumption components are used in the RIS and if required, energy can also be obtained from the surrounding signals. (Alexandropoulos et al., 2023)

2. Techniques to Adapt RIS for Different Applications

Signal adaptation: For different applications, the RIS can be adapted in such a way that it can change the direction and intensity of the signal. For example, the signal can be focused in areas requiring high data transmission and connectivity can also be provided in areas with low network by changing the direction of the signal.

Real-time optimization: RIS in smart cities can be optimized in real-time so that the RIS can change its settings according to the changing traffic and network demands. This helps in effectively managing the signal according to the traffic and user needs.

Compatibility for devices and networks: RIS needs to be synced with IoT devices, 5G networks, and other communication technologies. For this, the settings of the RIS are optimized in such a way that it can work with different protocols and devices.

Coverage optimization: In urban environments, RIS can be optimized in such a way that it can cover the maximum area. For this, the orientation and position of the RIS surfaces are optimized in such a way that it maximizes the signal coverage and delivers the signal to all the important locations.

3. Solution of Limitations in Hardware, Software and Scalability

Hardware limitations: The hardware of RIS may have limitations such as its energy consumption, its stability in outdoor environments, and signal degradation due to other obstructions. To solve these limitations, more energy-efficient components can be used and the surface can be clad to make it more weather-resistant.

Software limitations: There are many challenges in the software to make RIS fully adaptable to urban environments. To solve these challenges, smart algorithms are used, which can change the direction, intensity, and phase of the signal in real-time. Also, the software of RIS is designed in such a way that it can be coordinated with IoT and 5G networks.

Scalability limitations: Scalability is a challenge for large-scale deployment of RIS in urban areas. To solve this, the design of RIS can be made modular so that it can be divided and installed in smaller parts and the capacity of RIS can be increased as per the need.

Network Optimization: RIS is optimized to work in urban areas keeping in mind the various requirements of the network. For this, intelligent network protocols and management systems are used in RIS so that it can handle large-scale networks and serve multiple users simultaneously.

The design and optimization of RIS is essential to provide better connectivity and energy efficiency in urban environments. From the design of the main components of RIS to its optimization techniques, it is tailored for various applications in multiple stages. The latest technologies are used to solve the limitations in hardware and software. Considering the scalability of RIS, it is deployed in IoT and other networks of smart cities, making it a better, efficient, and more connected urban communication system. RIS technology plays a vital role in making smart cities more prosperous and environmentally friendly.

13.7 REAL-TIME RECONFIGURATION ALGORITHMS FOR RIS

Reconfigurable Intelligent Surfaces (RIS) need to be reconfigured in real-time to be efficient in urban environments. This means that RIS has to change its settings instantly according to different conditions to meet the changing needs of the network. In this chapter, we will understand what kind of algorithms are used to reconfigure RIS in real-time, what is the contribution of Artificial Intelligence (AI) and Machine Learning (ML) in this, and what are the challenges and future prospects in this process. (Salim ALRikabi et al., 2024)

1. Overview of Algorithms to Optimize RIS in Real-time

Basic Algorithms: The algorithms commonly used in RIS work to change the direction, intensity, and phase of the signal. These algorithms optimize the signal according to the user's condition and need. For example, if there is more network pressure in an area, the algorithm can divert the signal in that direction to provide more connectivity there.

Real-time optimization algorithms: Optimization algorithms are used to reconfigure the RIS in real-time. These algorithms constantly monitor the network status and adjust immediately according to any changes. As soon as the user's location changes or the network traffic changes, these algorithms immediately update the RIS's settings.

Optimization algorithms for reconfiguration: These algorithms read the data and reconfigure the signal's direction, strength, and phase. The function of these algorithms is to make the RIS cover the maximum area and provide the highest quality of service while using the least amount of energy.

2. Role of AI and Machine Learning in RIS Reconfiguration

Automated optimization by AI: Artificial intelligence (AI) can be used for automated optimization in RIS. AI algorithms constantly read various data and predict in which direction the signal should be sent. Thus, with the help of AI, it becomes possible to optimize RIS in real-time. (Ahmed et al., 2024)
Prediction by Machine Learning: Using machine learning, RIS can be enabled to learn about different network patterns. For example, when network traffic is high at a location, a machine learning model can recognize that pattern and direct the RIS to focus on that location when network pressure comes in the future.
Deep Learning Algorithms: Deep learning algorithms can be used in RIS to process more complex data. Like urban environments have lots of buildings and obstructions, deep learning helps in sending the signal in the right direction despite all these obstructions.
Data Processing and Prediction: AI and ML are also used for data processing so that RIS can predict what the network condition might be at the next moment and how it should be optimized. This prediction capability makes RIS more efficient and provides better service to users. (Faisal & Choi, 2021)

3. Challenges and Future Prospects in RIS Control

Challenges in real-time data processing: Data processing in real-time requires fast processors and high volumes of data to manage. Processing data from a large number of users in urban areas can be challenging, making it difficult for RIS to work efficiently.
Compatibility with devices and systems: Integrating RIS with various IoT devices and network systems in urban environments is a major challenge. RIS must be designed in such a way that it can work with all types of devices and networks.
High energy consumption: Real-time optimization requires constant data processing and signal adjustments, which can consume more energy. In the future, energy-efficient equipment and circuitry will need to be developed so that RIS can save energy and function sustainably.
Improving RIS optimization in the future: More efficient AI and ML models can be used in RIS in the future, which can make faster decisions and recognize changing network patterns. Apart from this, advanced energy-harvesting techniques can also be used to make RIS more energy-efficient.

Adapt to every situation: In the future, RIS will be developed in such a way that it can adapt itself according to any type of environment or situation. This means whether it is a crowded area or a quiet place, RIS will be able to change its setting according to the situation.

In this chapter, we saw what kind of algorithms are used to reconfigure RIS in real-time and how the reconfiguration capability of RIS can be enhanced with the help of AI and machine learning. Although adapting RIS in real-time is a challenging task, this process can be made simple and efficient through AI and ML. There are possibilities to further improve these capabilities of RIS in the future, which will lead to better connectivity and energy efficiency in smart cities. (Ren et al., 2023)

13.8 COMPATIBILITY WITH EXISTING COMMUNICATION INFRASTRUCTURE

The successful integration of reconfigurable intelligent surfaces (RIS) with existing and future communication networks such as 5G and upcoming 6G is very important for the development of smart cities. RIS technology, which is capable of adapting electromagnetic waves, can be effectively used only when it is fully compatible with the existing communication infrastructure. In this chapter we will see how RIS can be integrated with 5G and 6G networks, what are the challenges in integrating it with traditional wireless systems, and how these challenges can be addressed to enable seamless integration of RIS in urban networks.

1. Integration of RIS with 5G and Future 6G Networks

Compatibility with 5G networks: 5G networks are designed for higher data speeds, lower latency, and more device connectivity. These properties can be further improved with the help of RIS. For example, RIS surfaces can serve more users simultaneously by changing the direction of the signal, thereby improving network performance. (Liu et al., 2022)

Compatibility for 6G networks: 6G technology will have greater capabilities than 5G, requiring maximum data transfer speeds and extremely low latency. RIS can be designed to meet the growing demand for 6G networks to further improve connectivity in smart cities. The integration of RIS into 6G networks will enable efficient support of hyper-connected environments, such as sensitive equipment and smart devices that require large amounts of data processing. (Bae et al., 2024)

Compatibility for high frequency bands: 5G and 6G networks use high frequency bands, which can lead to low signal penetration. Using RIS, the direction and strength of the signal at these frequencies can be controlled to enable better connectivity.

2. Interoperability Challenges With Traditional Wireless Systems

Compatibility with different network protocols: There are differences between traditional wireless networks and modern RIS technology. Integrating RIS with different network protocols and standards is a challenging task. For example, 4G or older network protocols have limited data processing speed, while RIS requires faster signal adaptation.
Difference in signal processing: Traditional network systems process signals statically, while RIS needs to change the signals in real-time. Therefore, bridging the signal processing gap between RIS and traditional systems is a major challenge.
Compatibility with underlying devices: RIS should be compatible with traditional network devices such as antennas, routers, and transmitters. These devices need to be adapted to make their settings compatible with RIS so that proper coordination can take place between the two systems.
Limited support of RIS with older network systems: Many traditional network systems are not designed to work with RIS. Therefore, the proper functioning of RIS may hinder older network systems, and hence this aspect should be taken care of especially during RIS integration. (Velez et al., 2023)

3. Possible Solutions for Seamless Integration of RIS

Development of Multi-Protocol Support: Designing multi-protocol support is an effective solution to make RIS compatible with various network protocols. It allows RIS to work easily in various network environments and work better with traditional systems.
Hybrid RIS Design: Some hybrid designs of RIS have compatibility with both traditional and advanced network systems. This design makes RIS useful in older networks as well as helps in working easily with new networks.
Advanced Signal Processing Techniques: Using advanced signal processing techniques in RIS can make it easier to make it compatible with traditional systems. For this, AI and machine learning based processing techniques can be applied in RIS, allowing it to overcome the limitations of traditional networks.

Unified Control System for All Network Systems: An integrated control system can be developed to integrate RIS and traditional networks. This control system can enhance the performance of the network by coordinating the signals and data of both systems.

Special Adapters for RIS with Older Networks: Special adapters or interfaces can be developed to accommodate RIS with older networks, which help in creating synergy between the two systems. These types of adapters play a complementary role in making RIS compatible with older networks.

In this, we understood how RIS can be integrated with existing and future networks such as 5G and 6G. There are several challenges in integrating RIS with traditional networks, such as compatibility with different network protocols, differences in signal processing, and synergy with underlying devices. To overcome these challenges, multi-protocol support, hybrid design, advanced signal processing, and special adapters need to be developed. Integration of RIS with these measures not only enhances its utility but also strengthens connectivity in urban areas, creating an efficient and stable network.

13.9 SECURITY AND PRIVACY IMPLICATIONS OF RIS TECHNOLOGY

With the advancement of reconfigurable intelligent surfaces (RIS) technology, the challenges of network security and data privacy in smart cities have also increased. While RIS technology improves connectivity and communication quality in urban areas, it also poses a number of security and privacy risks. In this chapter, we will look at how RIS creates security risks in data transmission and network control, how privacy-related issues arise in IoT and other urban applications, and what policy and regulatory aspects are necessary to address these challenges. (Naeem et al., 2023)

1. RIS-Related Security Risks in Data Transmission and Network Control

Threat of Data Theft: The purpose of RIS is to control and optimize wireless signals, but in this process, RIS surfaces and the data transmitted under their control may become vulnerable. Cyber attackers can disrupt data transmission or steal sensitive information using RIS surfaces.

Signal Jamming and Spoofing Attacks: The process of directing signals may be illegally interrupted due to RIS. Some cyber attackers can bring the network to a standstill using signal jamming, causing disruption to critical systems in the city. In spoofing attacks, the direction and identity of the signal being sent by the RIS can be manipulated, leading to incorrect data reaching the network.

Risk of malware and viruses: The control system or processing unit connected in the RIS system can become the target of malicious software (malware) affecting the operation of the RIS. If viruses or other harmful software enter the control of the RIS, they can endanger the security of the entire network.

Intrusion into the RIS control circuitry: The control circuitry in the RIS system is used to control each unit cell. If cyber attackers intrude into this control circuitry, they can arbitrarily control the operation of the RIS and disrupt the functionality of the system.

2. Privacy Concerns in RIS-Optimized IoT and Urban Applications

Privacy of user data: RIS technology is being widely used in smart cities, such as camera networks, traffic systems, and IoT device-connected systems. A large amount of network data can be collected through RIS, which may also include user's personal information. This may threaten user's privacy.

Security of IoT device: Security of IoT device connected to RIS is also a big challenge. If IoT devices are unable to handle the data collected through RIS securely, users' confidential data may be exposed. This may affect not only personal information but also the privacy of other important persons in the smart city.

Violation of privacy: RIS technology is used to change the direction of signals at different locations, allowing a large amount of data to be collected. If RIS is misused, it may access personal and sensitive data, leading to privacy violation.

Difficulties in data encryption: RIS redirects signals in real time, increasing the challenge of keeping data secure continuously. Traditional encryption techniques can be difficult to adapt to RIS, increasing the risk of data being intercepted.

3. Policy and Regulatory Aspects for Secure RIS Deployment

Strict cybersecurity standards: The adoption of stringent cybersecurity standards is essential for the secure use of RIS technology. These standards should include strong security protocols for RIS devices, regular security audits, and development of signal protection technologies.

Compliance with data protection laws: RIS-based smart cities need to comply with data protection laws to ensure user privacy. Governments and organizations should ensure that data collected from RIS is secure and is used only on a legal basis.

Policies in RIS control and monitoring: It is also important to create policies for the operation and control of RIS technology. This can determine who will control the RIS systems and what steps can be taken in case of their misuse.

Authentication for RIS-based IoT devices: The authentication process is necessary to secure the communication between RIS and IoT devices. All IoT devices must be authenticated so that only authorized devices can connect to the RIS network and data remains secure.

Encryption Policy: Encryption policies should be created to keep RIS data secure. These policies should detail the encryption technique of the data, the time frame of encryption, and other security aspects to keep RIS data secure.

RIS technology has the opportunity to take communications in smart cities to new heights, but it also comes with several security and privacy challenges. RIS can pose security risks in data transmission and network control, affecting privacy in IoT and other urban applications. Adherence to cybersecurity standards, compliance with data protection laws, and adoption of policies in RIS control are essential for the secure deployment of RIS. The measures presented in this chapter can play a vital role in the secure use of RIS and the development of smart cities, ensuring user privacy and data security.

13.10 FUTURE DIRECTIONS FOR RIS RESEARCH AND DEVELOPMENT

Reconfigurable intelligent surfaces (RIS) are helpful in meeting the new needs of smart cities and still require a lot of research and development to fully unlock its potential. In this chapter, we will discuss the key areas of research, new trends, and potential innovations in RIS technology in the future. Also, we will see how RIS technology can impact smart cities in the future. (Kazim et al., 2023)

1. Key Areas of Further Research in RIS Technology

High-performance metasurface design: The core part of RIS is the metasurface, which is designed to control electromagnetic waves. An important area of research focuses on enhancing the capability of this metasurface. The communication efficiency and power of RIS can be further enhanced by the use of advanced metasurface design and effective materials.

Algorithms for real-time reconfiguration: RIS requires powerful algorithms to reconfigure quickly and efficiently. Further research can develop algorithms that can adapt RIS to the immediate user needs using machine learning and artificial intelligence (AI).

Low energy consumption designs: Another research area of RIS technology is to ensure that RIS consumes as little energy as possible. For this, work can be done on energy storage, wireless energy harvesting and RIS designs with minimal energy use.

Security and privacy measures for RIS: As the use of RIS increases, security and privacy will also become a priority. A major focus of researchers should be on developing mechanisms that keep data secure and make RIS operations possible with maximum privacy.

2. Emerging Trends, New Innovations and Potential Transformational Innovations

AI-integrated RIS systems: With the integration of artificial intelligence (AI) in RIS in the future, this technology can become highly effective. Using AI, RIS can be given the ability to learn automatically, allowing it to adapt itself to the changing environment.

Integration of RIS with 5G and 6G networks: A major trend for RIS is its integration with 5G and upcoming 6G networks. RIS can play a key role in future 6G technology, enabling higher speeds, better connectivity, and new types of applications.

Expansion of RIS in smart cities infrastructure: RIS can be integrated into many parts of smart cities, such as traffic lights, building facades, and smart road systems. This technology can make every part of urban infrastructure more communication-centric in the coming years.

Incorporation of quantum technology in RIS: Quantum technology can play a big role in the development of RIS. Quantum-based RIS can take signal processing and data security to a new level, making communications in smart cities even more secure and efficient.

3. Future of RIS and its Impact on Smart Cities

Better connectivity and communication: In the future, RIS will be present everywhere in smart cities, improving communication quality, connectivity, and coverage. It can solve the problems of network congestion and internet speed in urban areas.
Green City Solutions: Another effective contribution of RIS will be in promoting energy efficiency in smart cities. With low energy consumption, RIS technology will have less impact on the environment and will take a step towards green cities.
Improving Security and Surveillance: In the future, RIS can be used in public security and surveillance systems. This technology can make surveillance more efficient and effective by transmitting images and signals in a better way, thereby taking the security system to new heights in smart cities.
Expansion of Internet of Things (IoT): RIS will be helpful in easily integrating IoT devices in smart cities. Through this, IoT devices will be able to work more efficiently and urban life will be enhanced.

In this chapter, we have focused on the major areas of RIS research and development in the future. Advanced metasurface design, AI-based algorithms, low energy consumption RIS, and its integration with 6G networks are some of the major research areas. The impact of RIS in smart cities can be wide-ranging, leading to better communications, achieving green city goals, improved security, and expansion of IoT. In the future, RIS technology can play a vital role in strengthening the digital and environmental infrastructure of cities, making our cities even smarter and more connected.

13.11 CASE STUDIES: RIS IMPLEMENTATION IN SMART CITIES AROUND THE WORLD

Reconfigurable Intelligent Surface (RIS) technology has opened up new possibilities in smart cities. RIS has been tested and implemented in various countries and cities, and these real-life examples give us valuable insights into the functionality, challenges, and future use of RIS. In this chapter, we will discuss some of the key case studies of RIS implementation in various urban settings to illustrate how RIS can be effective in smart cities.

1. Real-life Examples of RIS in Various Urban Settings

Singapore: Use of RIS Technology in Smart Cities

Reconfigurable Intelligent Surface (RIS) technology has been used in smart cities like Singapore to improve urban connectivity. The Singapore government integrated RIS technology with 5G networks to improve traffic monitoring, smart street lighting, and environmental monitoring systems. The main purpose of RIS technology was to provide connectivity to various IoT devices (such as smart sensors, cameras, and meters) and enhance network coverage. (Cavada, Tight, & Foss Rogers, 2019)

Function of RIS:

Traffic Monitoring: RIS has been used to improve traffic monitoring systems in Singapore. RIS connects traffic lights and other monitoring devices, allowing traffic to be controlled more effectively. RIS technology helps to optimize traffic signals from time to time, improve traffic flow, and reduce congestion.

Smart Street Lighting: RIS technology has also contributed significantly to Singapore's smart street lighting system. RIS automatically controls street lights, ensuring energy savings and better lighting. It can change the status of street lights in different parts of the city in real time, reducing energy consumption at night.

Environmental monitoring: RIS technology has also been used in environmental monitoring systems. RIS is used to monitor air quality, temperature and other environmental conditions in Singapore. This data helps make city planning and policies more effective.

Connectivity with IoT devices: RIS technology has helped connect various IoT devices in Singapore to each other. It has ensured strong and stable connectivity between the city's Wi-Fi hotspots, smart buildings and other IoT devices. Data can be exchanged through RIS in a very fast and cost-effective way.

Combination of RIS and 5G networks: Singapore has integrated RIS technology with 5G networks, improving network speed and coverage. 5G networks provide high speeds and low latency, and RIS makes it even better. RIS can make 5G networks smart and energy-efficient, improving the quality of connectivity in the city.

The example of Singapore proves how RIS technology can be applied in smart cities to make urban life better and more efficient. The use of RIS in traffic control, smart lighting and environmental monitoring not only improves connectivity but also ensures energy savings and environmental sustainability. Singapore has set a role model for smart cities using RIS technology, which can also become an inspiration for other cities in the future.

China: RIS Used to Expand 5G Network

In China, Reconfigurable Intelligent Surface (RIS) has been used to expand 5G networks in large cities. Specifically, RIS technology was tested in densely populated cities like Beijing and Shanghai. The main purpose of this technology was to increase network coverage and improve data speed, thereby improving connectivity and communication quality in these cities. (Baark, 2022)

Function and Benefits of RIS:

Improve network coverage: Network coverage is a big challenge in large and densely populated cities like Beijing and Shanghai. RIS technology was used in these cities to strengthen the network signal and increase the coverage area. RIS controls the direction and strength of the network signal, which helps in getting a good signal even in places where the network was previously weak. This improved the stability and reach of the 5G network in all parts of the city.

Improve data speed: RIS technology also helped in increasing the speed of data transmission. It is difficult to use the full potential of 5G networks without high data speeds, and RIS improves this. With the help of RIS, 5G networks can transfer data faster, giving users a more accurate and faster Internet experience.

Improved communication between IoT devices: The use of IoT (Internet of Things) devices is increasing in China's big cities, such as smart cars, smart home devices, and smart meters. RIS technology has also improved communication between these devices. Data exchange between these IoT devices through RIS is faster and more effective, making smart city services smarter and more efficient.

Improved smart city services: RIS has been used to improve various services in smart cities such as traffic management, energy saving, and environmental monitoring. Traffic signals are controlled smartly through RIS, reducing traffic jams and pollution. In addition, the connectivity provided by RIS makes smart lighting, smart parking, and other services more effective and energy-efficient.

Combination of RIS and 5G networks: In China, RIS has been integrated with 5G networks to maximize network speed, efficiency, and reliability by combining these two technologies. The high speed and low latency of 5G networks are combined with the flexibility and adaptability of RIS, ensuring that the highest quality connectivity can be provided in urban areas.

The use of RIS in China has proven to be very effective in expanding 5G networks. In cities such as Beijing and Shanghai, RIS has played an important role in expanding network coverage, improving data speeds, and improving communication between IoT devices. RIS technology has also improved connectivity and communication as well as smart services in China's smart cities. This technology may also become an important tool for other cities in the future, helping to make smart cities more effective.

Bengaluru, India: RIS Used in Energy Efficiency and Smart City Strategies

In Indian cities like Bengaluru, Reconfigurable Intelligent Surface (RIS) technology is being used to implement smart city strategies and improve energy efficiency. The local government here has adopted RIS technology to improve urban energy management by combining solar panels and smart metering systems.

Functions and Benefits of RIS:

Improving Energy Efficiency: RIS technology is being used in Bengaluru to better manage energy. RIS integrates with solar panels to make them more efficient. RIS can optimize solar panels according to the direction and intensity of the sun, allowing for maximum energy use. It helps reduce energy consumption in a smart way.

Smart Metering System: RIS technology has also helped improve Bengaluru's smart metering system. Smart meters track energy consumption in homes and commercial spaces. Through RIS, the connectivity of these meters has been made stronger and more effective, providing real-time energy consumption information. This system helps users to control and save their energy consumption.

Smart City Strategies: RIS has also been used to pursue smart city plans in Bengaluru. Through this, effective use of energy in the city has been promoted, which can reduce the energy crisis in the city. The energy system of the city has been integrated by RIS technology, making urban life more sustainable and smart.

Urban Energy Management: RIS technology has played an important role in improving the urban energy management of Bengaluru. This technology helps to manage the distribution, consumption and supply of energy in the city in a more efficient way. Data from smart meters and solar panels is exchanged through RIS, which monitors and optimizes energy use across the city.

RIS technology is being used in Bengaluru to promote smart city strategies and improve energy efficiency. The combination of RIS with solar panels and smart metering systems makes urban energy management more effective. This technology not only reduces energy consumption but also ensures better connectivity and smart energy management in all parts of the city. The example of Bengaluru shows how the use of RIS technology can make smart cities more sustainable and energy-efficient. (Paksok et al., 2023)

2. Lessons Learned From RIS Implementation

Improve connectivity: The first implementations of RIS have made it clear that this technology can bring significant improvements to urban connectivity. Although, in the early stages, some RIS systems proved effective only in limited locations, as research and testing increased, its implementation became possible on a larger scale.
Compatibility with infrastructure: During the implementation of RIS, it was found that there were challenges in integrating it with the existing urban infrastructure. New construction and design approaches were needed to effectively integrate RIS with older buildings and structures. **User-centric design**: During the implementation of RIS, it was found to be important to keep the users' needs in mind while using the technology. Ensuring that the impact of RIS is equally beneficial to all users was a key lesson.

3. Ways to Scale RIS in Different Cities and Structures

Scaling to larger cities: Scaling RIS to larger cities requires more robust and advanced infrastructure. For example, in large cities like Singapore and Shanghai, RIS is integrated with smart traffic signals and sensors. In large cities, RIS is deployed in places with high population density and traffic problems.
Scaling to medium-sized cities: In medium-sized cities, RIS is used for urban connectivity, energy management, and sensing in public spaces. Deploying RIS in these cities requires minimal infrastructure, and this technology helps provide better network coverage and data speeds.
Scaling to smaller cities: RIS in smaller cities is commonly used for smart lighting, city sensor networks, and climate monitoring. When implementing RIS in smaller cities, it is important to keep in mind that RIS can be used effectively while staying within the cost limits. In smaller cities, RIS can be used as a simple, low-cost solution rather than more advanced or high-cost equipment.

RIS technology can revolutionize smart cities, as evidenced by its implementation in many cities around the world. In cities such as Singapore, Shanghai, and Bengaluru, RIS has improved network connectivity, data speeds, and urban infrastructure. Important lessons learned from RIS implementation include user-centered design, compatibility with infrastructure, and effective use in high-density areas. Strategies have been developed to scale RIS to cities of varying sizes, so that this technology can work effectively in large, medium and small cities.

This chapter shares experiences gained from the implementation of RIS in smart cities and shows how this technology can improve connectivity and functionality in various urban settings.

13.11 CONCLUSION AND ROADMAP FOR RIS IN SMART CITY TRANSFORMATION

Summary of the Impact and Opportunities of RIS in Smart Cities

Reconfigurable intelligent surfaces (RIS) are playing a vital role in the creation of smart cities. These technologies have improved communication, energy management, and connectivity in urban areas. Through RIS, we are seeing better network coverage, energy efficiency, and more intelligent IoT (Internet of Things) systems in smart cities.

The biggest strength of RIS is that it makes wireless communication flexible and adaptive in urban environments. It helps reduce network traffic, increase data speed, and control energy consumption. Additionally, this technology can also integrate with smart city applications such as smart traffic systems, smart metering, and solar panels, making urban life more efficient and sustainable.

The future of RIS in smart cities looks even brighter, as it has the potential to improve almost every aspect of urban life. Through this, it may be possible to create a smart, safe, and eco-friendly environment in smart cities.

Suggested Roadmap for Researchers, Policy Makers and Urban Planners

Some important steps must be taken to fully implement RIS technology in smart cities.

Research and development: More research is needed on RIS to fully understand its potential. More focus should be given to its working models, software algorithms and more accurate data collection. Researchers need to understand how RIS technology works in different urban environments and how it can be adapted.

Policy and regulation: Policy makers should create a strong legal and policy framework for the development of RIS. This framework is necessary to implement RIS technology in a safe, reliable and effective manner. It should focus on data privacy, security and energy efficiency.

Guidelines for urban planners: Urban planners need clear guidelines to integrate RIS technology in the architecture and design of cities. RIS can be adapted based on the size, structure and population density of cities. Planners should ensure that there is no disruption in communication and network connectivity with the RIS system.

Sensitivity and acceptability: Sensitivity and acceptability of city dwellers play a vital role in the adoption of new technologies like RIS. It is essential that common citizens are made aware about the benefits of RIS and how it works so that they can perceive its impact positively.

The Crucial Role of Interdepartmental Collaboration and Innovation

Interdepartmental collaboration is essential for the effective implementation of RIS. This technology is a joint effort of experts from various fields like wireless communication, electronics, energy management and urban planning. It is only through collaboration between these fields that the potential benefits of RIS technology can be properly implemented.

Apart from this, innovation also plays a vital role. Using technologies like Artificial Intelligence (AI), Machine Learning (ML) and Blockchain in the development of RIS can make it more smart and effective. Integrating RIS with these technologies can make it even more powerful and show a new direction in smart cities.

RIS technology plays an important role in changing the structure of smart cities. Through this, communication, energy management and connectivity can be improved in cities. For its successful implementation, researchers, policy makers and urban planners will have to work together. Along with this, innovation and inter-departmental collaboration can make RIS more effective and sustainable, which will help in the development of smart cities in the future.

REFERENCES

Ahmed, M., Xu, F., Wahid, A., Ali, K., Mirza, M. A., Khan, W., Dev, K., Hassan, S. A., & Hans, Z. (2024). *A Comprehensive Survey of Artificial Intelligence Advances in RIS-Assisted Wireless Networks*. doi:DOI: 10.36227/techrxiv.172349533.38675045/v1

Alexandropoulos, G. C., Phan-Huy, D.-T., Katsanos, K. D., Crozzoli, M., Wymeersch, H., Popovski, P., Ratajczak, P., Bénédic, Y., Hamon, M.-H., Gonzalez, S. H., Mursia, P., Rossanese, M., Sciancalepore, V., Gros, J.-B., Terranova, S., Gradoni, G., Di Lorenzo, P., Rahal, M., Denis, B., & Strinati, E. C. (2023). RIS-enabled smart wireless environments: Deployment scenarios, network architecture, bandwidth and area of influence. *EURASIP Journal on Wireless Communications and Networking*, *2023*(1), 103. DOI: 10.1186/s13638-023-02295-8

Baark, E. (2022). China's New Digital Infrastructure: Expanding 5G Mobile Communications. *East Asian Policy*, *14*(2), 124–136. DOI: 10.1142/S1793930522000162

Bae, J. S., Khalid, W., Lee, A., Lee, H., Noh, S., & Yu, H. (2024). Overview of RIS-enabled secure transmission in 6G wireless networks. *Digital Communications and Networks*, *10*(6), 1553–1565. DOI: 10.1016/j.dcan.2024.02.005

Basar, E., Di Renzo, M., De Rosny, J., Debbah, M., Alouini, M. S., & Zhang, R. (2019). Wireless Communications Through Reconfigurable Intelligent Surfaces. *IEEE Access : Practical Innovations, Open Solutions*, *7*, 116753–116773. DOI: 10.1109/ACCESS.2019.2935192

Cavada, M., & Rogers, C. D. F. (2019). A smart city case study of Singapore—Is Singapore truly smart? In *Smart City Emergence* (pp. 295-314). doi:DOI: 10.1016/B978-0-12-816169-2.00014-6

Faisal, M., & Choi, W. (2021). A Study on Machine Learning-based Approaches for Reconfigurable Intelligent Surface. In *2021 International Conference on Information and Communication Technology Convergence (ICTC)*.

Haider, T. H. (2024). Reconfigurable Intelligent Surfaces Between the Reality and Imagination. *Wasit Journal of Computer and Mathematics Science*, *3*(2), 42–50. DOI: 10.31185/wjcms.266

Hameed, I., & Koo, I. (2024). Enhancing Throughput in IoT Networks: The Impact of Active RIS on Wireless Powered Communication Systems. *Electronics (Basel)*, *13*(7), 1402. DOI: 10.3390/electronics13071402

Huang, C. (2020). Reconfigurable Intelligent Surfaces for Energy Efficiency in Wireless Communication. *IEEE Transactions on Wireless Communications, 19*(10), 6654–6669.

Kazim, J. R., Rains, J., Imran, M. A., & Abbasi, Q. H. (2023). Application and Future Direction of RIS. In *Intelligent Reconfigurable Surfaces (IRS)* (pp. 171–188). Wiley-IEEE. doi:DOI: 10.1002/9781119875284.ch9

Kisseleff, S., Martins, W. A., Al-Hraishawi, H., Chatzinotas, S., & Ottersten, B. (2020). Reconfigurable Intelligent Surfaces for Smart Cities: Research Challenges and Opportunities. *IEEE Open Journal of the Communications Society*, *1*, 1781–1797. DOI: 10.1109/OJCOMS.2020.3036839

Liu, R., Alexandropoulos, G. C., Wu, Q., Jian, M., & Liu, Y. (2022). 2022 IEEE/CIC International Conference on Communications in China (ICCC Workshops), 11-13 August. doi:DOI: 10.1109/ICCCWorkshops55477.2022.9896658

Liu, Y., Liu, X., Mu, X., Hou, T., Xu, J., Di Renzo, M., & Al-Dhahir, N. (2020). *Reconfigurable Intelligent Surfaces: Principles and Opportunities*. doi:DOI: 10.48550

Naeem, F., Ali, M., Kaddoum, G., Huang, C., & Yuen, C. (2023). Security and Privacy for Reconfigurable Intelligent Surface in 6G: A Review of Prospective Applications and Challenges. *IEEE Open Journal of the Communications Society*, *4*, 1196–1217. DOI: 10.1109/OJCOMS.2023.3273507

Paksok, D., Riba, J., Powrel, D., & Kiron, B. (2023). Strategies for advancing bengaluru as smart city: A case study" In *Sustainable Civil Infrastructure Development: Case Studies* (pp.114-131). 4D.

Qian, J., Yu, L., Liu, C., Lv, X., Wang, Y., & Wang, Z. (2023, June). Optimization design of RIS-assisted high-capacity visible light communications based on HDMA. *Physical Communication*, *58*, 102056. DOI: 10.1016/j.phycom.2023.102056

Rafique, N., Ul Hassan, M., Zubair, I. H., Naqvi, M. Q., Di Renzo, M., Debbah, M., & Yuen, C. (2023). Reconfigurable Intelligent Surfaces: Interplay of Unit Cell and Surface-Level Design and Performance under Quantifiable Benchmarks. *IEEE Open Journal of the Communications Society PP*, (99), 1–1. DOI: 10.1109/OJCOMS.2023.3292357

Rajaan, R., Baishya, B. K., Rao, T. V., Pattanaik, B., Tripathi, M. A., & R, A. (2024). Efficient Usage of Energy Infrastructure in Smart City Using Machine Learning. *EAI Endorsed Transactions on Internet of Things, 10*. doi:. DOI: 10.4108/eetiot.5363

Reconfigurable Intelligent Surfaces (RIS) ETSI Industry Specification Group (ISG). (2023). *Reconfigurable Intelligent Surfaces (RIS): Technological challenges, architecture and impact on standardization.* ETSI, Industry Specification Group (ISG) Report GR RIS 002 V1.1.1, 08.

Ren, Y., Zhou, R., Teng, X., Meng, S., Zhou, M., & Tan, W. (2023, October). On Deployment Position of RIS in Wireless Communication Systems: Analysis and Experimental Results. *IEEE Wireless Communications Letters*, *12*(1756-1760), 1756–1760. DOI: 10.1109/LWC.2023.3292125

Selvaraj, M., Vijay, R., Anbazhagan, R., & Rengarajan, A. (2023). Reconfigurable Metasurface: Enabling Tunable Reflection in 6G Wireless Communications. [PubMed]. *Sensors (Basel)*, *23*(22), 9166. DOI: 10.3390/s23229166

Singh, A. K., Maurya, A. K., Prakash, R., Thakur, P., & Tiwari, B. B. (2022). Reconfigurable intelligent surface with 6G for industrial revolution: Potential applications and research challenges, *Journal of Behavioral Robotics*. doi:. DOI: 10.1515/pjbr-2022-0114

Velez, V., Pavia, J. P., Souto, N., Sebastião, P., & Correia, A. (2023). Performance Assessment of a RIS-Empowered Post-5G/6G Network Operating at the mmWave/THz Bands. *IEEE Access : Practical Innovations, Open Solutions*, (99), 1–1. DOI: 10.1109/ACCESS.2023.3277388

Wu, Q., & Zhang, R. (2019). Towards Smart and Reconfigurable Environment: Intelligent Reflecting Surface Aided Wireless Network. *IEEE Communications Magazine*, *58*(1), 106–112. DOI: 10.1109/MCOM.001.1900107

Chapter 14
Reconfigurable Intelligent Surface-Assisted Device-to-Device Communications: Applications and Research Challenges

K. Kavitha
https://orcid.org/0000-0002-9159-9690
Velammal College of Engineering and Technology, India

Raj Kumar S
https://orcid.org/0009-0002-0499-9619
Velammal College of Engineering and Technology, India

K. Karthika
https://orcid.org/0000-0001-9968-5027
Kumaraguru College of Technology, India

ABSTRACT

Reconfigurable Intelligent Surfaces (RIS) are emerging as a pivotal technology in next-generation wireless networks, particularly in enhancing Device-to-Device (D2D) communications. This chapter explores the applications, challenges, and future directions of RIS in D2D contexts. By leveraging programmable meta-materials, RIS can significantly improve spectrum efficiency, enhance reliability, and optimize energy consumption in D2D networks. However, challenges such as beamforming optimization, channel estimation, and practical deployment remain. Moreover, the

DOI: 10.4018/979-8-3693-8099-4.ch014

integration of machine learning (ML) and artificial intelligence (AI) is crucial for optimizing RIS configurations and addressing dynamic network conditions. As we approach the era of 6G, RIS-assisted D2D communications promise to revolutionize how devices interact, enabling ultra-reliable and low-latency communication for various applications, including IoT, smart cities, and autonomous systems.

INTRODUCTION

The rapid evolution of wireless communication technologies has driven the relentless pursuit of innovative solutions to meet the ever-growing demands for high-speed, low-latency, and energy-efficient connectivity. Among the various technological advancements shaping the future of wireless networks, Reconfigurable Intelligent Surfaces (RIS) have emerged as a transformative paradigm, offering unparalleled flexibility in shaping wireless propagation environments. By intelligently altering the phase, amplitude, and polarization of incoming signals, RIS enables significant improvements in spectral efficiency, energy efficiency, and communication reliability, making it a cornerstone of next-generation networks (Cui et al., 2014; Smith et al., 2004; Di Renzo et al., 2020; Liu, Guo, & Zhang, 2020).

The primary motivation behind RIS is its ability to create a "smart radio environment," where the wireless channel becomes an active participant in optimizing communication performance. Unlike traditional techniques that rely solely on active components such as base stations and relays, RIS leverages passive, software-controlled metasurfaces, which operate with minimal energy consumption (Zappone et al., 2020; Huang et al., 2020; Luo & Yuen, 2021). These properties make RIS a promising solution for addressing challenges in dense urban deployments, IoT networks, and beyond. Recent works have demonstrated its potential in diverse applications, including indoor signal focusing, beamforming optimization, and power control, showcasing its versatility and adaptability (Bjornson, Sanguinetti, & Debbah, 2019; Wang et al., 2020; Huang, Hu, & Alexandropoulos, 2020).

In parallel with RIS, Device-to-Device (D2D) communication has gained significant traction as a disruptive communication paradigm. D2D enables direct communication between devices without relying on cellular infrastructure, thereby reducing latency, improving spectral efficiency, and alleviating network congestion (Asadi, Wang, & Mancuso, 2014; Xu et al., 2015; Wang et al., 2015). Early studies have explored various aspects of D2D communication, such as feasibility, resource allocation, and power control mechanisms (Zhou et al., 2018; Yu, Xu, & Zhang, 2020). By facilitating proximity-based communication, D2D has become integral to applications like public safety, vehicular communication, and content sharing, especially in scenarios where network coverage is limited or unavailable.

Despite its advantages, D2D communication faces significant challenges, including interference management, resource allocation, and security concerns. To address these challenges, researchers have investigated advanced techniques such as coalitional games, queueing theory, and optimization algorithms (Smith et al., 2004; Wang et al., 2015; Xu et al., 2015). These methods have laid the groundwork for integrating D2D with other emerging technologies, such as RIS, to further enhance its performance and reliability (Liu, Guo, & Zhang, 2020; Chen, Chen, & Li, 2021).

The integration of RIS with D2D communication represents a paradigm shift in wireless network design, offering a synergistic approach to overcoming the limitations of conventional systems. By leveraging the reconfigurability of RIS, it is possible to dynamically manipulate the propagation environment, mitigate interference, and optimize resource utilization in D2D networks (Di Renzo et al., 2020; Basharat et al., 2022; Zappone et al., 2020). For instance, RIS can create highly directional signal paths that reduce interference and enhance signal quality, thereby improving the overall performance of D2D communication (Wang et al., 2020; Liu, Zeng, & Zhang, 2020).

Moreover, the combination of RIS with artificial intelligence (AI) and machine learning (ML) techniques has opened new avenues for optimizing system performance in complex and dynamic environments. Recent studies have explored the use of deep reinforcement learning, beamforming optimization, and predictive analytics to enhance the efficiency of RIS-assisted D2D communication (Liu, Guo, & Zhang, 2020; Elbir, Papazafeiropoulos, & Ioannidis, 2020). These approaches enable real-time adaptation to channel state information (CSI) uncertainty, ensuring robust and reliable communication even in challenging scenarios (Ding et al., 2024; Bjornson et al., 2019).

The integration of RIS with D2D communication is particularly relevant in the context of 6G, where the emphasis is on creating ultra-reliable, low-latency communication systems that can support a wide range of applications. As highlighted by recent studies, RIS can play a pivotal role in enabling these capabilities by providing flexible and energy-efficient solutions for managing interference, enhancing connectivity, and optimizing resource allocation (Gong, Lu, & Khoshnevisan, 2020; Liang, Zhang, & Di Renzo, 2020). Furthermore, RIS-assisted D2D communication can facilitate new use cases, such as smart cities, autonomous vehicles, and industrial IoT, by enabling seamless connectivity and efficient data exchange (Liu, Zeng, & Zhang, 2020; Di Renzo et al., 2020).

This chapter provides a comprehensive survey of the current state-of-the-art in RIS-assisted D2D communication, focusing on theoretical advancements, practical applications, and open challenges. Key contributions from the literature are discussed in detail, highlighting the role of RIS in creating a smart and reconfigurable wireless environment. By addressing these challenges, researchers can unlock the full potential of RIS and D2D communication in shaping the future of wireless networks.

RIS-ASSISTED D2D COMMUNICATIONS

Reconfigurable Intelligent Surfaces (RIS) have emerged as one of the most promising innovations in wireless communication, especially as we advance toward the era of *6G networks*. RIS comprises engineered surfaces embedded with programmable meta-materials that can dynamically manipulate electromagnetic waves. Controlled by software, these surfaces can adjust key wave properties, such as phase, amplitude, and polarization, enabling precise control over signal propagation. This capability makes RIS a transformative technology for optimizing communication environments, particularly for applications demanding high efficiency, reliability, and adaptability.

In the context of *Device-to-Device (D2D) communication*, RIS offers an unprecedented opportunity to overcome critical challenges. D2D communication allows devices to exchange data directly, bypassing the conventional reliance on centralized network infrastructure like base stations or access points. This decentralized approach is highly advantageous for reducing latency, minimizing network congestion, and supporting localized data exchanges in applications such as IoT, autonomous systems, and emergency communications. However, the performance of D2D networks is often compromised by issues like *path loss*, *multipath fading*, *signal attenuation*, and *interference*, particularly in complex propagation environments such as urban areas or dense indoor settings.

RIS can address these limitations by dynamically creating a more favorable propagation environment. By improving *line-of-sight (LoS)* connectivity and effectively redirecting or reflecting signals, RIS minimizes the impact of path loss and fading. This enhances signal strength and reliability, even in scenarios where direct LoS is naturally obstructed by physical barriers. Furthermore, RIS can mitigate interference by precisely controlling the directionality of signals, ensuring that devices operate with minimal disruption from other transmissions in the same frequency spectrum. This capability is particularly critical in dense networks, where spectrum scarcity and interference pose significant challenges.

Another major advantage of RIS in D2D systems is its potential to improve *spectrum utilization* and *energy efficiency*. Traditional methods of enhancing D2D communication often involve increased transmission power or reliance on additional spectrum, which can lead to inefficiencies and environmental concerns. RIS, on the other hand, optimizes the existing spectral and energy resources by passively reflecting and steering electromagnetic waves without requiring active power-consuming components like amplifiers or signal generators. This not only reduces the energy footprint of D2D networks but also aligns with the sustainability goals of future wireless technologies.

Moreover, the integration of RIS with advanced control mechanisms, powered by *Machine Learning (ML)* and *Artificial Intelligence (AI)*, enables dynamic adaptation to varying network conditions. For example, RIS can autonomously adjust its configurations based on real-time environmental changes, traffic patterns, or user mobility, ensuring consistent performance under diverse scenarios. This adaptability further enhances the potential of RIS-assisted D2D communications in applications requiring ultra-reliable and low-latency communication (URLLC), such as industrial automation, connected vehicles, and smart city infrastructure.

Moreover, RIS-assisted D2D communications represent a revolutionary advancement in wireless technology. By leveraging the unique properties of RIS, these systems can overcome longstanding challenges such as interference, signal degradation, and spectrum scarcity, while simultaneously enhancing reliability, energy efficiency, and spectral utilization. As we move closer to the widespread deployment of 6G networks, RIS is poised to become an integral component of D2D communication systems, unlocking new possibilities for seamless, efficient, and sustainable wireless connectivity. Figure 1 shows the flowchart of the RIS Deployment Process in D2D Networks and Figure 2 shows the RIS-assisted D2D communication system.

Figure 1. Flowchart of the RIS Deployment Process in D2D Networks

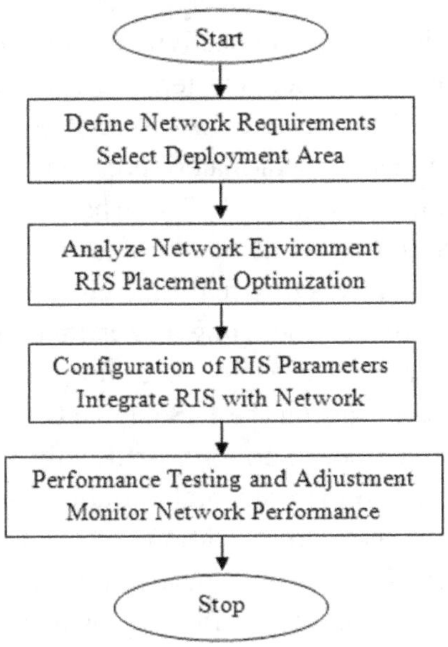

Figure 2. RIS-assisted D2D communication system

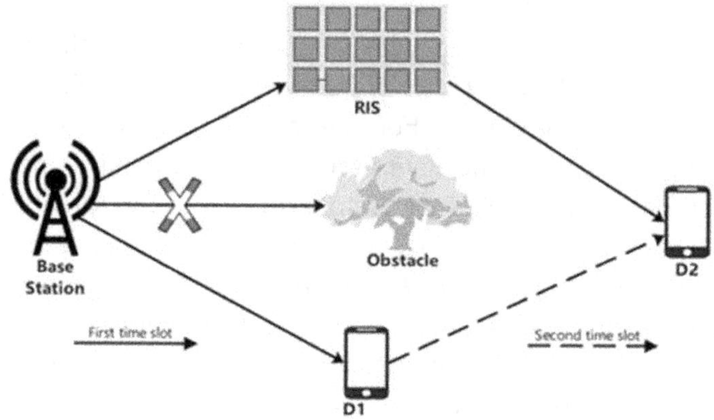

(Courtesy: Ding et al. [48])

Integration of RIS in D2D Networks

Reconfigurable Intelligent Surfaces (RIS) are emerging as a game-changing technology in Device-to-Device (D2D) communication, addressing persistent challenges like interference, signal attenuation, and spectrum inefficiency in densely populated communication scenarios. RIS enhances D2D communication by dynamically improving line-of-sight (LoS) connectivity through intelligent signal steering and optimizing signal paths to reduce interference. Its passive operation makes it highly energy-efficient, a critical feature for sustainable communication systems.

RIS comprises programmable meta-materials capable of dynamically adjusting the phase and direction of incoming signals. These panels are strategically deployed in complex environments such as urban areas, indoor spaces, and regions with high propagation complexity. By leveraging real-time channel state information (CSI) collected through precise channel modeling and estimation, RIS enables adaptive signal control and optimization. Machine Learning (ML) algorithms calculate the optimal phase shifts, configuring the RIS elements for efficient beamforming toward specific D2D receivers. Furthermore, Artificial Intelligence (AI) enhances RIS efficiency by monitoring environmental variables, user mobility, and traffic patterns to adapt to dynamic network conditions.

Through its integration, RIS not only ensures ultra-reliable, low-latency communication but also improves spectrum efficiency by managing signal propagation intelligently, reducing spectrum collisions, and maximizing resource reuse. Its ability to integrate seamlessly with existing D2D communication protocols makes RIS a cornerstone for next-generation 6G networks. This enables low-latency, interference-free, and energy-efficient communication that meets the requirements of applications such as the Internet of Things (IoT), smart cities, and autonomous systems.

Challenges in D2D Networks with RIS Integration

Despite its transformative potential, integrating RIS into D2D networks presents challenges, particularly in highly dense and dynamic environments. Signal interference, attenuation, and resource allocation complexities often hinder optimal performance. However, RIS mitigates these issues by dynamically controlling beamforming and enabling precise phase adjustments. It also significantly improves spectrum efficiency by reducing dependency on centralized base stations, thereby minimizing network congestion.

The implementation of RIS is vital for mission-critical applications, such as disaster recovery, emergency response, and industrial IoT, where ultra-reliable and low-latency communication is essential. In environments with dense obstacles, RIS can provide alternate communication paths, ensuring stable connectivity even in scenarios where direct transmission is obstructed. Ding et al. (2020) highlighted the effectiveness of RIS in overcoming these challenges, emphasizing its ability to sustain high performance in urban and indoor settings.

Interference Management and Spectrum Efficiency

RIS significantly improves spectrum efficiency in D2D communications by allowing devices to bypass centralized base stations. By adjusting the reflection properties of its programmable elements, RIS minimizes interference and optimizes signal strength for multiple users operating within the same spectrum. Research by Li et al. (2021) demonstrated how RIS reduces signal collisions, leading to enhanced spectrum reuse and improved spectral efficiency.

Enhancing Reliability in Mission-Critical Applications

In mission-critical scenarios like emergency response and disaster recovery, stable and reliable communication is crucial. RIS plays a pivotal role in ensuring robust D2D communication by creating alternative transmission paths when direct communication is disrupted. This capability is particularly beneficial in urban environments with numerous obstacles. Ding et al. (2020) demonstrated that RIS enhances communication reliability by mitigating signal blockages, enabling critical applications to maintain seamless connectivity.

Energy Efficiency

Energy efficiency is a crucial factor in D2D networks, especially for battery-operated devices such as IoT sensors and portable communication systems. RIS significantly reduces energy consumption by optimizing signal propagation and minimizing the need for high transmission power. Studies by Tang et al. (2021) revealed that RIS-driven communication systems effectively lower power requirements, enhancing the sustainability of D2D networks. By intelligently reflecting signals toward target devices, RIS reduces energy wastage, prolonging device battery life and making D2D communication more energy-efficient and environmentally friendly.

Like this, RIS is revolutionizing D2D networks by addressing interference, spectrum inefficiency, and energy concerns while paving the way for next-generation communication systems. Its integration with AI and ML-driven optimization strategies ensures ultra-reliable, low-latency, and energy-efficient communication, making it indispensable for 6G networks. With its ability to adapt dynamically to evolving network conditions, RIS holds the promise of transforming the future of wireless communication.

Figure 3. RIS-assisted D2D system and conventional D2D system versus SNR

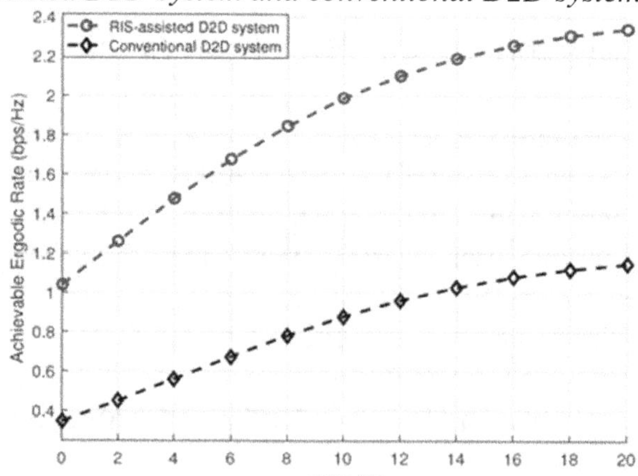

(Courtesy: Ding et al. [48])

RESEARCH CHALLENGES IN RIS-ASSISTED D2D COMMUNICATIONS

Reconfigurable Intelligent Surfaces (RIS) have shown great promise in revolutionizing Device-to-Device (D2D) communications, yet their integration and optimization pose several research challenges that need to be addressed for realizing their full potential. These challenges span multiple domains, including beamforming, channel estimation, practical deployment, scalability, and security, and require ongoing advancements in algorithms, hardware design, and system architectures.

Beamforming and Phase Shift Optimization

A central aspect of RIS operation lies in the optimization of beamforming and phase shifts, which are crucial for enhancing signal strength, improving spectral efficiency, and minimizing interference in D2D communications. RIS can modify the propagation environment by adjusting the phase, amplitude, and direction of signals, but achieving optimal configurations in real-time remains a significant challenge, especially in environments with multiple devices and rapidly changing conditions. In particular, when RIS is deployed in multi-user environments, beamforming optimization becomes even more intricate due to diverse channel characteristics and varying user mobility. Zhou et al. (2021) proposed a Singular Value Decomposition (SVD)-based phase shift design algorithm that improves the Signal-to-Noise Ratio (SNR) by optimizing the reflection parameters in RIS-D2D systems. While this approach shows promising results, real-time implementation is hindered by the high computational complexity of the algorithm. Furthermore, the adaptability of beamforming strategies in the face of fluctuating network conditions and the need for low-latency operation adds another layer of difficulty. To address these issues, advanced machine learning algorithms and optimization techniques that can perform adaptive beamforming while minimizing computation are being explored, but their practical deployment is still a topic of active research.

Channel Estimation

Channel State Information (CSI) is essential for optimizing the performance of RIS-assisted D2D communications. Accurate CSI allows for effective beamforming, interference management, and overall system optimization. However, obtaining reliable CSI in real-time, particularly in dynamic environments with high mobility or challenging propagation conditions, remains a major challenge. Abeywickrama et al. (2020) noted that phase shift estimation in RIS systems is significantly affected by mobility and environmental changes, which makes it difficult to predict and control signal paths accurately. Traditional CSI acquisition methods, which rely on explicit measurements and feedback from the network, struggle to maintain accuracy in highly dynamic or rapidly changing scenarios. To overcome this, recent studies have proposed the use of deep learning and artificial intelligence techniques for real-time CSI prediction, which can enhance the accuracy and speed of channel estimation. However, these methods require extensive training data and computational resources, which can be prohibitive in practical deployments. Moreover, the challenges in predicting CSI in scenarios with high mobility, non-line-of-sight conditions, or complex interference patterns still require further refinement. To address these challenges, new channel modeling techniques that account for the unique characteristics of RIS-assisted systems need to be developed, along with more efficient algorithms for real-time CSI acquisition.

Practical Deployment and Scalability

Despite the theoretical benefits, deploying RIS on a large scale in real-world D2D networks presents several practical challenges. One of the most significant concerns is scalability—how to integrate RIS in large, dense networks while maintaining cost-effectiveness and energy efficiency. RIS hardware, which typically consists of programmable meta-material elements, must be lightweight, affordable, and energy-efficient to make widespread deployment feasible. Basar et al. (2019) discussed the hardware limitations of RIS, such as the need for low-cost, energy-efficient reconfigurable elements and lightweight designs that can be seamlessly integrated into existing network infrastructures. The process of scaling up RIS to support numerous D2D devices, especially in highly urbanized environments or in massive Internet of Things (IoT) networks, presents challenges in terms of both hardware and software. Additionally, the large-scale deployment of RIS requires compatibility with existing communication protocols and network architectures. Developing seamless integration methods for RIS into current wireless systems, without disrupting network performance or increasing operational costs, is critical for realizing the full potential of RIS in D2D communications. Future research must also focus on addressing issues related to the maintenance and reliability of RIS in large-scale deployments, including ensuring robustness against environmental changes and wear-and-tear of hardware components.

Security Concerns

As RIS technology enhances the performance of D2D communications, it also introduces new security risks that must be carefully managed. The main concern is the possibility of RIS being exploited for malicious purposes, such as eavesdropping or jamming attacks. RIS can manipulate electromagnetic waves to create secure transmission paths, but if compromised, they can also be used to intercept sensitive data or inject malicious interference into the communication network. Xu et al. (2021) emphasized the importance of designing secure RIS-assisted systems that can withstand potential attacks, particularly in the context of future 6G networks, where privacy and data protection will be paramount. Securing RIS not only involves protecting the physical devices but also ensuring the security of the communication protocols and algorithms used to control them. The integration of RIS with advanced encryption methods, secure key exchange protocols, and real-time threat detection systems is crucial to mitigate security risks. Furthermore, as D2D communications are often deployed in decentralized environments, ensuring secure end-to-end communication without relying on a central trusted authority adds another layer of complexity. Developing robust security frameworks that can prevent, detect, and respond to potential threats in real-time will be critical for the widespread adoption of RIS in D2D networks.

Interference Management and Resource Allocation

Managing interference is one of the most significant challenges in RIS-assisted D2D communications, particularly in dense and highly congested environments. RIS can play a crucial role in mitigating interference by intelligently redirecting signals and adjusting reflection angles to avoid collisions between transmissions. However, the dynamic nature of D2D communication environments means that interference patterns can vary rapidly, making it difficult to maintain an optimal configuration. Effective interference management requires continuous monitoring and adaptation of RIS elements in response to changes in the network, user mobility, and environmental conditions. In addition, efficient resource allocation is essential to ensure that the limited spectrum and energy resources are utilized effectively. Optimizing the assignment of RIS resources, including the allocation of beamforming parameters and reflection coefficients, is a key area of ongoing research. Machine learning and optimization algorithms can help improve interference management and resource allocation in RIS-assisted D2D systems, but the trade-offs between system complexity and performance must be carefully balanced.

The integration of RIS into D2D communications offers substantial improvements in spectrum efficiency, reliability, and energy efficiency. However, several significant research challenges remain in the areas of beamforming and phase shift optimization, channel estimation, practical deployment and scalability, security, and interference management. Addressing these challenges requires a multidisciplinary approach, combining advances in hardware design, optimization algorithms, machine learning, and cybersecurity. As research in RIS technology and D2D communication systems continues to evolve, overcoming these obstacles will pave the way for more efficient, reliable, and secure communication networks, particularly as we transition toward the 6G era.

APPLICATIONS OF RIS IN D2D COMMUNICATION

Reconfigurable Intelligent Surfaces (RIS) have emerged as a transformative technology for enhancing Device-to-Device (D2D) communication networks, especially in the context of next-generation wireless networks like 5G and 6G. RIS can significantly enhance network performance by manipulating the wireless environment and offering solutions to many challenges faced in modern communications, including interference, energy consumption, signal blockage, and coverage. Below, we explore the key applications of RIS in D2D communication, illustrating how this technology improves performance, energy efficiency, reliability, and coverage.

Figure 4. Emerging applications of RIS for 6G wireless networks

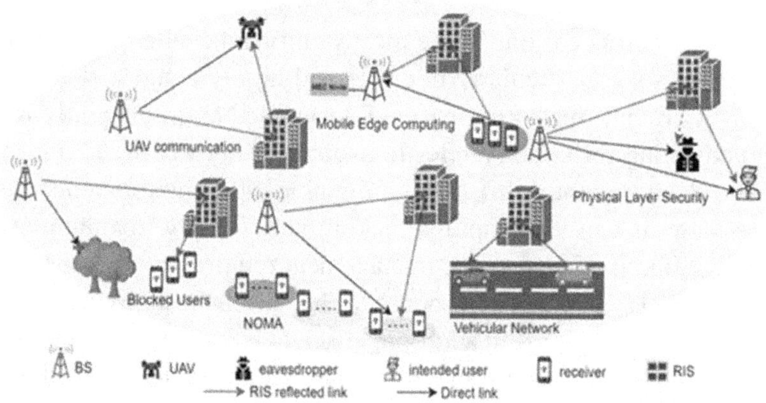

(Courtesy: Basharatet al. [47])

Spectrum Efficiency and Coverage Enhancement

One of the most impactful applications of RIS in D2D communication is its ability to significantly enhance spectrum efficiency and extend network coverage. By intelligently reflecting and refracting signals toward users, RIS can provide superior coverage in areas that suffer from signal blockage, such as urban canyons or indoors (Zhang et al., 2021; Khalid et al., 2021). In densely populated environments, where traditional communication systems often face challenges related to interference and low signal-to-noise ratios (SNR), RIS can mitigate these issues by improving the propagation conditions. This leads to more reliable and efficient communication between devices, reducing the need for additional infrastructure such as base stations.

RIS can also transform hostile environments into favorable ones by leveraging controlled reflections. These surfaces can dynamically adjust their properties to focus energy where it is needed most, effectively bypassing obstacles or interference from physical structures (Zeng et al., 2021). A practical example of this capability was demonstrated by Tang et al. (2021), who performed path loss modeling experiments for RIS-enhanced D2D networks. Their results showed that RIS deployment could significantly improve communication performance in complex propagation environments, such as in urban areas with significant obstructions.

The large-scale deployment of RIS can, therefore, drastically improve the coverage of D2D communication systems by enhancing signal quality, extending network reach, and reducing dependency on conventional relay stations.

Energy Efficiency

Energy efficiency is another attractive feature of RIS in D2D communications. Traditional relay systems, such as repeaters and base stations, rely on active power sources and can consume considerable amounts of energy. In contrast, RIS operates passively, requiring very little energy to function while still providing significant improvements to the signal quality and strength (Han et al., 2021). This makes RIS a highly energy-efficient alternative to conventional systems in energy-sensitive applications like D2D communications.

Research by Han et al. (2021) explored the design of RIS-empowered D2D setups, demonstrating that the use of RIS could significantly reduce energy consumption while maintaining an acceptable level of communication quality. By reducing the need for power-hungry infrastructure, RIS also contributes to the overall sustainability of communication networks. This energy-efficient aspect is particularly important in D2D communication, where multiple devices might be operating simultaneously, often in remote or off-grid environments.

Additionally, Yadav et al. (2021) highlighted that RIS-assisted systems could outperform traditional D2D communication schemes in terms of energy efficiency. By passively reflecting signals rather than requiring energy-intensive signal amplification or retransmission, RIS offers a greener, more cost-effective solution for enhancing communication networks, making it an ideal technology for future wireless systems that prioritize sustainability.

Improved Reliability for Mission-Critical Applications

In mission-critical applications, such as disaster recovery, emergency response, and public safety, the reliability of communication links is of utmost importance. In these scenarios, the availability of base stations or other infrastructure may be limited or entirely unavailable. This is where RIS can play a crucial role by dynamically adapting to changing channel conditions, ensuring that communication remains reliable even in the absence of traditional network infrastructure.

RIS can improve the robustness of D2D communication links by reshaping the wireless environment to maintain high-quality signal transmission. For example, in disaster recovery scenarios, where infrastructure may be damaged, RIS can help establish ad hoc communication links between devices, such as between emergency responders or citizens. By reflecting and refracting signals, RIS can create reliable communication paths in areas where conventional infrastructure is compromised (Ding et al., 2020).

RIS also helps mitigate channel fading, a common problem in mobile communication systems, especially in urban or rural areas with a lack of direct line-of-sight between devices. Through adaptive and intelligent signal manipulation, RIS can continuously adjust its properties to combat the challenges posed by environmental factors, such as physical obstructions or interference from other devices. As a result, RIS enhances the reliability of communication links in mission-critical situations, enabling continuous communication even under challenging circumstances.

User-Centric Communication

RIS can also support user-centric D2D communication, where devices interact with each other directly without involving a base station. This type of communication is becoming increasingly important as devices become more interconnected, and the demand for low-latency, high-throughput communication increases. RIS can dynamically optimize the communication environment for each device, ensuring that users receive the best possible connection based on their location, signal quality, and network conditions.

By adapting to the real-time needs of users, RIS can facilitate highly efficient, peer-to-peer communication, enhancing the performance of D2D communication systems in both urban and rural environments. For example, in IoT networks, where many low-power devices may communicate directly with one another, RIS can optimize the wireless channels to ensure seamless and efficient data transmission between devices, thereby improving the overall performance of the IoT ecosystem.

Integration with Edge Computing and Mobile Networks

As mobile networks evolve, especially with the advent of 5G and 6G, there is a growing need for edge computing to handle data processing closer to the source of data generation. RIS can play a pivotal role in this context by improving communication between mobile devices and edge computing nodes in D2D networks. With RIS-enabled D2D communication, mobile devices can be provided with reliable and high-speed connections to local edge servers, reducing the latency associated with cloud-based data processing.

In scenarios where edge computing is used to offload heavy computational tasks, RIS can help in maintaining low-latency communication channels between devices and edge nodes, enhancing the overall efficiency of mobile networks. Additionally, RIS can assist in optimizing the data flow between the mobile network and edge nodes, ensuring that communication is not only fast but also efficient, with minimal energy consumption.

Integration with 6G Networks

The transition to 6G networks promises even greater integration of RIS, enhancing D2D communications further. 6G will likely feature extremely dense networks with billions of devices connected, creating challenges related to network congestion, signal interference, and energy consumption. RIS offers a scalable solution to address these challenges, as it allows for fine-grained control over the wireless environment, enabling more efficient and dynamic D2D communication.

In 6G, RIS could be integrated into a network architecture that combines traditional cell towers, edge computing, and D2D communication systems to create a seamless, high-performance wireless network. As such, RIS will play a vital role in enhancing network performance and making D2D communication a core component of 6G's vision.

The applications of RIS in D2D communication are wide-ranging and offer transformative potential for next-generation wireless networks. By enhancing spectrum efficiency, improving coverage, optimizing energy consumption, and increasing reliability in mission-critical applications, RIS offers a scalable and sustainable solution for the challenges faced by D2D networks. As we move toward 5G and 6G, RIS will play an increasingly crucial role in improving network performance, supporting new communication paradigms, and enabling seamless connectivity in highly dynamic environments. As such, RIS is poised to be a key enabler of the future wireless landscape, providing the infrastructure necessary to meet the growing demands of modern communication systems.

ROLE OF AI, ML, AND DL IN RIS-ASSISTED D2D COMMUNICATION1

The integration of **Reconfigurable Intelligent Surfaces** (RIS) into **Device-to-Device (D2D)** communication networks represents a paradigm shift towards creating highly adaptive, efficient, and dynamic communication systems. The power of **Artificial Intelligence (AI)**, **Machine Learning (ML)**, and **Deep Learning (DL)** further enhances the performance of RIS in D2D communication, addressing the challenges associated with dynamic environments, interference, energy efficiency, and security.

These intelligent technologies can provide continuous learning and adaptation, making RIS systems not only reactive but predictive, allowing for optimal communication under varying network conditions.

Channel Estimation and Prediction

- **Traditional Approach**: In conventional communication systems, the Channel State Information (CSI) is usually estimated through pilot signals, and these estimates are fixed unless recalibrated. The accuracy of these estimates in traditional systems depends on periodic updates.
- **RIS-Assisted Approach**: RIS systems offer dynamic and adaptive channel conditions by controlling signal reflections and refractions. However, this dynamic nature complicates traditional CSI estimation.
- **AI/ML/DL Approach**: Advanced AI techniques, particularly **Deep Learning (DL)**, can predict future channel states by learning from past interactions between devices. For example, **Convolutional Neural Networks (CNNs)** and **Recurrent Neural Networks (RNNs)** can be used to identify channel patterns and forecast future states. These techniques significantly improve the accuracy of CSI estimations, enabling RIS to adjust its configuration in real time for optimal performance. By leveraging large amounts of historical data, **Machine Learning (ML)** algorithms like **Support Vector Machines (SVM)** or **K-Nearest Neighbors (KNN)** can provide accurate channel predictions with minimal overhead.

Beamforming and Reflection Pattern Design

- **Traditional Approach**: In traditional wireless communication, beamforming is typically handled using fixed antennas with predefined patterns. This approach lacks the flexibility needed for dynamic environments.

- **RIS-Assisted Approach**: RIS utilizes an array of passive reflective elements to manipulate the phase and amplitude of incoming signals. These elements are designed to adjust the propagation path of signals, improving overall communication quality.
- **AI/ML/DL Approach**: AI plays a pivotal role in dynamically optimizing RIS's beamforming capabilities. **Reinforcement Learning (RL)**, in particular, is well-suited for this task as it allows RIS to continuously learn the best configurations for reflection patterns based on the ongoing network conditions. In RL, agents (RIS) interact with their environment (the wireless channel) and adjust their parameters (beamforming patterns) to maximize a predefined reward, such as signal strength or capacity. Moreover, **Supervised Learning** can be used to pre-train models that help predict the best beamforming configurations given a set of environmental parameters.

Interference Management

- **Traditional Approach**: Interference management in traditional D2D systems is often handled through frequency planning or by allocating specific time slots for each device. However, as the number of devices increases, interference management becomes a significant challenge.
- **RIS-Assisted Approach**: RIS can actively mitigate interference by dynamically reshaping the wireless environment, rerouting signals to avoid interference zones and optimizing resource usage. The challenge lies in determining the best way to configure the RIS elements to minimize co-channel interference.
- **AI/ML/DL Approach**: **AI-based optimization techniques**, such as **Reinforcement Learning (RL)** or **Deep Q Networks (DQN)**, are emerging as promising methods to control interference. These methods allow RIS to predict and adaptively avoid interference by learning from environmental changes. ML-based algorithms can classify sources of interference and predict future interference patterns, allowing RIS to dynamically adjust its configuration to avoid conflicts. These adaptive techniques ensure minimal signal overlap between devices and maximize throughput.

Energy Efficiency Optimization

- **Traditional Approach**: Traditional relay systems often require active power sources, which increase energy consumption, especially in dense communication scenarios. The energy consumption in conventional systems is typically high due to the need for signal amplification, transmission power, and retransmissions.

- **RIS-Assisted Approach**: One of the key advantages of RIS is its passive operation. Unlike traditional relay systems that require energy to amplify or retransmit signals, RIS only modifies the direction of signal propagation, consuming significantly less power.
- **AI/ML/DL Approach**: AI and ML techniques can further optimize energy usage in RIS-assisted systems. By using **Supervised Learning**, RIS can learn optimal configurations based on real-time energy usage data and communication requirements. **Reinforcement Learning** can also be used to dynamically adjust RIS settings, such as the number of reflective elements in operation or the intensity of reflections, based on network load and available energy resources. This results in a substantial reduction in energy consumption while maintaining or improving communication quality.

Security and Privacy

- **Traditional Approach**: In traditional communication systems, security mechanisms such as encryption are often applied to protect the transmitted data. However, these methods may not be sufficient in a RIS-enabled D2D communication system, where the control over the signal propagation introduces potential vulnerabilities.
- **RIS-Assisted Approach**: The ability of RIS to control signal propagation can be exploited by malicious entities for eavesdropping, creating security vulnerabilities. RIS could potentially be used to intentionally direct signals to unauthorized recipients, violating privacy.
- **AI/ML/DL Approach**: AI plays an important role in ensuring that RIS systems are secure and private. **Anomaly detection** algorithms based on **Unsupervised Learning** can identify abnormal behaviors in the system, such as unauthorized signal redirection. **Deep Learning** models can be trained to detect such security breaches by learning from network activity patterns. Additionally, **Federated Learning** can enable RIS to perform secure data analysis without disclosing sensitive information, ensuring privacy protection for the users.

Resource Allocation and Network Slicing

- **Traditional Approach**: Resource allocation in traditional D2D communication systems is usually static, with fixed allocations of spectrum, time slots, or power levels. This approach does not adapt well to changing network demands and can lead to inefficient utilization of resources.

- **RIS-Assisted Approach**: RIS can provide adaptive resource allocation by shaping the wireless environment to suit specific communication needs. It can dynamically adjust the network parameters to ensure optimal performance and resource utilization.
- **AI/ML/DL Approach**: **Reinforcement Learning (RL)** can be used to optimize resource allocation in real-time. By interacting with the network environment, RIS can learn the best configurations that minimize resource waste and maximize throughput. Moreover, **Deep Q Networks (DQN)** can be used to determine optimal network slicing strategies, ensuring that different applications (e.g., emergency services, video streaming, etc.) receive appropriate resources without interference.

Adaptation to Dynamic Environments

- **Traditional Approach**: Traditional D2D communication systems often struggle to adapt to rapidly changing environments, such as when devices are moving, or network conditions fluctuate.
- **RIS-Assisted Approach**: RIS systems offer more flexibility, as they can dynamically reconfigure to accommodate changing channel conditions. However, the reconfiguration process needs to be managed effectively to avoid performance degradation.
- **AI/ML/DL Approach**: AI and ML techniques can continuously monitor the network environment and adapt RIS configurations accordingly. **Deep Learning** models, such as **Recurrent Neural Networks (RNNs)**, can be employed to track the state of the network over time, learning the best adaptation strategies for RIS. Additionally, **Reinforcement Learning (RL)** can enable RIS to learn from its environment and autonomously adjust to optimize communication in real-time.

Table 1. Efficiency parameters in RIS-Assisted D2D communication

Parameter	Traditional Systems	RIS-Assisted Systems	AI, ML, and DL Enhanced RIS Systems
Channel Estimation Accuracy	Moderate, Static	Improved with RIS	High, Adaptive, Predictive (via DL)
Beamforming Efficiency	Fixed, Static	Dynamic with RIS	Highly Dynamic and Optimized (via ML/RL)
Interference Mitigation	Limited to Fixed Methods	Mitigated via RIS	Optimized in Real-time (via AI/ML)
Energy Consumption	High, due to active relays	Low, passive operation	Optimized dynamically (via AI/ML)
Reliability in Complex Environments	Reduced in urban/ Rural areas	Enhanced via RIS	Significantly Enhanced (via AI-driven optimization)
Security and Privacy	Vulnerable to Eavesdropping	Potential Vulnerabilities	Secure, AI-based detection & encryption
Real-time Adaptation	Limited adaptability	Moderate adaptability	High adaptability (via AI/ML)
Network Capacity	Limited by static resources	Improved with RIS placement	Highly Optimized, Predictive (via AI/ML)
Network Coverage	Restricted by physical infrastructure	Extended via RIS reflection	Enhanced in Dynamic Conditions (via ML/DL)
Resource Allocation	Static, based on network design	Static but better than traditional	Highly Adaptive, Optimized (via RL)

AI, ML, and DL techniques are crucial for unlocking the full potential of RIS in D2D communication systems. By incorporating these technologies, RIS systems can dynamically adapt to changing environments, optimize energy consumption, manage interference, and ensure secure communications, all while maximizing resource efficiency. The integration of intelligent systems like AI/ML and DL into RIS-assisted D2D networks will pave the way for smarter, more robust communication infrastructures capable of meeting the demands of modern wireless applications.

Table 2 outlines the main challenges that need to be addressed and some potential solutions to advance RIS-assisted D2D communications.

Table 2. Key challenges and potential solutions

Challenge	Description	Potential Solutions
Channel State Information (CSI) Estimation	Accurate and real-time CSI is essential for RIS configurations, but obtaining it in dynamic D2D environments is challenging.	1.**AI/ML-Based CSI Estimation**: Leveraging AI/ML algorithms for more accurate and faster CSI estimation. 2.**Low-ComplexityAlgorithms**: Developing algorithms with reduced computational overhead for CSI estimation in real-time.
Beamforming and Phase Shift Optimization	Optimizing the beamforming and phase shifts in RIS configurations can be computationally expensive, especially in large-scale networks.	1.**AI-Powered Beamforming**: Using machine learning techniques to optimize beamforming in real-time with minimal resources. 2.**Distributed Optimization**: Designing distributed beamforming algorithms to reduce centralized computation load.
Energy Efficiency of RIS Devices	RIS devices are passive, but their large-scale deployment may still incur significant energy consumption for signal reflection and control.	1.**Low-Power RIS Design**: Developing RIS devices with lower power consumption. 2. **Energy-Aware Algorithms**: Using energy-efficient algorithms for RIS configuration and management.
Security and Privacy Concerns	The reconfigurable nature of RIS poses risks in terms of security and privacy, especially in sensitive D2D applications like healthcare and financial systems.	1.**RIS-Based Encryption**: Implementing encryption techniques directly on RIS to enhance communication security. 2.**AI for Intrusion Detection**: Using AI for detecting anomalous behaviors in RIS-assisted networks.
Scalability in Dense Urban Environments	Large-scale deployment of RIS in densely populated urban environments faces challenges related to signal blockage, interference, and large-scale coordination.	1.**Distributed RIS Networks**: Implementing distributed RIS networks for better coverage and interference mitigation. 2.**Autonomous Coordination**: Using AI to autonomously coordinate RIS devices for scalable deployment.
Integration with Legacy Systems (4G/5G)	Integrating RIS technology with existing 4G and 5G infrastructures in D2D communications while maintaining backward compatibility is challenging.	1.**Backward-Compatible RIS Designs**: Developing RIS architectures that can work alongside existing communication systems. 2.**Hybrid Networks:** Leveraging hybrid networks that combine 5G/6G with RIS for seamless integration.
Latency and Real-Time Communication	Achieving ultra-low latency is critical for D2D communication in mission-critical applications (e.g., autonomous driving, healthcare) but challenging in RIS networks.	1.**Low-Latency AI Algorithms**: Implementing AI algorithms optimized for low-latency communications. 2.**Edge Computing**: Integrating RIS with edge computing for reduced latency in communication.

continued on following page

Table 2. Continued

Challenge	Description	Potential Solutions
Dynamic Network Topology	The network topology in RIS-assisted D2D communication can change rapidly due to mobility, making it difficult to maintain optimal performance.	1.**AI for Topology Adaptation**: Using AI and machine learning to dynamically adapt to changing network topologies. 2.**Mobility-Aware Algorithms**: Designing algorithms that can predict and adapt to mobile devices' movement patterns.
RIS Deployment and Cost	Deploying RIS at scale can be costly, both in terms of hardware and the complexity of deployment processes.	1.**Cost-Effective RIS Designs**: Developing low-cost, scalable RIS hardware. 2.**Deployment Optimization Algorithms**: Using optimization algorithms for cost-effective RIS deployment strategies.
Interference Management	Interference between RIS-assisted and non-RIS-assisted D2D communications or other networks could degrade performance.	1.**RIS-Assisted InterferenceMitigation**: Developing RIS configurations that minimize interference. 2.**AI for Interference Prediction**: Using AI to predict and mitigate interference in real-time.

AI and Machine Learning (ML) are indispensable in addressing challenges such as Channel State Information (CSI) estimation, beamforming optimization, and security in RIS-assisted D2D communications. These technologies enable adaptive, real-time solutions by dynamically analyzing network conditions and optimizing resources effectively, thus enhancing overall system performance. Energy efficiency and scalability remain critical challenges for large-scale deployment in urban environments. The development of low-power RIS devices and energy-efficient algorithms will be crucial for sustainable and scalable network implementations.

Security and privacy concerns continue to be a major focus in RIS-assisted networks due to their wireless and open nature. Future advancements could involve encryption mechanisms embedded directly at the RIS level and the deployment of AI-based intrusion detection systems to safeguard networks against potential threats. Furthermore, latency and real-time communication are essential for mission-critical applications such as disaster recovery and autonomous systems. AI-driven latency reduction techniques, combined with edge computing, can ensure rapid response and improved real-time communication in such critical scenarios.

FUTURE DIRECTIONS FOR RIS-ASSISTED D2D COMMUNICATIONS

The future of RIS-assisted D2D communications offers immense potential to revolutionize wireless networks. One critical area for advancement is the development of advanced CSI estimation techniques. Accurate and real-time CSI estimation is essential for optimizing RIS performance, and future research should focus on AI-powered algorithms that utilize historical data and real-time feedback to improve CSI accuracy, especially in dynamic environments. Similarly, integrating AI with edge computing will play a vital role in minimizing latency and enabling rapid decision-making by processing RIS configurations locally at edge nodes.

Energy efficiency will remain a key focus area, particularly for large-scale deployments. Innovations such as passive and semi-passive RIS designs and energy-harvesting technologies like solar panels could provide sustainable solutions. Additionally, the design of energy-efficient control algorithms will optimize RIS operations and reduce power consumption. Security and privacy will continue to be major priorities. Future work could explore the use of encryption protocols embedded in RIS hardware and AI-based anomaly detection models to enhance network integrity and safeguard against cyber threats.

RIS systems will also need to be interoperable with emerging technologies such as Massive MIMO, millimeter-wave communication, and Terahertz (THz) frequencies. Research should focus on designing RIS systems that can seamlessly operate across these paradigms while maintaining compatibility with evolving wireless standards. Another promising direction is the use of RIS in dynamic and mission-critical scenarios. Applications such as disaster recovery, autonomous vehicles, and industrial IoT demand ultra-reliable and adaptive communication, and future research could focus on self-healing RIS networks that dynamically adjust to maintain communication links under adverse conditions.

Scalability is a significant challenge, particularly in dense urban environments. Research could focus on designing cost-effective, high-performance RIS networks with minimal deployment barriers. Multi-panel RIS systems that collaborate to provide optimized coverage in dense areas could significantly enhance urban deployments. Standardization and protocol development are equally important for widespread adoption. Efforts should be made to define operational guidelines, spectrum-sharing mechanisms, and regulatory frameworks to facilitate the global adoption of RIS-assisted D2D networks.

Performance optimization for 6G networks is another critical research direction. RIS is expected to be a cornerstone technology for 6G, and optimizing its capabilities for ultra-high bandwidth, low latency, and massive connectivity requirements will be crucial. Techniques like joint optimization of RIS and D2D protocols could unlock unprecedented performance levels. Furthermore, RIS has great potential in IoT and smart city applications. Deploying RIS in smart city infrastructure could enhance connectivity, reduce energy consumption, and enable seamless communication between IoT devices, paving the way for intelligent urban ecosystems.

Integration with 6G and Beyond

As the telecommunications industry prepares for 6G networks, RIS technology is expected to become foundational for achieving ultra-high-speed, low-latency services. RIS will play a crucial role in enabling smart radio environments where D2D communication can occur seamlessly, with almost no latency. According to Saad et al. (2020) and Zhang et al. (2021), RIS is seen as a key enabler for realizing 6G's vision by enhancing signal propagation, increasing network capacity, and reducing interference.

The integration of RIS with D2D communications will be particularly important for applications requiring real-time data exchange, such as autonomous systems, smart cities, and advanced Internet of Things (IoT) systems. The massive deployment of RIS in 6G will offer unprecedented levels of network performance, supporting the vision of ubiquitous, ultra-low-latency D2D communication.

AI-Driven RIS for 6G D2D Networks and their Optimization

As we transition to 6G networks, AI-driven RIS technology will be integral to creating ultra-reliable, low-latency communication systems. Saad et al. (2020) indicated that AI-based optimization approaches would help achieve the desired performance levels for D2D communication in 6G, particularly in unpredictable environments. Machine learning models can predict traffic patterns, optimize resource allocation, and help RIS adjust in real-time to changing conditions, thus improving overall system performance.

Shen et al. (2021) also discussed how predictive modeling powered by AI could assist in forecasting network demands and adjusting RIS configurations accordingly, further improving the reliability of D2D communication.

The integration of Artificial Intelligence (AI), particularly machine learning (ML) and reinforcement learning (RL), into RIS will be one of the major advancements in D2D communications. AI-driven algorithms can dynamically optimize beamforming and phase shifts, which are essential for improving the performance of D2D networks. Traditional approaches to RIS configuration often struggle with complex and dynamic environments, but AI allows for adaptive real-time optimization.

Taha et al. (2020) proposed leveraging deep learning for optimizing large-scale RIS configurations, a concept that can significantly enhance the flexibility and effectiveness of RIS in D2D communications. By applying AI, RIS can autonomously adjust to environmental changes, interference, and network congestion, improving communication reliability and energy efficiency in real-time. This level of optimization will make D2D communications more robust, scalable, and cost-effective.

RIS in Millimeter-Wave (mmWave) D2D Networks

The deployment of millimeter-wave (mmWave) frequencies in 5G and 6G networks presents a significant opportunity to enhance data transmission due to the large available bandwidth. However, mmWave signals are highly susceptible to blockage from obstacles, such as buildings or even people. RIS technology can overcome this challenge by providing alternative paths for signal propagation through its reconfigurable surfaces.

Zhao et al. (2020) highlighted the potential of RIS in improving D2D communication at mmWave frequencies, helping to mitigate the negative effects of physical obstructions on signal coverage. Future research will likely focus on optimizing RIS configurations specifically for mmWave and even higher frequencies like terahertz (THz) communications, which are expected to be critical for future high-speed networks.

RESEARCH OPENINGS IN FUTURE DIRECTIONS FOR RIS-ASSISTED D2D COMMUNICATIONS

As Reconfigurable Intelligent Surfaces (RIS) continue to evolve, there are several promising research openings across key future directions that will shape the advancement of RIS-assisted Device-to-Device (D2D) communications. Below, we outline the specific research opportunities within each identified future direction:

For AI-Driven RIS for 6G D2D Networks

- **AI-Powered Predictive Models**: AI can be used to predict network traffic, interference, and user mobility patterns. Developing predictive models for RIS deployment in 6G D2D communication networks will help improve performance by preemptively adjusting configurations before issues arise.
- **AI for QoS and Resource Management**: A research opening exists in using AI to optimize Quality of Service (QoS) and resource allocation, ensuring

that RIS is used optimally in terms of power consumption and computational load while maintaining high performance for D2D communication.

- **AI-Based Security Models for RIS Networks**: As AI algorithms play a larger role in RIS optimization, securing the AI-driven systems from attacks, such as model poisoning or adversarial inputs, will be a significant area of research. Ensuring AI robustness in RIS-controlled environments is critical for the security of the network.

In AI-Driven Optimization

- **Real-Time Optimization Algorithms**: There is a need for developing AI-based algorithms that can dynamically adjust RIS configurations in real-time. Research could focus on creating reinforcement learning or deep learning models for large-scale and complex environments where traditional methods fail to adapt quickly enough.
- **Hybrid AI Models**: Exploring hybrid AI models that combine multiple approaches (e.g., combining supervised learning for initial configurations and reinforcement learning for continuous optimization) could yield better performance in RIS-assisted D2D communications.
- **Low-Complexity AI Algorithms**: Many current AI optimization methods are computationally expensive. Research into developing lightweight AI algorithms that maintain high optimization efficiency without heavy computational requirements will be crucial for practical deployment.
- **Cross-Layer Optimization**: AI-based optimization could be applied across multiple layers of the network (physical layer, MAC layer, network layer) to ensure global optimization, not just focusing on the RIS configurations alone but also on traffic management and routing.

Integration with 6G and Beyond

- **RIS Integration with Advanced 6G Technologies**: Research on how RIS can be seamlessly integrated with 6G technologies, such as terahertz (THz) communication and quantum communication, is needed. This includes understanding how RIS can be adapted to high-frequency bands and their impact on signal propagation at these frequencies.
- **Interoperability with Legacy Systems**: While focusing on 6G, researchers should investigate how RIS technology can work in conjunction with existing 4G and 5G systems, providing backward compatibility and seamless communication for devices transitioning to 6G.

- **RIS for Ultra-Reliable Low Latency Communication (URLLC)**: A key focus for 6G is ultra-reliable low-latency communications. Research into how RIS can improve reliability and reduce latency in D2D communications, especially in high-demand and critical use cases (e.g., autonomous vehicles), is critical.

For RIS in Millimeter-Wave (mmWave) D2D Networks

- **RIS for mmWave Propagation in Urban Environments**: Given that mmWave signals are prone to blockages in urban settings, there is an opportunity for research into the deployment of RIS in such environments to enhance line-of-sight and overcome blockages.
- **RIS for Dynamic Environments**: Future research could explore RIS deployment strategies for dynamic environments, such as moving vehicles or drones, where traditional beamforming methods struggle to maintain consistent coverage.
- **Optimizing RIS at mmWave Frequencies**: Investigating novel RIS materials and design principles optimized specifically for mmWave frequencies, to reduce losses and improve signal propagation over shorter ranges, would be valuable for D2D communications.

Enhanced RIS Design for Scalability

- **Low-Cost RIS Manufacturing**: Research into developing cost-effective materials and scalable manufacturing techniques for RIS will be critical for widespread deployment in large-scale D2D networks. Innovations in fabrication techniques, such as printable RIS or low-cost meta-materials, offer significant opportunities.
- **Scalable Architecture Design**: There is an opening for research into scalable RIS designs that can be implemented across large geographical areas without compromising performance. This could involve modular designs that can be easily expanded or reconfigured based on demand.
- **Energy-Efficient RIS Solutions**: Since RIS can enhance network performance while also reducing energy consumption, researchers should focus on designing energy-efficient RIS technologies that require less power to operate, further improving their viability for large-scale deployment.

For Integration with Other 6G Technologies

- **RIS and Quantum Computing for Secure D2D Communication**: Quantum computing offers the potential for ultra-secure communication. Research into how RIS and quantum computing can complement each other in D2D networks, especially in terms of encryption and secure key management, presents a unique research area.
- **Synergy between RIS and Terahertz Communications**: As the demand for high-speed, high-capacity wireless communications increases, terahertz bands are becoming a crucial frequency range. Exploring the role of RIS in improving the propagation of terahertz signals in D2D communications is an exciting research avenue.
- **RIS-Assisted Federated Learning**: The combination of RIS and Federated Learning (FL), a decentralized machine learning model, could significantly improve the adaptability and performance of D2D communications by enabling distributed, privacy-preserving optimization of RIS configurations across the network.

Table 3. Research openings with respect to future directions

Future Direction	Research Openings
AI-Driven Optimization	• **Real-Time Optimization Algorithms**: Development of AI models for dynamic RIS configuration in real-time. • **Hybrid AI Models**: Combining multiple AI approaches (e.g., supervised learning and reinforcement learning). • **Low-Complexity AI Algorithms**: Design lightweight AI models that ensure high optimization efficiency with minimal computational cost. • **Cross-Layer Optimization**: AI for optimizing across multiple network layers (physical, MAC, network layers).
Integration with 6G and Beyond	• **RIS Integration with Advanced 6G Technologies**: Exploring RIS's role in THz communication, quantum communication, and high-frequency bands. • **Interoperability with Legacy Systems**: Ensuring RIS works with 4G and 5G networks. • **RIS for Ultra-Reliable Low Latency Communication (URLLC)**: Investigating how RIS improves reliability and latency for critical 6G applications (e.g., autonomous systems).
RIS in Millimeter-Wave (mmWave) D2D Networks	**RIS for mmWave Propagation in Urban Environments**: Enhancing signal coverage in urban settings with RIS. **RIS for Dynamic Environments**: RIS deployment strategies for moving vehicles, drones, and other dynamic use cases. **Optimizing RIS at mmWave Frequencies**: Developing RIS materials and designs optimized for mmWave frequencies to reduce losses and improve signal propagation.

continued on following page

Table 3. Continued

Future Direction	Research Openings
AI-Driven RIS for 6G D2D Networks	**AI-Powered Predictive Models**: Creating AI-based models to predict network traffic, interference, and mobility patterns. **AI for QoS and Resource Management**: Optimizing Quality of Service (QoS) and resource allocation using AI. **AI-Based Security Models**: Developing security measures for AI-driven RIS systems, ensuring robustness against attacks (e.g., model poisoning, adversarial attacks).
Enhanced RIS Design for Scalability	**Low-Cost RIS Manufacturing**: Researching cost-effective materials and scalable production techniques for RIS. **Scalable Architecture Design**: Designing scalable RIS solutions for large D2D networks. **Energy-Efficient RIS Solutions**: Developing energy-efficient RIS technologies that consume minimal power and optimize deployment.
Integration with Other 6G Technologies	**RIS and Quantum Computing for Secure D2D Communication**: Investigating the integration of RIS with quantum computing for secure communication. **Synergy between RIS and Terahertz Communications**: Exploring RIS to improve signal propagation for terahertz frequencies in D2D networks. **RIS-Assisted Federated Learning**: Investigating the use of federated learning for distributed optimization of RIS configurations across D2D networks.

The future directions for RIS-assisted D2D communications are expansive, offering numerous opportunities for groundbreaking research and technological advancements. Integrating Artificial Intelligence (AI) into RIS systems presents a promising avenue, as AI-driven algorithms can optimize real-time configurations, adapt to dynamic network conditions, and enable efficient resource management. This synergy between RIS and AI is expected to significantly enhance the performance and reliability of D2D communications. Advancing the integration of RIS with emerging 6G technologies is also a key direction. This involves exploring how RIS can complement other 6G enablers such as terahertz communication, massive MIMO, and edge computing to deliver ultra-reliable, low-latency, and energy-efficient D2D networks. These efforts related to the future scope of RIS research will shape the future of D2D interactions across diverse applications, including IoT, smart cities, and autonomous systems.

CONCLUSION

Reconfigurable Intelligent Surfaces (RIS) are a groundbreaking innovation in wireless communications, especially in Device-to-Device (D2D) interactions. By leveraging programmable meta-materials, RIS can enhance spectrum efficiency, improve reliability, and reduce energy consumption in D2D networks. Its ability to manipulate electromagnetic waves ensures robust communication links even in high-interference, high-path-loss environments, significantly boosting overall network performance. However, challenges like complex beamforming, phase shift optimization, accurate channel state information (CSI) in dynamic conditions, scalability, and security concerns must be addressed for practical deployment. Integrating machine learning (ML) and artificial intelligence (AI) with RIS-assisted D2D networks can significantly improve system performance and operational efficiency. AI algorithms enable real-time optimization of RIS configurations, dynamic adaptation to network variations, and efficient resource allocation. As telecommunications advance toward 6G and beyond, the synergy of RIS, D2D communication, and AI opens doors to transformative applications in IoT, smart cities, and autonomous systems. RIS-assisted D2D networks promise enhanced performance, reliability, and innovative services, standing at the forefront of future wireless networks. Continued research and development will deepen theoretical understanding, overcome technical barriers and accelerate practical implementations, revolutionizing device communication in an increasingly connected & intelligent world.

REFERENCES

Abeywickrama, S., Zhang, R., & Yuen, C. (2020). Intelligent reflecting surface: Practical phase shift model and beamforming optimization. *IEEE Transactions on Communications*, 68(9), 5844–5857. DOI: 10.1109/TCOMM.2020.3001125

Asadi, A., Wang, Q., & Mancuso, V. (2014). A survey on device-to-device communication in cellular networks. *IEEE Communications Surveys and Tutorials*, 16(4), 1801–1819. DOI: 10.1109/COMST.2014.2319555

Basar, E., Renzo, M. D., Rosny, J. D., Debbah, M., Alouini, M. S., & Zhang, R. (2019). Wireless communications through reconfigurable intelligent surfaces. *IEEE Access : Practical Innovations, Open Solutions*, 7, 116753–116773. DOI: 10.1109/ACCESS.2019.2935192

Basharat, S., Khan, M., Iqbal, M., Hashmi, U. S., Zaidi, S. A. R., & Robertson, I. (2022). Exploring reconfigurable intelligent surfaces for 6G: State-of-the-art and the road ahead. *IET Communications*, 16(13), 1458–1474. DOI: 10.1049/cmu2.12364

Bjornson, E., Sanguinetti, L., & Debbah, M. (2019). A primer on massive MIMO. *Foundations and Trends® in Signal Processing*, 11(3-4), 154-355. DOI: 10.1049/cmu2.12364

Chen, Y., Chen, X., & Li, L. (2021). Artificial intelligence meets intelligent reflecting surface: The age of machine learning-assisted wireless communications. *IEEE Wireless Communications*, 28(2), 138–144.

Cui, T., Qi, M., Wan, X., Zhao, J., & Cheng, Q. (2014). Coding metamaterials, digital metamaterials and programmable metamaterials. *Light, Science & Applications*, 3(10), e218. DOI: 10.1038/lsa.2014.99

Di Renzo, M., Zappone, A., Debbah, M., Alouini, M. S., Yuen, C., & Tretyakov, S. A. (2020). Smart radio environments empowered by reconfigurable AI meta-surfaces: An idea whose time has come. *EURASIP Journal on Wireless Communications and Networking*, 2020(1), 129. DOI: 10.1186/s13638-019-1438-9

Ding, G., Zhang, Z., Liu, T., Wu, Q., Yu, F. R., & Song, Z. (2020). Risk-based routing optimization for mission-critical disaster recovery with intelligent reflective surfaces. *IEEE Transactions on Vehicular Technology*, 69(10), 11603–11618.

Elbir, A. M., Papazafeiropoulos, A., & Ioannidis, M. (2020). Deep learning for beamforming optimization in reconfigurable intelligent surfaces-based multi-user MIMO systems. *IEEE Transactions on Communications*, 68(8), 5186–5195.

Fodor, G., Dahlman, E., Mildh, G., Parkvall, S., Reider, N., Miklós, G., & Turányi, Z. (2012). Design aspects of network assisted device-to-device communications. *IEEE Communications Magazine*, 50(3), 170–177. DOI: 10.1109/MCOM.2012.6163598

Gong, S., Lu, X., & Khoshnevisan, M. (2020). Toward smart wireless communications via intelligent reflecting surfaces: A contemporary survey. *IEEE Communications Surveys and Tutorials*, 22(3), 1270–1298. DOI: 10.1109/COMST.2020.3004197

Han, Y., Huang, C., & Debbah, M. (2020). Reconfigurable intelligent surface assisted wireless communications: Antenna design, prototyping, and experimental results. *IEEE Access : Practical Innovations, Open Solutions*, 8, 120073–120084.

Huang, C., Alexandropoulos, G. C., Yuen, C., & Debbah, M. (2020). Indoor signal focusing with reconfigurable intelligent surfaces. *IEEE Wireless Communications Letters*, 9(12), 1985–1989.

Huang, C., Hu, S., Alexandropoulos, G. C., Zappone, A., Yuen, C., Zhang, R., & Debbah, M. (2020). Holographic MIMO surfaces for 6G wireless networks: Opportunities, challenges, and trends. *IEEE Wireless Communications*, 27(5), 118–125. DOI: 10.1109/MWC.001.1900534

Khalid, Z., Ding, M., & Zhang, R. (2021). Wireless communications with large intelligent surfaces: Opportunities and challenges. *IEEE Transactions on Communications*, 69(5), 3125–3136.

Li, X., He, Z., & Wu, Q. (2021). Reconfigurable intelligent surface aided wireless communication: How many reflecting elements are needed? *IEEE Wireless Communications Letters*, 10(1), 68–71.

Liang, J., Zhang, H., & Di Renzo, M. (2020). RIS-assisted wireless communications: Prototyping, challenges, and future directions. *IEEE Wireless Communications*, 27(6), 93–99.

Lin, X., Andrews, J. G., Ghosh, A., & Ratasuk, R. (2014). An overview of 3GPP device-to-device proximity services. *IEEE Communications Magazine*, 52(4), 40–48. DOI: 10.1109/MCOM.2014.6807945

Liu, F., Guo, H., & Zhang, Z. (2020). RIS-enhanced communications: Performance evaluation and beamforming optimization. *IEEE Transactions on Wireless Communications*, 19(11), 7340–7353.

Liu, Y., Zeng, M., & Zhang, R. (2020). Deep reinforcement learning for intelligent reflecting surface-assisted device-to-device communications. *IEEE Transactions on Wireless Communications*, 19(8), 5378–5387.

Luo, H., & Yuen, C. (2021). Optimizing power and beamforming in RIS-assisted MISO systems. *IEEE Transactions on Communications*, 69(3), 1653–1665.

Saad, W., Bennis, M., & Chen, M. (2020). A vision of 6G wireless systems: Applications, trends, technologies, and open research problems. *IEEE Network*, 34(3), 134–142. DOI: 10.1109/MNET.001.1900287

Shen, H., Zhao, T., & Zhang, R. (2021). Phase-shift design for intelligent reflecting surfaces in wireless communications: A new approach with SVD-based optimization. *IEEE Transactions on Communications*, 69(5), 3393–3407.

Smith, D. R., Padilla, W. J., Vier, D. C., Nemat-Nasser, S. C., & Schultz, S. (2004). Composite medium with simultaneously negative permeability and permittivity. *Physical Review Letters*, 84(18), 4184–4187. DOI: 10.1103/PhysRevLett.84.4184 PMID: 10990641

Taha, A., Alrabeiah, M., & Alkhateeb, A. (2020). Enabling large intelligent surfaces with compressive sensing and deep learning. *IEEE Access : Practical Innovations, Open Solutions*, 8, 79682–79699.

Tang, W., Chen, M., & Zeng, Y. (2021). Wireless communications with reconfigurable intelligent surfaces: Path loss modeling and experimental measurement. *IEEE Transactions on Wireless Communications*, 20(1), 421–439. DOI: 10.1109/TWC.2020.3024887

Tang, W., Dai, J. Y., Shen, H., & Zeng, Y. (2020). Programmable metasurface-based RF chain-free 8PSK wireless transmitter. *Electronics Letters*, 56(11), 564–567.

Tang, W., Zeng, Y., & Wu, Q. (2021). Intelligent reflecting surface-enhanced wireless networks: From theory to practical implementation. *IEEE Communications Magazine*, 59(6), 60–66.

Wang, B., Zhang, L., Zhang, J., & Liu, Z. (2015). Resource allocation for D2D communication underlaying OFDMA cellular networks using coalitional games. *IEEE Transactions on Wireless Communications*, 14(6), 3259–3272.

Wang, B., Zhang, L., Zhang, J., & Liu, Z. (2015). Resource allocation for D2D communication underlaying OFDMA cellular networks using coalitional games. *IEEE Transactions on Wireless Communications*, 14(6), 3259–3272.

Wang, J., Feng, W., & Niu, Z. (2020). Beamforming optimization for intelligent reflecting surface-assisted communications with channel state information uncertainty. *IEEE Transactions on Wireless Communications*, 19(10), 6719–6733.

Wu, Q., & Zhang, R. (2019). Towards smart and reconfigurable environment: Intelligent reflecting surface aided wireless network. *IEEE Communications Magazine*, 58(1), 106–112. DOI: 10.1109/MCOM.001.1900107

Xu, J., Wang, W., & He, S. (2021). Reconfigurable intelligent surfaces for secure communications in 6G: Challenges, opportunities, and research directions. *IEEE Communications Magazine*, 59(7), 82–88.

Xu, Y., Liu, L., Zhou, S., & Niu, Z. (2015). Opportunistic device-to-device communications: A queueing perspective. *IEEE Journal on Selected Areas in Communications*, 33(1), 68–80.

Xu, Y., Liu, L., Zhou, S., & Niu, Z. (2015). Opportunistic device-to-device communications: A queueing perspective. *IEEE Journal on Selected Areas in Communications*, 33(1), 68–80.

Yadav, A., Singh, R. K., & Li, J. (2021). Meta-material inspired reconfigurable intelligent surfaces for wireless communication: A comprehensive survey. *IEEE Access : Practical Innovations, Open Solutions*, 9, 103558–103579.

Yu, X., Xu, D., & Zhang, R. (2020). Smart and reconfigurable environment with intelligent reflecting surfaces: A tutorial on modeling and analysis. *IEEE Communications Surveys and Tutorials*, 23(2), 942–973.

Zappone, A., Di Renzo, M., & Debbah, M. (2020). Wireless networks design in the era of deep learning: Model-based, AI-based, or both? *IEEE Transactions on Communications*, 67(10), 7331–7376. DOI: 10.1109/TCOMM.2019.2924010

Zeng, Y., Zhang, R., & Wu, Q. (2021). Reconfigurable intelligent surfaces in wireless communication: Recent advancements and future trends. *IEEE Communications Magazine*, 59(7), 20–25.

Zhang, H., Wang, L., & Huang, C. (2021). Towards 6G wireless networks: Prototyping of reconfigurable intelligent surfaces. *IEEE Wireless Communications*, 28(2), 86–92.

Zhang, Q., Zhang, R., & Wu, Q. (2021). Beyond the era of massive MIMO: Intelligent reflecting surfaces for 6G wireless networks. *IEEE Transactions on Wireless Communications*, 20(4), 2457–2471.

Zhang, S., Zhang, H., & Di Renzo, M. (2020). Capacity analysis of reconfigurable intelligent surfaces aided MIMO systems. *IEEE Journal on Selected Areas in Communications*, 38(8), 1824–1838. DOI: 10.1109/JSAC.2020.3000814

Zhao, H., Liang, Y., & Niu, Z. (2020). Intelligent reflecting surfaces assisted device-to-device communications in millimeter-wave wireless networks. *IEEE Transactions on Vehicular Technology*, 69(9), 10401–10410.

Zhou, G., Yuan, W., Yu, X., & Zhang, R. (2021). Intelligent reflecting surface: Practical phase shift model and efficient beamforming optimization. *IEEE Communications Letters*, 25(1), 118–122.

Zhou, S., Wang, J., Niu, Z., & Zhang, Y. (2018). Joint resource allocation and power control for D2D communications underlaying cellular networks. *IEEE Transactions on Wireless Communications*, 17(11), 7595–7609.

Chapter 15
Design and Implementation of Self-Powered Reconfigurable Intelligent Surfaces: Applications and Future Trends

C. Manonmani
https://orcid.org/0009-0001-2725-097X
V.S.B. Engineering College, India

Ashok Vajravelu
University Tun Hussein Onn Malaysia, India

S. Vidhya
V.S.B. Engineering College, India

Sheshang Degadwala
Sigma Institute of Engineering, India

ABSTRACT

RISs could improve radio broadcasts by adding virtual channels, for example. This would make the sound better and less noisy. If you change how electromagnetic waves move and behave, you can make things that gather energy, take pictures, and find things better. People can stay safe and private with RISs because they can let or stop certain bands of electromagnetic energy. This article gives a brief history of RIS buildings. It also talks about how the buildings at RIS are planned and put together. Why and how RIS technology can be used, like how it can make Wi Fi links faster, save energy, improve tracking and picture quality, and make monitoring and

DOI: 10.4018/979-8-3693-8099-4.ch015

security stronger. Finally, we discuss some businesspeople issues and attempt to guess how RIS technology will be utilized. We think much progress will be made in this area because RISs can improve many tools and apps.

1. INTRODUCTION

A lot of work went into making smart radios with parts that can be switched out. They can connect to networks that are connected and are far away. Changing any surface in RIS to fit an intelligent radio K. Chen and Y. J. Feng (2022). EMF waves can change direction when they hit something, which is one way to say this. It is in a controlled device block because the channel can be changed. Everything will work better after this is fixed. RIS is an artificial surface made of electromagnetic (EM) material managed by electronics built in. They now have a new, cheap way to send radio data that uses less power and charge. These parts of the surface might be able to send radio waves. These concepts are used in a new study on RIS to find the best possible signal-to-noise ratio (SNR), signal-to-interference ratio (SINR) maximization, physical layer security solutions, cognitive radio applications, and AI solutions (like deep learning) Özdogan, Ö,(2019) Many new studies and thoughts have been put together on how to make RISs work better. There are various ways to write about RISs Ali Araghi, Mohsen Khalily (2022).

They've been called “reconfigurable intelligent surfaces,” “intelligent reflecting surfaces,” and other names. Scientists have also looked into VLC, FSO (open space optics), and mmWave intelligent surfaces. The body parts must be kept safe. News stories also discussed the first RISs were used with OFDM, SM, and SSK W. Saad, M. Bennis, (2020). They have been looking into channel models for several years, improving the signal-to-interference ratio (SINR) and the best mix of active and passive beamforming Nadeem, Q.-U.-A (2020). As was already said, studies have also been done on the uplink frequency effectiveness, the average data rate, and the failure rate of RISs when they are used for transmission and reception in new areas such as mmWave, FSO, VLC systems, and uncrewed aerial vehicles (UAV). Some experts have also looked into how machine learning tools, physical layer security solutions, and intelligent surfaces could be helpful to for mmWave and terahertz uses over the past few years [D. M. Pozar,(1997), W. T. Chen, A. Y. Zhu,(2018)].

It works better for RIS so that you can share data quickly with MIMO. mmWave lets us talk in new ways. When electricity waves hit an RIS, they can change direction. After this, radio waves might move differently. People are discussing RIS because it's better to speak over FSO and mixed RF/FSO systems than normal MIMO systems. When RIS is mixed with RF/FSO, it can gain much more. It costs less and is easier to set up than networks with more sources that boost and forward. This plug-in lets you get both VLC and VLC-RF. You can send more, do more, and get around faster with them. With this mix, I can see well in places that change a lot, like next to cars. This could make your Wi-Fi apps work again. In some parts of the network [Bito, Ryan Bahr,(2017), C.J. Vaca-Rubio, (2020)], there are some tricky links, but it doesn't slow down. A new study says that RIS can make it easier for UAVs to talk to each other and use less power. They do this when they need to fix the engine or more than one thing simultaneously. It was written Long, H.; Chen, M.; (2020) to keep it safe and use less power. They sent more power, changed the RIS's phase shift, and simultaneously turned the UAV in a different direction to make this happen.

Many people have different thoughts on the new RIS and metasurface tools. It's good that this study shows all of them. The study also looks at new ways to use RIS, like active RISs and their signal models, and rates how well they work compared to a form that isn't active. People talk about how the link budget changes to show how different they are. Some bad things happen when RIS is used. Let us show you what would be good if these bad things were fixed. Everything will work better after this. A new type of RIS that can be controlled differently is made with field-programmable gate arrays (FPGAs). J. Ye, S. Guo, and M.-S. Alouini, (2020) That's all there is to say about tools that let people talk to each other. They are put into different groups based on how well they do various jobs. This has already been looked into, and more will be done tomorrow.

2. LITERATURE REVIEW

This part talks about what has already been said about the IRS's technology, terms, and the good and bad points of a radio interview that the IRS helped with.

2.1. Historical Concept of Intelligent Reflecting Surface Technology

Look at Table 1 and Figure 1 to get a sense of what took place. By turning FSS on and off, you can change the EMF waves. After this, they might do other things. Instead of FSS, two-dimensional metamaterials that could work at different frequencies could be made. They talked about many ways to use 2D metasurfaces. They also discussed how waveguides can control metasurfaces to make them work better at terahertz frequencies and how terahertz devices can hold and move electromagnetic energy between two metasurfaces. Radio waves can pass through only two dimensions of metasurfaces Ntontin, K.;.(2020) People also talk about the radio bands they work on. These surfaces don't take up as much room as 3D surfaces, so it's not as bad for them. The SMMs can be tuned based on metasurfaces because they are on the walls. You can use this to increase the volume or make the sound bigger. Metamaterials (MM) can now change the phases 0 and π to a binary 0 and 1.

These can be used to change how the EM works. This works better now that the metamaterials tool is used. It also talked about a PIN diode-based metasurface that could be changed. This was said in 2016, and parts of changed reflect arrays that aren't used were made public. With a big bright surface, you can go even farther than MMO. Sixteen Many people talked about echo, transfer, and how sound moves when mirrored. LISA, an enormous bright screen on the ground, was also shown. O. Abari (2017) A hype surface was shown that can be used in IoT devices to talk to each other and be managed without a LoS. Some people believe that the direction of the electromagnetic field (EM) could be changed by changing the phases of surfaces that reflect light.

Figure 1. IRS timeline.

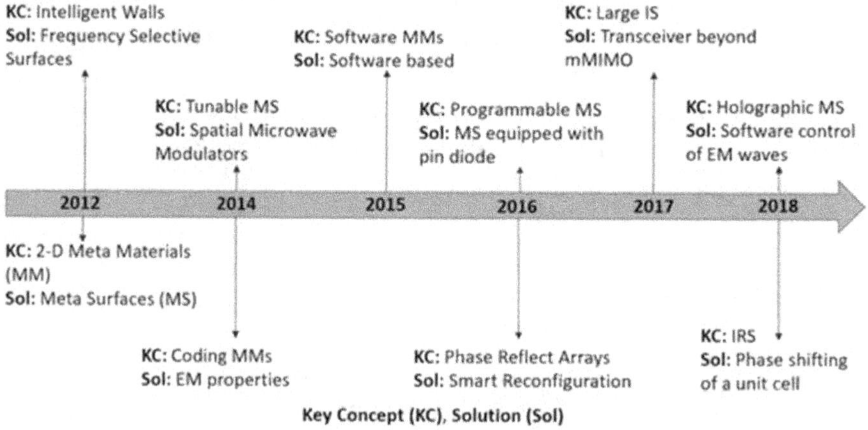

Table 1. IRS history.

Important Ideas	Answer:	Year
Smart walls	Flat surfaces that can pick up frequencies	2012
Metamaterials that are flat and round	"Meta-Surface"	2012
Tunable MS	Modulators for microwaves in space	2014
Coding MMs	Properties of EM	2014
Software for MMs	Based on software	2015
MS with programming	MS with a pin diode	2016

continued on following page

Table 1. Continued

Important Ideas	Answer:	Year
Arrays of Phase Reflections	Smart changes	2016
Big IS	That is more than MMO	2017
The Holographic MS	Software is in charge of EM waves.	2018
IRS	Moving the unit cell's phase	2018

2.2. Nomenclature and Literature Terminologies

Even though the term "IRS" is used, the main idea can be applied to any passively intelligent or software-controlled metasurface that can change how the signal gets to the user. This lets everyone change how their Wi-Fi works. (J. Choi, G. Kwon 2021) There are various ways to write "IRS." Some of them are talked about in a new study paper. Tax experts use these words, like those in Surface that are smart and can be changed (RIS): Software can change radio waves. It looks like wallpaper and is very thin and cheap. Limes bright surface (LIMS): Dang, J.; Zhang, Z.;(2020) Besides these, many other low-cost metamaterial antennas can change the signal phase to reflect them without changing them. When people look at the stream, it gets bigger. SDMS stands for " software-defined metasurface." The EM waves that hit it are very different now. Their length, direction, power, and phase are a few examples. There shouldn't be any lines in the world. The big smart surface (LIS): Alluhaibi, O.(2016) It, along with a big active radio array and special RF lines, helps the system use less power.

Live radio bands are not required. It's called a PIS, which stands for passive intelligent surface. It gets texts and sends them back in a soft voice. Something fresh and green. It can work for a long time with five networks. There are groups of good thoughts: T. J. Cui, (2014) One way for the IRS to fix the issue of signal blocking in mmWave internal communications is to move the incident signal toward the user's goal. This will make it easy for businesses to talk to each other. A lot of these look like the IRS. HMIMO stands for a picture that isn't being used. Why do we need MIMO? With the help of the dielectric particles, it is possible to get what you want and make Wi-Fi better Wi-Fi. The surface is cheap and flat. A lot of lines can talk to each other quickly and easily. There is no difference between LIS, RIS, LIMS, SDMS, and IRS. These less expensive green tools should help with EE and SE. Signs don't work with these. They change the signs' order to reach the right person (Hemadeh, I.A.; El-Hajjar, M.;et 2016). This is the right way to do it. Most of the time, they speed things up without making them louder.

2.3. Benefits of IRS-Aided Wireless Communication

EE and SE think the IRS will make 5G and other future wireless networks work well. This is the best way to change the reflection coefficient, another name for the phase shift. You can share and receive data without radio frequency (RF) lines when you have an IRS. R. J. Mailloux, (2005) How an IRS grows can be changed by moving the mirror pieces around. This makes a ray out of the signal so it can be picked up. It gets stronger (MU) so other people can't mess with it.

The main point of this piece is to explain how an IRS can be used in future wireless networks (FWNs), why it's essential to have a smart radio environment, and the idea of an ongoing link. An IRS that uses radio waves already in use can do this. When they wrote the study, they were excited about what a RIS could do. They talked about its past and how it's different from other tech. Also, the mathematical top level of how well the IRS can do its job was talked about. Someone needs to answer a fundamental study question.

They also talked about other ways it could be used besides networks and 5G in more depth. Setting up a radio network that works with the IRS is simple. It also talks about the IRS and how to deal with them. Besides that, it talks about the pros and cons of the IRS compared to other methods in the same field. Things that can go wrong with plans are also talked about. The HMIMO surface was also talked about. These are bits of copper or metal that work with the IRS. The waves are too short to see. The story said that the IRS can be changed. The new ways it can be used and the speed tests showing how much better IRS-assisted Wi-Fi networks are were also discussed. Its problems and how they will impact 5G and other intelligent networks in the future were also discussed. With the IRS's help, businesses should do all of these things. It lasts a long time and is easy to set up: An IRS is a flat hypersurface made up of cheap parts that don't do anything. One metasurface lets you see a lot of things. Building walls, roofs, homes, and caves with a clear line of sight (LoS) to the base station (BS) is easy.

Since the IRS doesn't have any RF bands, it doesn't need a lot of power. With passive beamforming, you can do and change the following: The phase change in each spreading part should be as small as possible so that the work doesn't get in the way. It's easy to send event and stop signs going the wrong way when you have a lot of them. This makes the Wi-Fi network work better in general. Many things are spread out: Many people use 5G is suitable for making it work better. mmWave bands don't go far enough to be beneficial, either. When there are a lot of base stations, though, it worsens. SINR (signal-to-noise plus interference) is lower now, so there is less flow (Andrews, J.G.; Buzzi, S.;2014). In this case, an IRS is beneficial because it can be used to beamform intelligently to make the system work better, make the signal more robust, and lower the noise power of the receiver. Setting it up doesn't cost much, either. Not as much loss at the cell's edge: People who live close to the edge of the cell get less service and hear more noise.

People close to the edge of a cell network can get a better signal from an IRS because it reduces interference. There are bits in MU that can help with the network facts. The signal is spread out by these bits. The sum rate works better; the IRS can provide better service, and less power is used. Help make new tech: Many IRS workers will use all of these new tools. The Internet of Things can also be used for virtual reality (VR) and picture chat. In this case, the IRS can make many good things happen by telling the right people. You can still share and receive messages even if you can't see each other. An intelligent way to move power without wires is also available. You can keep your network safe and stop other people from messing with it, among other things. Table 2 shows a plan for the book study.

Table 2. IRS technology review related works.

Contribution
Why and how an IRS could be used in 5G and FWNs
Bright surfaces that can be managed and EM at different frequency bands can be made with 2D MS.
The idea behind and creation of an IRS is to create an intelligent radio setting, and there are more talks on how to set it up in FWNs.
Talk about some basic study questions that need to be answered and explain how they could be used in FWNs. Use statistical methods to find the possible performance limit of an IRS.
It works with metals or dielectric particles shorter than a wavelength and is similar to IRS. We will discuss HMIMO surfaces.
Describe briefly the IRS, what makes it better than other technologies that do the same thing, any problems that came up while it was being made, and how it can help FWNs.

2.4. Comparison of IRS with Other Related Technologies

Some things make the Internal Revenue Service (IRS) unique. Here is a list of some of those on the list: These metasurfaces are called IRS, which stands for "internal reflection surfaces." They don't need a single power source for quiet presentations. Antenna-to-digital converters (ADCs), digital-to-analogue converters (DACs), or power amplifiers are unnecessary to boost signals that IRSs copy or add noise. They don't need a lot of power to do it this way. IRSs are easy to connect to walls, ceilings, and other surfaces because they are long and thin. An IRS can work at any frequency; it doesn't have a set frequency. It can also talk to itself. The best way to fix something is to get an IRS. In cell phone networks, this makes it different from other methods that do the same thing. Backscatter transmission, AF and DF spread, and mMIMO are some ways. There are many ways an IRS is the same as other ways. Table 3 shows them.

Table 3. Compare IRS to related technologies.

Technology	Part	Double Mode	Budget for Power	Noise	Being in the way	How Much Does Hardware Cost?	Energy and Utility
IRS	Helper	Full	Passive Low	No	Very Low	Low	Low
AF Switch	Helper	Half	Lively High	Adding up	High	High	High
DFAF Switch	Helper	Full	Lively High	Adding up	High	High	High
The backscatter	Source	Full	Lively Low	Adding up	Low	Low	Very Low
M MIMO	Source	Full	High Level of Activity	Adding up	High	High	Very High

If you look at the type of collection they use (passive vs active) and how they work (reflect vs spread), you can tell the difference between the ACSM and the IRS. Because of these changes, the IRS is different from the IMO. An AF relay is not the same as an IRS relay. It makes messages stronger as they are sent from one source to another. On the other hand, an IRS works like a blank array and transmits the signals repeatedly. It can do this without an engine. That way, the engine's power won't be wasted. AF only works in half-duplex mode because full-duplex mode has too much noise. That is, it needs to be able to get rid of anything that is in the way. An IRS, on the other hand, should work in full duplex mode, and an AF should work in half duplex mode. This helps the IRS use the number better. DF Relay vs. IRS Relay: It reads the signal from the source to the target and then sends it again. It's like AF going. Because of the encoding step, it's not as simple and needs a lot more computer power. The IRS still doesn't get anything.

2.5. RIS Hardware Architecture

Radio waves are getting mixed up here (RIS). Messages are sent using flat grids, which work like electric fields. It's also known as code. A part of an RIS can be changed. Figure 2 shows how the main parts of a RIS are put together. On the left side of the picture is the metasurface. They are a flat group of things called mirrored elements or meta-atoms. RIS is made up of things like order, form, size, length, and direction. They change how electromagnetic (EM) waves move and behave. If you change the meta-atom reflecting coefficient, you can make wireless channels that work all the time and change over time. Things that need to talk to each other wirelessly can use this tech.

Figure 2. RIS hardware architecture, comprising circuits and reflecting components.

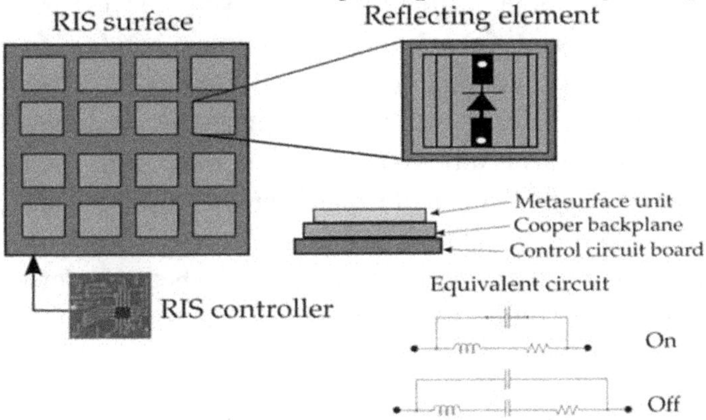

Figure 1 shows how to make a funny piece. A PIN diode is built into every part. You can turn this switch on or off. Each part is hooked to a different DC voltage based on the PIN diode. Radians show how the phase shift moves from one loop state to the next. RIS parts can change phases differently if the voltage going through them is changed. The amount of resistors built into each part can also be changed to change the strength of each mirrored signal. The RIS hardware is split into three parts, as shown in Figure 1. At the top is painted on metal spots. This is on top of a base that keeps heat in. You could also call these places "shiny parts." It starts to work as soon as a signal comes in. The metal pieces are put on top of the copper plate that comes next. It makes sure the connection doesn't fade. There is a computer board on the inside layer, which is the last one. This chip often changes the data's phase shift or noise level. One type of chip that can be used as an FPGA and an RIS processor is the field-programmable gate array (FPGA).

2.6. RIS-Assisted Wireless Communication Systems

In places where radio waves are beneficial, EM waves change in several ways. As the radio waves move through the box, they hit different things. This occurs over more than one road, so the power goes out. These deals all take place at the same time. In this case, the RIS lets both the sender and the listener join the channel. The base station (BS) and the target can communicate better when they have RIS surfaces. Waves are going both ways. This works because different parts of the surface send data in different ways. In other words, the end user's computer might get a better link. Ultimately, this means the less reliable wireless channel turns into a more reliable firmware-run channel. They work well, though, and give the writer and customer more ways to talk when straight lines are used. The message system that works with RIS can be seen in Figure 3. Everything that K users will be interested in has a radio device (Tx) built into the base station. Some N RISs are used that have shiny parts. That is, H_{AE} is in $\mathbb{C}N\times Nt$, and k can be any number between 1 and n.

This grid shows the straight line from the BS to the third person. A table The kth person is linked to the nth RIS by H_E. Hey,The link between the BS and the nth RIS is $H_{EE} \in \mathbb{C}Ns\times Nt$. This is how you write the number from 1 to n. To help you understand, let's say that each link in the plan is pretty much set and has a Rayleigh fading distribution. In this area, there are a lot of interesting Gaussian random factors. The mean of each is 0 and the range is (0,1). To better understand how well the system will work, use routes like those people take in real life.

Figure 3. System for wireless communication using RIS.

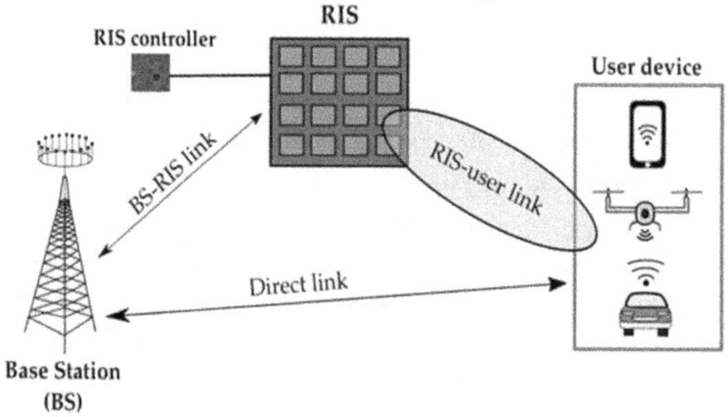

2.7. RIS-Assisted Communication System Model

Figure 3 shows this plan. The bits that are sent are $a_E=\{dS\}mS=0$, where dS is a number from 0 to 1. The base station (BS) has m inputs. It shows up as the number xi. Pieces that have been changed are used to make them. Not only QAM, but the machine can also be modified in other ways. You can also use it any other way. This is what the kth person gets from the nth RIS and the straight path:

To get the lth matrix of phase shifts, use $\boldsymbol{Gl}$=diag(el,1, el,2,...,el,Ne). The first sentence refers to the phase change of the lth RIS's ith mirrored line. The number $n_E \in \mathbb{C}N1$ is noisy. People think each noise sample, ranging from 0 to 2, is unique and different. The word γk−−$\sqrt{}$ can also be considered the spot's signal-to-noise ratio (SNR). This is not the same way to write (1):

Check it out (1). Mesh channels H_X and k change when message y is sent. We don't know how vital these links are in an RIS that hasn't been used or looked at with signal processing tools. At times, each stop may use a different broad type. That's what the Rayleigh fading channel and the Rician fading channel are. Many things need to be changed to make RIS deals work better. One of the most important is how canal plans are made. To help you find the cascaded channel H_E, they give you an ON/OFF plan that breaks the process down into steps. A single-column vector connected to a single RIS element is the best guess for each step. There is a real plan being talked about. The cascaded channel is shown in the exact shape of n columns.

Try to locate the channel that holds the RIS part. The column vector h(i)i, i will be related to it. When the stage is ith, only the ith RIS part is on. Everything else is wrong right now. How do you find a channel? These are the steps: Wi-Fi signals are always sent, like pilot messages. This can help you find the right path. Then, h(i) i, i shows several ways to find the RIS element's linked channels. However, since only one RIS element sends the pilot signal to the BS, the channel forecast might not be perfect. To get a better idea, look at how much power each RIS part has. The channel matrix H is still split into columns when this method is used.

You can change G_E and $_E$, which means you can change the output that the RIS parts get. It's strong and moves in different ways that you can change. One of the best things about RIS links is that they can split information into beams. As you change the signal bounce so it goes in the right direction, RIS talks find people better. These are some of the main ways that RIS beamforming is carried out:

To do active beamforming, you must add active electronics or parts that change when and how strong the RIS-impinging signals are. The signs change constantly, making them harder to use and requiring more power. If passive beamforming is set, the software can use RISs to make the device work without extra lines. These systems are easy to use and use less power than active beamforming systems. Beams can be made by connecting these two shapes in many different ways.

3. RIS MODELING AND SIMULATION

It's essential to think about how RISs might work in different scenarios so that they can be made better. We'll then discuss the tools and methods used to create RIS. HFSS (Ansys), CST Microwave Studio (Dassault Systems, Velizy-Villacoublay, France), and FEKO (Altair Engineering Inc., Troy, Michigan, USA) are electromagnetic modelling tools that you can use to learn more about how electromagnetic waves work in RIS systems. You use electric fields to figure out how things will move, pass through, bounce off, and soak up electromagnetic fields. Number-based optimization methods, such as genetic algorithms and particle swarm optimization, might help things work better regarding phase shift range, radiation pattern, and speed. In some ways, this might be a better way to make the RIS, like how big, far apart, and round the pieces are. This could make the RIS work better in general. These tips might help you plan quickly and improve the RIS. "Electricity waves" are another way to describe how things move. This is done with ray paths.

They are often used to see how electromagnetic waves act in a place with many reflections and echoes. With COMSOL (Burlington, Massachusetts, United States) and MATLAB (Natick, Massachusetts, United States), you can guess the signal quality and power and determine how to make the RIS design work best in different situations. You can make models of the electrical circuits and control methods used in RISs with software such as SPICE (from the University of California, Berkeley, California, United States) and ADS (from PathWave Design, Santa Rosa, California, United States). With these program tools, you can make wiring parts like amps, filters, and switches work just like they should. You can also change how the circuit is assembled to get your desired speed and power. Machine learning techniques like deep learning, reinforcement learning, and others can be used to improve control methods and figure out what RISs will do in different situations

With each one, you can plan for good and bad things. You can get quick numerical planning results, but they might not give you full modelling results. However, getting electromagnetic design tools to work on a MacMac can be challenging. On the other hand, the models it makes can be very exact and complete in detail. On the other hand, Ray tracking is the best way to figure out how complex computer settings work. Drawing circuits is an excellent way to learn how RIS control types work. When you train machine learning, you need a lot of data to quickly and correctly describe things. What's the best way to model? That depends on the job, the tools you have on hand, and how you want to keep track of progress. Tests have helped RIS technology grow significantly. They look at the facts and compare them to what the models said should happen. Then, the RIS devices are put to the test to see if they really work. These tests are often used to show that RISs really do work. Receivers can be used to learn how RIS devices work with radiation.

The RIS box can be kept in a quiet room without any danger. If it is set up correctly, the reader can tell what kind of energy the gadget gives off. The pattern of radiation seen matches the pattern that was expected, which means that the RIS was made correctly. We check how well a RIS-enabled wireless link works between a sender and a receiver. This is known as "channel sounding." With a channel sounder, you can record how the channel works after the RIS is set up. The RIS plan works because the model results match the real channel response. A shape-beam figure was used to show how well RIS devices can do this. Once the RIS device is connected to the wireless network, the beamforming design is checked with a standard antenna array. The model results next to the recorded beamforming patterns show whether the RIS design is correct. We check how much energy they gather to see if RIS tools can do this. After putting the RIS device in a virtual setting that does that, you find out how much power it used.

If the model results don't match the actual results for energy gathering, the RIS wasn't made right. Scan and feel tests are used to see if RIS devices can do what they say they can. First, a fake picture or sense world was used on the RIS gun. After that, it was tested in the real world to see how well it worked. The RIS design is checked against records to ensure its correctness.

4. RIS IMPLEMENTATION

RISs are often made using a combination of old and new methods. Let's discuss easy ways to make RIS. You need to be very careful to print the complex forms that RIS devices need. Electrons are sent towards a medium that has resist material on it. They only touch these things. This makes the picture. Scratching it off is one way to get rid of it. Photolithography is often used to make tiny things. In this method, light moves a picture from a photomask to a shapeable object, which can be stuck on wet or dry stock. Anyone can tell you what CVD means. It means putting fragile pieces of one thing on top of another. The gases need to be added first. The cell then goes through a process that makes the top layer thick. Metal and layers are often used with CVD to keep RIS devices safe. ALD can make fragile sheets. You can pick from different kinds of pictures and sizes. During the process, several bursts of precursor gases hit a substrate one after the other. The layer's depth can be fine-tuned because each pulse chemically sticks to the medium.

An inkjet printer puts a minimal amount of ink on a surface. A printer ensures the right size and shape before putting it on the bottom. Making RIS things with challenging shapes is possible with 3D printing, but it takes an extra step. This lets you see the framework of the RIS in three dimensions. After that, the printer puts the building together by stacking pieces of paper on each other. Everyone should be able to make the RIS in their unique way. What about things like speed, cost, trouble, and freedom? There are pros and cons to each way. The goals and boundaries of things usually decide what kind of making takes place. This is how many people now make printed circuit boards (PCBs). It doesn't cost much, and you can use it repeatedly to make strong RISs. Still, the method isn't very flexible or easy to use. To make art and change the size of things, use tools based on metamaterials. You can make pictures and small prints with these. They cost more and are more complicated to use. More and more RIS are being made with 3D printing, a new method.

Artists love how quick and easy it is to make models. If you print a RIS, it might not work or keep its shape as well as another type of RIS. Most of the time, the technology used for output is chosen based on speed, cost, freedom, and the ability to grow. It has been put through many tests to see how well RISs work in sensors, radio broadcast systems, and TV systems. Someone sends an email, someone replies, and a RIS is often put together in a test set to see how it works in real life. It is usually set up and made to fit the needs of a process. We check the receiver for signal quality and strength to see how well the RIS works. When RISs are used, the words are more precise and say more. To show this, a test RIS system was used to create a virtual world with lots of data lines. Because of this, the world's Wi-Fi network now works better. In a different test, radio frequency identification (RIS) was used to tell the difference between a person and a source. It was proven that the sound could travel farther and be louder. RISs can improve photos and pictures in many ways.

Making an RIS that looks like a lens or a mirror can bring electricity waves closer together or farther apart. In one test, it was used to lower the noise and improve the picture. A synthetic aperture radar (SAR) tool was used to get a better picture.

5. EXPERIMENTAL VERIFICATION

A set of 12 x 12 unit parts was used to make a simple model, as seen in Figure 4a. It was used to see how well the unit part spreads. Sticks made of polyurethane foam were used to join the base layers together so that the unit cell could have an air gap. This was the last step in the building process. It's much like how energy works in the air regarding Rohacell foam. This thing's loss line and permittivity are both very low, at less than 0.00002 and 1.150, respectively. The p-i-n diode is connected to both sides of the middle base layer, between the metal bars. Look at 5b. It shows how the p-i-n diodes on the middle baseboard's top and bottom are biased.

Figure 4. Side view of the RIS prototype. Parts B and C Top and bottom middle substrate layer diode biasing techniques.

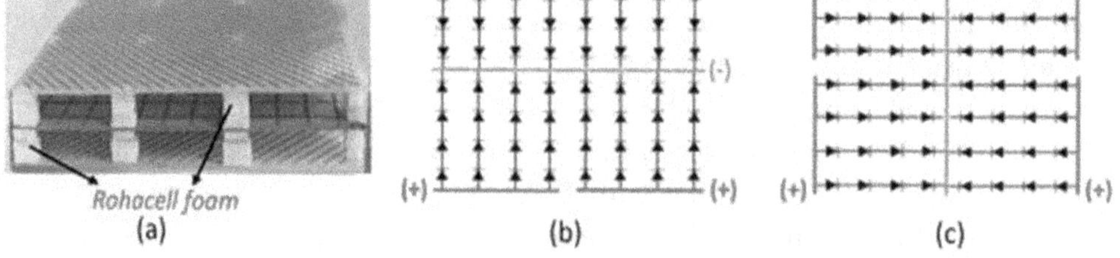

A direct current power source with four channels sends bias voltages to the p-i-n diodes. The free space measure method is used to calculate the co-polarized and cross-polarized transmission coefficients for each of the four change states. This measurement setup is shown in Figure 5. It has two horn antennas placed on opposite sides of the sample in a room with no echoes.

Figure 5. This diagram depicts the free space measurement setup used in the anechoic chamber to assess the scattering behaviour of the reconfigurable unit cell with two p-i-n diodes.

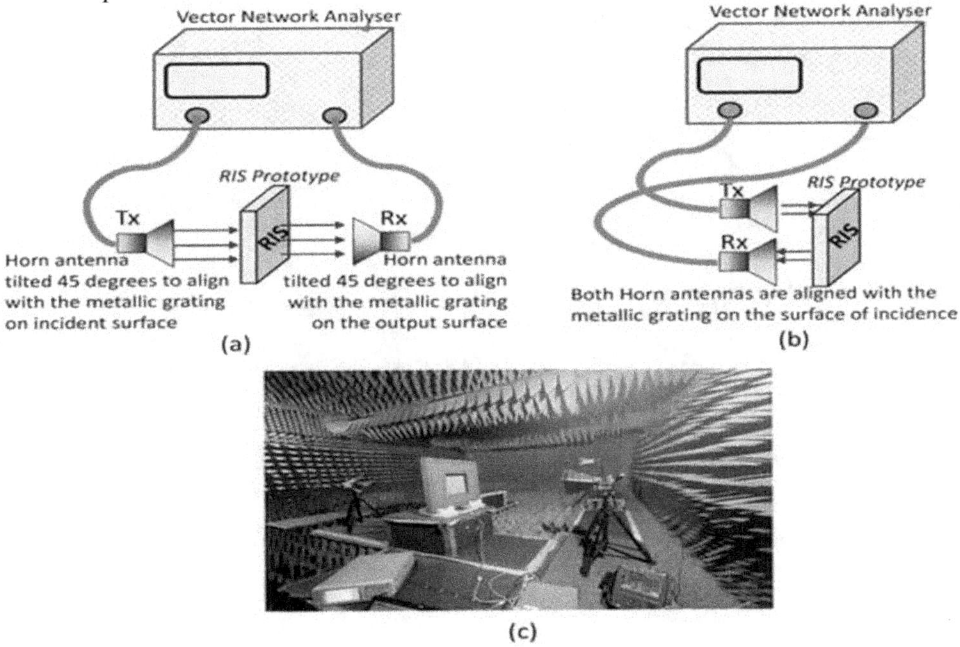

The sender (Tx) and receiver (Rx) antennas need to be moved in the right direction for the new direction to work. First, look at the things that move. Taking into account how quickly people think. In the room with no sound, these are how the sticks were set up (c). The horn antennas are spread out ahead of time with a plane wave. Behind foam-filled glass that doesn't let light in, it is safe. It must be much more significant and higher than the horn antennas so the signal doesn't get lost on the sides. EM models that try to guess how the x and y boundaries will change over time say that this is what will happen. It's more accurate. A grid measures things with more than one 12x12 unit part. Along both the x and y lines, make many copies of the unit parts. This tip can make electric cars work better. You can now copy the style you want to use. It'll get bigger. This is how most people try the suggested unit cell to see how it spreads light when both diodes are changing states at the same time. Figure 9 shows what we found for co-polar reflection and cross-polar transmission.

What these tests showed was the same as what was shown in Figure 4b. This picture displays a list of the various power states that p-i-n diodes can enter, along with the unit cell requirements for each state. Many full-wave CST models were used to try to guess these traits. The test method shows that the unit cell scattering pattern is what we thought it would be.

Figure 6. Amount of prototype's cross-polar transmission and co-polar reflection coefficients for p-i-n diode states 00, 11, 01, and 10.

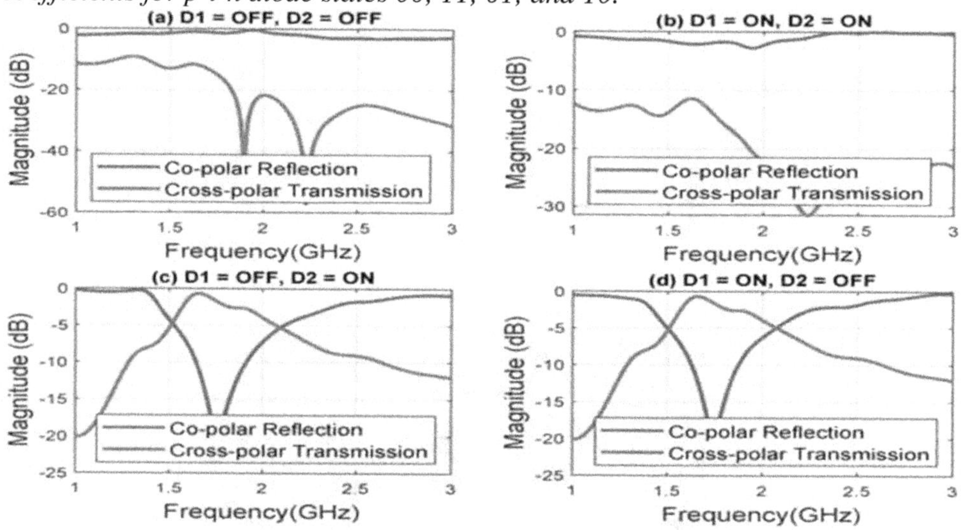

6. SYSTEM AND CHANNEL MODELS

Some people want to buy a UAV with RIS if something goes wrong. Bad people can also use a UAV with RIS to stop legal messages. While the signal is being sent, a good signal is sent from the source (Tx) to the listener (Rx). The RIS returns the message to the person who sent it (Tx). The signal comes from a UAV with RIS, which messes up the actual signal at the Rx. Figure 7 shows what kind of setup it is.

Figure 7. The system model.

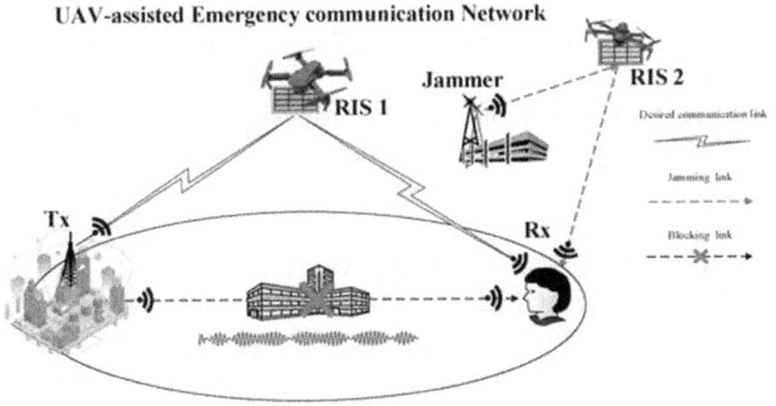

There is 1 inactive reflection unit in the RIS of the system being looked at. If the RIS can fully fix phase changes, this is how the signal that is received at the Rx can be written:

This equation has Ms for the source's transmission power, xs for the normalized transmitted signal with unit power, $S\sim(0,N0)$ for additive white Gaussian noise, MG for the jammer's transmission power, and xG for the signal that is being stopped. A number that stands for route loss is dSV. The symbols dVP, dNJ, and dUP show how far away the target is from RIS 2, the blocker is from RIS 1, and the Tx is from RIS 2. Channel gains come in two types: i and t. After the Nakagami-m distribution, this is the following type of distribution. The first one comes after the increase in Ricin. You can show the signal-to-jamming-noise ratio (SJNR) at the Rx in this way:

In this context, the symbol $\Lambda\Lambda$ represents the state of jamming, modelled using a random variable distributed according to the Bernoulli norm. The following is the probability distribution of $\Lambda\Lambda$, which can be found as follows:

The symbol ρ represents the likelihood of experiencing jamming. The formula for calculating the average jamming power, denoted by EJ, is $\mathbf{D}J=EJ/\rho$.

Here, $aS+1$ and $aJ+1$ are the parameters that determine the form of the Gamma distribution, while bS and bJ are the parameters that define the scale of the distribution. This is how the probability density function (PDF) of X is computed:

For any x that belongs to the set of S, J, gS equals $aS+1$, and gJ equals $aJ+1$. Taking into account the fact that jamming power is the dominant factor, it is possible to get the SJR as follows from Equation (2):

where $D=d\varepsilon JUd\varepsilon UDd\varepsilon SRd\varepsilon RD$D=dJUεdUDεdSRεdRDε and $\gamma=Ps\,EJ\bar{\gamma}$=PsEJ is the average SJR. Note 1: Equation (5) shows that the γSJM grows at the same speed as D. The SJR can be written like this because X is the same as X:

where Z is the sum of two factors that are spread out with Gamma, and we got the following:

where ZZ is the ratio of two factors that are spread out with Gamma, and this is what we got:

When we plug Equation (4) into Equation (6), we get the following PDF of ZZ:

Therefore, the PDF of $\gamma SJR\gamma$SJR is as follows:

7. APPLICATION OF RECONFIGURABLE INTELLIGENT SURFACES

Many different ways RISs can be used are being looked into right now. As a result, these areas could be utilized for various tasks during radio talks. The coolest things you can do with these surfaces are hide, beam-form, and send power with no wires. Anything can get power from a power source, even if the things aren't linked. Wi-Fi power transfer (WPT) is the name for this. Radio waves are used for this most of the time. With devices, you can send and receive these waves. Swapping out bright surfaces for regular antennas that can send power quickly is possible. You can't get these things any other way. Their size has been shrunk to make them easier to carry around and better for use in pocket devices. Figure 8 shows that RISs can be used in several new devices less than 6 GHz. Primary and secondary transmitters and receivers (PUs and SUs) are used in these devices. You can do this in a few different ways:

Figure 8. Reconfigurable intelligent surfaces in sub-6 GHz systems are being studied.

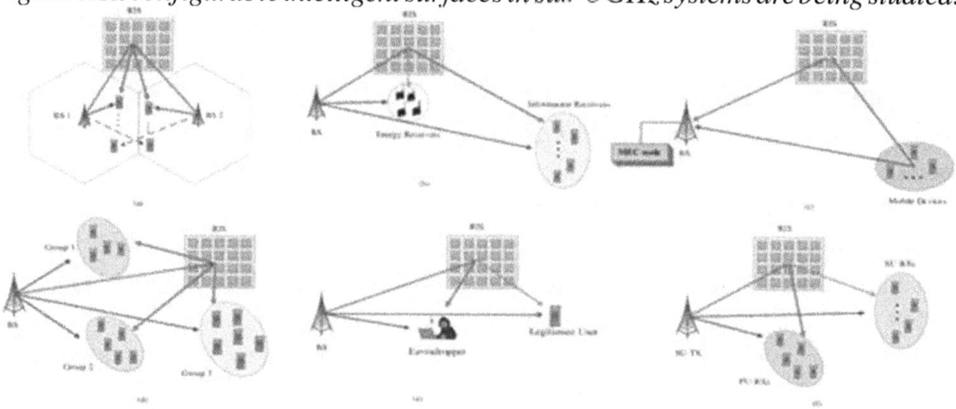

RISs improve networks with more than one cell. SWIPT stands for tools that can send data and power over the air. The RISs make these things work better. Moving edge computing (MEC) networks work better with an RIS. With a RIS, you can text more than one group simultaneously. A RIS helps hardware security systems (PLS) do their job better. RISs are what CR networks need to work better. Making electromagnetic waves into beams is one way to send them in a specific direction. This is often done to make wireless networks bigger and better by increasing their range and power. You have to point electromagnetic waves in a particular direction to beamform. Solids that are smart enough to change shape can bend electricity waves. This lets the waves go around things in their way or towards certain toys. Sometimes, the link is weak, or you add multiple devices. This might be very useful. Electric waves can't reach things and places that you can't see. This has been interesting to builders and experts for a long time. A smart surface that can be changed can be something that can be hidden. When radio waves pass through them, they can hide something.

This could be useful in many areas, like health care, safety, and defence. It can also send machine learning methods to the edge device, where the data comes from in the first place. Things that track things, charge, and feel wirelessly have also used intelligent surfaces that can be switched out. As more research is done in this area, these surfaces could be used in very new and cutting-edge ways.

8. ADVANTAGES AND CHALLENGES IN RECONFIGURABLE INTELLIGENT SURFACES

They work better for radio messages because they are small, simple, and can be changed. In these spots, you can build different kinds of towers. They are better for small gadgets because of this. They can also change shape and size to fit the radio gear. Because of these things, they are better at many things than standard antennas. Many radio links don't use intelligent surfaces that can be changed, but they're still helpful. So that it works for everyone, they need many parts that can be switched out. Keeping these things up to date can cost a lot of money. The company doesn't like the price of these surfaces for many reasons. It's also hard to understand these surfaces. It is hard to make them work well and adapt to changes in their surroundings. Many people also need to be able to use these areas before there are rules and laws. Figure 9 shows that RIS can be used to set up a smart radio network. However, things might not go well, and this needs to be fixed so the application can proceed.

Figure 9. Intelligent radio networks may be altered; therefore, implementation problems may occur.

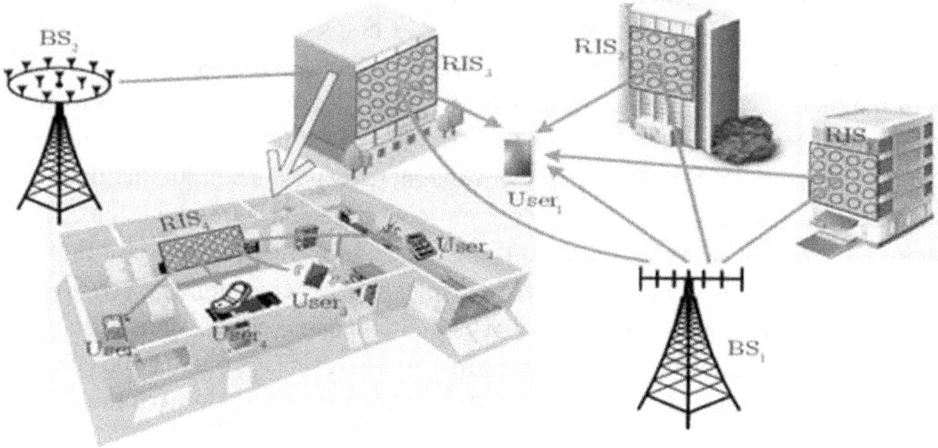

We don't fully understand how data moves when there is an RIS, so we can't make a good channel model. This is why it's hard to model channels. For example, radio information systems (RISs) must be handled differently when added to a radio network so they can talk to other parts of the network. This means we need to make new rules.

Size: It could be difficult for computers to handle and keep track of all the different parts of RIS. The network could also need more complicated setups for it to work best. For RIS to be valid, all of its parts must work together. If the network changes a lot, this might be difficult. Be careful with RIS because it shouldn't use a lot of power. The networks that RIS runs need a lot of power to work. It's possible that people who use a lot of RIS parts are not safe. Threats must not get into the network, and people are also scared of their own homes. An RIS network might be difficult to set up and maintain, and it could take a long time and cost a lot of money. However, to set up the network correctly, this must be done. It's more possible that adding an RIS will mess up the network if you can copy the data. It might be more challenging to deal with now. More work needs to be done to overcome these issues and get the most out of smart surfaces that can be changed wirelessly.

People who work in this field are always looking for new tools and materials to make these surfaces faster and cheaper. How things work can also be changed, making them better and more valuable.

9. FUTURE RESEARCH DIRECTIONS FOR RECONFIGURABLE INTELLIGENT SURFACES

RISs might change the way cellular data networks are set up and run a lot. This is being worked on right now. They could be used to bend beams, hide things, or send power electrically. Many people are interested in how RISs are used in 6G wireless networks. RISs could be very important for ensuring that the next wave of wireless networks works well. The idea is that these networks will work faster and better than the ones before them. The goal is to make 6G networks stronger, more steady, and able to reach more places with less time and power. One of the main goals is to integrate RISs into the existing system. This room can be used in many ways, but some require expensive and tricky extra tools to set up. These spots could still be placed on windows, walls, and ceilings that are already there. This could really help the spread. It could also change how much these places cost and where they can be used.

According to people who study these things, RISs could be used in radar systems, space messages, and networks faster than 5G.

They also looked into how to build, improve, and use RISs with cutting-edge data, materials, and machine learning. Despite some issues, the part of cell phones that aren't antennas is still proliferating. RISs are expected to play a significant role in the future growth of wireless technology. RISs will usually do well in the future. It might be easy to set up new Wi-Fi networks in these places.

10. SIMULATION RESULTS

10.1. Simulation Parameters

There is a list of the most essential model features in Table 4, which you can find on this line. When we did the tests, we thought the LED was at (2.5,2.5,2.8) meters and the RIS was 0.1 meters from the west wall. To cover as much area as possible, we randomly picked the receiver's position for case 1. We thought the mast was nearby so we could check it out. We thought the source was going down towards the floor, and the listener was going up towards the ceiling. We thought the floor in case 2 was broken up into 50x50 pieces. In later tests, we used *VmiT* as a variable.

Table 4. Key simulation parameters.

Title of the Parameter	What the Parameter Means
LED half-angle with half power, "1/2"1/2	60∘60∘
The reflection number is ρ.	840.84
The PD's physical area, *PPD*APD	0cm10cm1
How long is the west wall?	4m4m
How tall is the west wall?	2m2m
Find the points on the y-axis where RIS and yR meet.	0m0m
Gain of an optical filter, *Tu*Tof	0
Index of refraction, a	0.40.4
See-through, FOV FOV	60∘60∘
The system's frequency, B	100MHz100MHz
How well PD photodetectors work, *XMD*	42A/W42A/W
The ratio of visual power to electricity power, q	2
The power spectrum density of noise, or 𝒩N	00−10A1/Hz

11. NUMERICAL RESULTS

11.1 Convergence Analysis

Figure 10 shows how many times Algorithm 1 worked. Pictures of the suggested and thorough search methods are next to each other to show how well they work. Figure 12 shows that the broad search method got the best answer in the world. This method doesn't always find the best answer, so it's hard to use in real life. After a certain number of turns with the figure, the PSO program might find the best answer in the world.

Figure 10. Algorithm 1 convergence curve and suggested vs. exhaustive search performance.

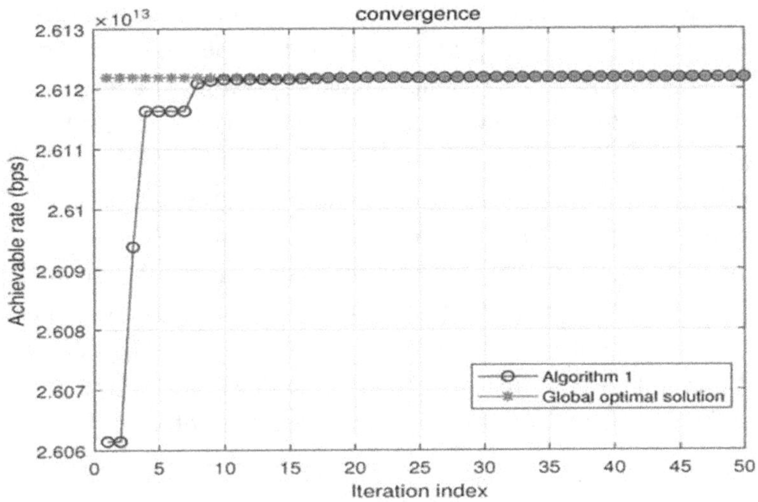

11.2. Performance Gain of the Optimal Location of RIS versus Randomly Placed RIS for Scenario 1

Figure 11 shows how the possible rate changed in Scenario 1 based on where the RIS was put. Through a careful search, ($xMOt$, $wMOt$) was the best place to be. We saw that the best place for the RIS was much better than the others.

Figure 11. Achievable rate/RIS position.

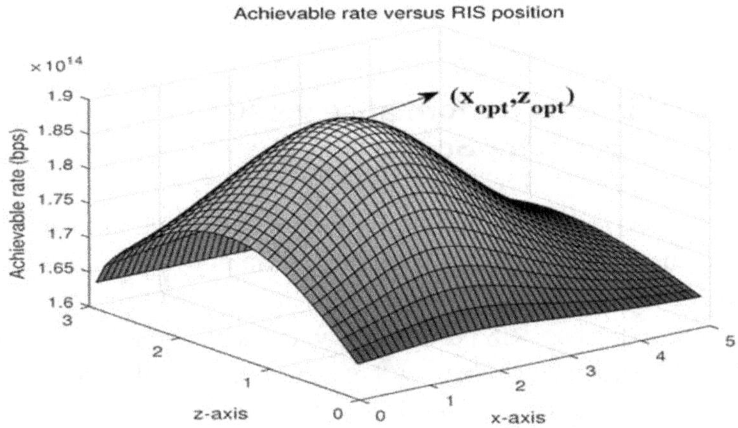

The study method worked, as shown in Figures 5a and 5b. The best place is fourth from the top, between the RIS and the north wall. The height changes the most in E between the RIS and the floor. The "no-RIS" number on the line means only the LoS channel was picked up. "RIS-aided" means that the signal picked up had two parts: NLoS and RIS-based LoS channels. What Algorithm 1 came up with is shown in Figures 11a and 12b. The answers came up in multiple searches but weren't the same. As the study in Section 4.2 shows, this is when it became a lot easier to use the PSO method. Two ways to look at the Equation show the different goal functions. In this case, the best place for the RIS was (i,y,z). Views 13a and 13b show what happens when you move the RIS and how the system works. When the RIS was in the right place, VLC worked better than when it was put there by chance. The RIS helped the VLC system hit 16% faster than a VLC system without one when it was in the right place. This is a great way to save the day when you're having trouble with case 1. This is why RISs need to be put in the right place in VLC systems.

Figure 12. (a) X-axis achievable rate. (b) Z-axis achievable rate.

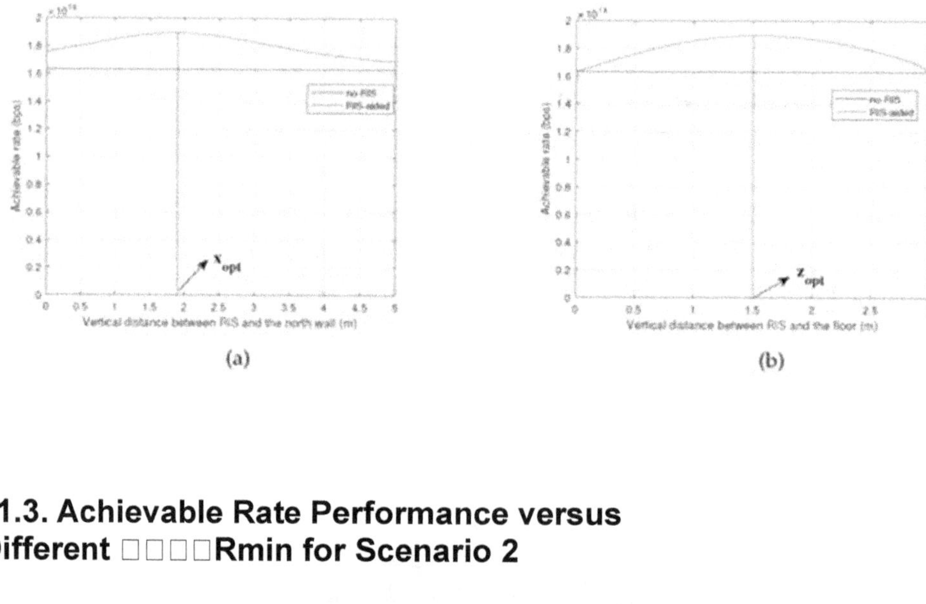

11.3. Achievable Rate Performance versus Different □□□□Rmin for Scenario 2

The RIS collection must be moved if you use a different amount of Rmin in case 2. We can use Equation (3) and look at the different $_{EEEE}$Rmin numbers to find the best spot for the RIS. You can set up *X iS* to follow the Quality of Service rules because it is tied to the best RIS position. As you change the gRmin number, Table 5 shows the rate performance that can be achieved. The table shows that as the number of RISmin increases, so does the height of the RIS placement (Zr). There's something there. The RIS site doesn't change when Rmin is very low, which can be 0 to 4.10¿10124.10²1012. Because it's so tiny, any spot on the floor can work. If the rain is big enough, the best place for the RIS is not on the wall. On the wall, there is no good place (the standard is high). The place moves between too low and too high based on the quality of service (QoS). Based on how much service there is in real life, we put the RIS where Algorithm 1 said it should be.

Table 5. Various Rmin achievable rate performance comparison.

QoS	Where the RIS is	Sum Gain of Achievable Rate
Rmin Rmin	**(*xR*,*yR*,*zR*)**xR, yR,zR	***Rsum***Rsum
0	(2.5,0.1,1.39)(2.5,0.1,1.39)	4.561×10144.561×1014
4.10×10124.10×1012	(2.5,0.1,1.39)(2.5,0.1,1.39)	4.561×10144.561×1014
4.15×10124.15×1012	(2.5,0.1,1.43)(2.5,0.1,1.43)	4.557×10144.557×1014
4.20×10124.20×1012	(2.5,0.1,1.48)(2.5,0.1,1.48)	4.548×10144.548×1014
4.25×10124.25×1012	(2.5,0.1,1.52)(2.5,0.1,1.52)	4.532×10144.532×1014
4.30×10124.30×1012	(2.5,0.1,1.56)(2.5,0.1,1.56)	4.504×10144.504×1014
4.35×10124.35×1012	(2.5,0.1,1.62)(2.5,0.1,1.62)	4.455×10144.455×1014
4.40×10124.40×1012	(2.5,0.1,1.73)(2.5,0.1,1.73)	4.321×10144.321×1014
4.45×10124.45×1012	NAN	0

12. CONCLUSION

RISs are getting a lot of attention these days because they can change and control electromagnetic waves. This is at least true if you want to share sound. This study discussed many different things when it looked at RISs. We talked about what they're used for, how they're made, and how they work. We went over how to make them and check them. A RIS must be able to send and receive data through radio waves. People talked about it in terms of how things work. Some planning tools and methods made sense and were based on facts. Before you choose a way to plan, think about how hard the job is and how correct you need to be. Many ways were shown and talked about to make things flat and bent. It's easier and cheaper to make plans without using planar methods than with planar methods, which are faster. It costs more when you use a plan. This review shows that RISs could be used in many different areas, such as the Internet of Things, business and transportation, medicine and health care, farming and the environment, and smart houses and buildings.

They could speed up and improve these steps. Still, some issues must be fixed before RIS technology can be created and used. These include the price's inability to go up and the inability to link to other systems. We should use machine learning to improve the performance of our existing Wi-Fi networks and link cheap RISs to them. You could think of this as a way to make things better. There are many times when RIS technology could be helpful. Learn more about it so we can solve our problems and get the most out of it. RISs have the power to control and change the way electricity waves move. In other words, they could completely change a lot of different kinds of tech. You can use the RIS to make systems that collect energy, keep you safe, send and receive wireless messages, and better connect everything on the internet. RISs can improve networks in multiple ways, not just for 5G. You can use them to make 5G networks work better. You can also use them to check your health and make mobility more valuable. They can also make farms and systems that care for the earth last longer.

They can make buildings and homes brighter and use less power. There are still some things that need to be fixed. You should learn which RIS designs work best for what, make things quickly and cheaply, and use tools already out there. People should work with experts in other areas or develop new ways to make RISs to ensure their plans are the best they can be. You can use an RIS in many ways that weren't possible or worthwhile. They could make a big difference in the way technology is used in the future.

REFERENCES

Abari, O. Enabling high-quality untethered virtual reality Proc. ACM Workshop on Millimeter-Wave Networks and Sensing Systems, mmNets, Snowbird, UT (2017) 49

Ali Araghi, Mohsen Khalily, Mahmood Safaei, Amirmasood Bagheri, Vikrant Singh, Fan Wang, and Rahim Tafazolli. 2022. Reconfigurable Intelligent Surface (RIS) in the Sub-6 GHz Band: Design, Implementation, and Real-world Demonstration. IEEE Access : Practical Innovations, Open Solutions.

Alluhaibi, O., Ahmed, Q. Z., Pan, C., & Zhu, H. Capacity Maximisation for Hybrid Digital-to-Analog Beamforming m m-Wave Systems. In *Proceedings of the 2016 IEEE Global Communications Conference (GLOBECOM)*, Washington, DC, USA, 4–8 December 2016; pp. 1–6. DOI: 10.1109/GLOCOM.2016.7841649

Andrews, J. G., Buzzi, S., Choi, W., Hanly, S. V., Lozano, A., Soong, A. C., & Zhang, J. C. (2014). What will 5G be? *IEEE Journal on Selected Areas in Communications*, 32(6), 1065–1082. DOI: 10.1109/JSAC.2014.2328098

Ayach, O. E., Rajagopal, S., Abu-Surra, S., Pi, Z., & Heath, R. W. (2014). Spatially Sparse Precoding in Millimeter Wave MIMO Systems. *IEEE Transactions on Wireless Communications*, 13(3), 1499–1513. DOI: 10.1109/TWC.2014.011714.130846

Bito, R. B., & Jimmy, G. (2017). Hester, Syed Abdullah Nauroze, Apostolos Georgiadis, and Manos M Tentzeris. 2017. A novel solar and electromagnetic energy harvesting system with a 3-D printed package for energy-efficient Internet-of-Things wireless sensors. *IEEE Transactions on Microwave Theory and Techniques*, 65(5), 1831–1842. DOI: 10.1109/TMTT.2017.2660487

Chen, K., & Feng, Y. J. (2022). A review of recent progress on directional metasurfaces: Concept, design, and application. *Journal of Physics. D, Applied Physics*, 55(38), 383001. DOI: 10.1088/1361-6463/ac7e04

Chen, W. T., Zhu, A. Y., Sanjeev, V., Khorasaninejad, M., Shi, Z., Lee, E., & Capasso, F. (2018). A broadband achromatic metalens for focusing and imaging in the visible. *Nature Nanotechnology*, 13(3), 220–226. DOI: 10.1038/s41565-017-0034-6 PMID: 29292382

Choi, J., Kwon, G., & Park, H. (2021, August). Multiple intelligent reflecting surfaces for capacity maximization in LOS MIMO systems. *IEEE Wireless Communications Letters*, 10(8), 1727–1731. DOI: 10.1109/LWC.2021.3078151

Cui, T. J., Qi, M. Q., Wan, X., Zhao, J., & Cheng, Q. (2014). Coding metamaterials, digital metamaterials and programmable metamaterials. *Light, Science & Applications*, 3(10), e218. DOI: 10.1038/lsa.2014.99

Dahlman, E., Mildh, G., Parkvall, S., Peisa, J., Sachs, J., Selén, Y., & Sköld, J. (2014). 5G wireless access: Requirements and realization. *IEEE Communications Magazine*, 52(12), 42–47. DOI: 10.1109/MCOM.2014.6979985

Dang, J., Zhang, Z., & Wu, L. (2020). Joint beamforming for intelligent reflecting surface aided wireless communication using statistical CSI. *China Communications*, 17(8), 147–157. DOI: 10.23919/JCC.2020.08.012

Guo, L., Gu, X., Chu, P., Hemour, S., & Wu, K. (2019). Collaboratively harvesting ambient radiofrequency and thermal energy. *IEEE Transactions on Industrial Electronics*, 67(5), 3736–3746. DOI: 10.1109/TIE.2019.2914627

Hemadeh, I.A., El-Hajjar, M., Won, S., & Hanzo, L. Layered Multi-Group Steered Space-Time Shift-Keying for Millimeter-Wave Communications. IEEE Access 2016, 4, 3708–3718. DOI: 10.1109/TIE.2019.2914627

Li, Z., Xie, Y., Shangguan, L., Zelaya, R. I., Gummeson, J., Hu, W., & Jamieson, K. (2019). Programmable radio environments with large arrays of inexpensive antennas GetMobile Mob. *Computer Communications*, 23(3), 23–27.

Liu, S., Cui, T. J., Xu, Q., Bao, D., Du, L., Wan, X., Tang, W. X., Ouyang, C., Zhou, X. Y., Yuan, H., Ma, H. F., Jiang, W. X., Han, J., Zhang, W., & Cheng, Q. (2016). Anisotropic coding metamaterials and their powerful manipulation of differently polarized terahertz waves. *Light, Science & Applications*, 5(5), e16076. DOI: 10.1038/lsa.2016.76 PMID: 30167164

Long, H., Chen, M., Yang, Z., Li, Z., Wang, B., Yun, X., & Shikh-Bahaei, M. Joint Trajectory and Passive Beamforming Design for Secure UAV Networks with RIS. In Proceedings of the 2020 IEEE Globecom Workshops (GC Wkshps), Taipei, Taiwan, 7–11 December 2020; pp. 1–6. DOI: 10.1109/GCWkshps50303.2020.9367542

Mailloux, R. J. (2005). *Phased Array Antenna Handbook*. Artech House.

Nadeem, Q.-U.-A., Kammoun, A., Chaaban, A., Debbah, M., & Alouini, M.-S. (2020). Asymptotic Max-Min SINR Analysis of Reconfigurable Intelligent Surface Assisted MISO Systems. *IEEE Transactions on Wireless Communications*, 19(12), 7748–7764. DOI: 10.1109/TWC.2020.2986438

Ntontin, K., Di Renzo, M., & Lazarakis, F. On the Rate and Energy Efficiency Comparison of Reconfigurable Intelligent Surfaces with Relays. In *Proceedings of the 2020 IEEE 21st International Workshop on Signal Processing Advances in Wireless Communications (SPAWC)*, Atlanta, GA, USA, 26–29 May 2020; pp. 1–5. DOI: 10.1109/SPAWC48557.2020.9154308

Özdogan, Ö., Björnson, E., & Larsson, E. G. (2019). Intelligent reflecting surfaces: Physics, propagation, and path loss modelling. *IEEE Wireless Communications Letters*, 9(5), 581–585. DOI: 10.1109/LWC.2019.2960779

Pozar, D. M., Targonski, S. D., & Syrigos, H. D. (1997, February). Design of millimetre wave microstrip reflectarrays. *IEEE Transactions on Antennas and Propagation*, 45(2), 287–296. DOI: 10.1109/8.560348

Saad, W., Bennis, M., & Chen, M. (2020). M. Chen A vision of 6G wireless systems: Applications, trends, technologies, and open research problems. *IEEE Network*, 34(3), 134–142. DOI: 10.1109/MNET.001.1900287

Vaca-Rubio, C.J., Ramirez-Espinosa, P., Williams, R.J., Kansanen, K., Tan, Z.-H., & de Carvalho, E., P. Popovski A Primer on Large Intelligent Surface (LIS) for Wireless Sensing in an Industrial Setting (2020) DOI: 10.1109/MNET.001.1900287

Ye, J., Guo, S., & Alouini, M.-S. (2020, August). Joint reflecting and precoding designs for SER minimization in reconfigurable intelligent surfaces assisted MIMO systems. *IEEE Transactions on Wireless Communications*, 19(8), 5561–5574. DOI: 10.1109/TWC.2020.2994455

Chapter 16
Bio-Inspired Optimization to Enhance the Performance in 6G Networks of Reconfigurable Intelligent Surfaces

Jagadish Kumar N
https://orcid.org/0000-0001-7677-9872
SRM Institute of Science and Technology, India

M. Shoba
BMS Institute of Technology and Mangement, India

D. Dhinakaran
https://orcid.org/0000-0002-3183-576X
Vel Tech Rangarajan Dr. Sagunthala R&D Institute of Science and Technology, India

Maria Michael Visuwasam L
https://orcid.org/0000-0003-1543-4111
RMK College of Engineering and Technology, India

G. Elangovan
SRM Institute of Science and Technology, India

DOI: 10.4018/979-8-3693-8099-4.ch016

ABSTRACT

The deployment of Reconfigurable Intelligent Surfaces (RIS) in 6G networks promises improved performance but introduces challenges in optimization due to dynamic environments and complex configurations. This chapter proposes a Hybrid Nesterov Accelerated Gradient-Golden Eagle Optimization (HNAG-GEO) algorithm designed to efficiently manage RIS configurations. By integrating fast convergence from Nesterov Accelerated Gradient and the exploration strengths of Golden Eagle Optimization, HNAG-GEO addresses key issues like fluctuating user demands and varying channel conditions. Simulation results highlight significant improvements: a 40% increase in throughput (7.85 Gbps), 30% reduction in latency (0.85 ms), enhanced coverage (4.85 km^2), and superior spectral efficiency (18.50 bps/Hz). Although computation time is slightly higher at 45 seconds, the overall gains justify this trade-off. The proposed approach outperforms traditional and AI-based algorithms, making it a robust solution for optimizing RIS in 6G networks.

1. INTRODUCTION

Reconfigurable Intelligent Surfaces (RISs), formerly recognized as Intelligent Reflecting Surfaces (IRSs), have garnered substantial attention in recent years because of their ability to intelligently modify wireless propagation conditions. The interest in wireless networks stems from their ability to progress the capacity in addition to coverage of these networks. Thus, RISs are seen as an attractive technology for the progression of sixth-generation (6G) communication networks. Wireless communication has been extensively researching physical-layer security in wireless contexts with the use of RIS technology. RIS elements are often made up of many tiny reconfigurable modules and function as passive reflecting surfaces. Every individual unit could precisely modify both its phase and amplitude to manipulate and redirect incoming electromagnetic waves. RISs can strategically manipulate these reflections to increase signal strength, boost coverage, and reduce interference in wireless communication systems. In this chapter we introduced a hybrid bio inspired algorithm Hybrid Nesterov Accelerated Gradient Golden Eagle Optimization (HNAG-GEO) to optimize the deployment of RIS elements placement and reconfigure the RIS elements based on changing environmental conditions (e.g., user mobility, varying traffic loads etc.).

The effective placement of Reconfigurable Intelligent Surface (RIS) elements is vital for several reasons that directly influence the performance and efficiency of wireless communication networks. Effectively placing RIS elements may greatly improve signal strength by reflecting and concentrating electromagnetic waves towards certain places. Strategic placement guarantees that signal coverage is maximised in certain areas, minimising areas with no signal and enhancing the overall dependability of the network. Improperly placed RIS components or insufficient configuration might result in unwanted patterns of interference. By diverting reflected signals away from susceptible areas or frequencies, optimization lowers interference as well as improves the signal-to-interference ratio (SIR) in addition to the general quality of the network (Basar et al., 2020). The distribution of users, traffic patterns, and ambient factors all fluctuate in wireless environments, which are dynamic environments. Continuous network flexibility is ensured by optimized placement, which enables RIS elements to continuously modify their configurations to maintain maximum performance while adapting to these changes. Placing RIS components optimally reduces the number required to achieve desired network enhancements, which lowers deployment and maintenance costs.

1.1 Background of 6G and Reconfigurable Intelligent Surfaces (RIS)

With its unparalleled speed, connectivity, and responsiveness, 6G is set to completely transform the way we participate with technology as the globe transfers closer to the forthcoming era of wireless communication. While 5G networks brought ultra-reliable low-latency communications along with enhanced mobile broadband, 6G is expected to drive these abilities uniform further, opening the door to genuinely immersive experiences like holographic communication, extended reality (XR), in addition to real-time AI-driven applications. Terahertz communication, ultra-massive MIMO, integrated sensing and communication (ISAC), and extreme connectedness are some of the major components of the 6G concept that are intended to create a smooth and intelligent network environment.

Reconfigurable Intelligent Surfaces (RIS) are a key technological innovation in the deployment of 6G. RIS has designed surfaces made of programmable components that may scatter, refract, and reflect electromagnetic waves to manipulate them. Unlike traditional passive surfaces, RIS are embedded with controllable metamaterials that allow dynamic adjustment of the signal propagation environment. By tuning the properties of each unit element, such as phase, amplitude, or polarization, RIS can actively control how signals are directed toward intended users or devices. In a 6G environment, RIS serve as a fundamental enabler for achieving smart and adaptive communication channels. They provide additional degrees of freedom in optimizing signal transmission, enhancing spectral and energy efficiency, and mitigating interference. The flexibility of RIS enables coverage extension in urban areas, overcoming blockages and dead zones by intelligently redirecting signals. This adaptability is particularly beneficial in complex environments like smart cities, dense urban areas, and indoor settings, where conventional line-of-sight communication is often challenging.

From an architectural perspective, RIS can be integrated into the 6G network as part of base stations, relay nodes, or standalone intelligent reflecting panels deployed across strategic locations. The interaction between RIS and the core 6G network introduces new dimensions of signal processing and resource management, where optimization algorithms can be applied to dynamically configure RIS based on real-time network conditions. Moreover, RIS plays a vital role in advancing key 6G objectives such as low-latency communication, enhanced connectivity density, and improved localization accuracy. By leveraging bio-inspired optimization techniques and intelligent control algorithms, RIS can be seamlessly coordinated to achieve system-wide performance goals, including reduced power consumption, optimized beamforming, and adaptive resource allocation. In this context, the integration of Hybrid Nesterov Accelerated Gradient-Golden Eagle Optimization (HNAG-GEO) algorithms further enhances the network's ability to optimize these configurations efficiently, ensuring a balanced trade-off between exploration and exploitation.

1.2 Importance of Optimization in 6G RIS Deployments

Reconfigurable Intelligent Surfaces (RIS) are a revolutionary technology that allows wireless propagation environments to be dynamically controlled in 6G networks. To fully utilize RIS, nevertheless, extensive optimization techniques must be used to overcome the difficulties brought on by the intricacy of 6G networks. The goal of 6G is to offer ultra-high data speeds, condensed latency, enhanced energy effectiveness, along with ubiquitous connectivity. In order to ensure operative resource utilization, optimal signal quality, and dependable network performance in a dynamic and diversified environment, optimization becomes increasingly important.

1.2.1 Complexity in 6G RIS Networks

6G networks are inherently complex, with densely interconnected devices, massive MIMO systems, and ultra-high-frequency bands such as the terahertz spectrum. In this context, RIS deployment introduces additional layers of complexity. Each RIS is composed of numerous reflecting elements, each with tunable properties like phase shifts and amplitudes. Coordinating these elements to enhance signal strength, minimize interference, and meet quality-of-service (QoS) requirements demands real-time, multi-dimensional optimization. The interaction between RIS and the surrounding environment is dynamic and context-aware. Factors such as user mobility, environmental changes, and varying traffic loads constantly affect network performance. Effective optimization algorithms must adapt to these fluctuations in real-time, ensuring that the RIS configurations remain optimal despite the changing conditions.

1.2.2 Resource Allocation and Network Efficiency

In a 6G RIS-empowered network, resource allocation plays a perilous role in balancing spectral efficiency, energy consumption, in addition to user experience. Optimization is crucial in determining how to allocate resources like bandwidth, power, and RIS configurations to maximize network throughput while minimizing latency and energy usage. Given the ultra-dense nature of 6G networks, with billions of connected devices and ultra-reliable low-latency communication (URLLC) demands, old-style optimization approaches may fall short in addressing the increasing complexity (Dhinakaran et al., 2023). Advanced algorithms that integrate bio-inspired techniques and machine learning are essential to navigating the large solution spaces and identifying near-optimal configurations. By incorporating techniques such as the Hybrid Nesterov Accelerated Gradient-Golden Eagle Optimization (HNAG-GEO), the network can intelligently adjust RIS settings to strike an ideal balance between conflicting objectives, such as maximizing coverage while minimizing power consumption.

1.2.3 Enhancing Signal Quality and Coverage

One of the key functions of RIS is to enhance signal quality by intelligently reflecting and directing signals toward intended receivers. However, optimizing signal quality involves multiple interdependent factors, including the optimal placement of RIS panels, selection of phase shifts, and coordination with base stations. Optimization in this context ensures that signals are delivered with minimal degradation and maximal strength, even in non-line-of-sight (NLOS) environments. In scenarios such as smart cities, indoor environments, and densely populated urban areas, RIS optimization can help overcome obstacles like physical blockages, ensuring seamless communication. The use of robust optimization frameworks ensures that RIS dynamically adapts its configuration to maintain strong and reliable communication links, contributing to improved user experience and network resilience.

1.2.4 Scalability and Real-Time Adaptation

As 6G networks scale up in terms of connected devices and RIS deployments, the optimization process must also scale. The computational complexity of optimizing RIS configurations across multiple surfaces and devices grows exponentially, requiring scalable optimization techniques that can efficiently handle large-scale systems. Real-time adaptation is another crucial requirement in 6G RIS deployments (Yu et al., 2019). The network environment can change rapidly due to factors like user mobility, environmental dynamics, and varying service demands. Optimization algorithms must be capable of performing swift, on-the-fly adjustments to RIS settings to ensure continuous and stable communication. Techniques such as HNAG-GEO bid the agility desirable for such real-time adaptation, empowering the network to rapidly converge on the way to optimal solutions while continuously exploring new configurations to handle unforeseen challenges.

1.2.5 Energy Efficiency and Sustainability

One of the core goals of 6G is to achieve sustainable and green communication. RIS technology is inherently energy-efficient since it relies on passive or low-power active elements to manipulate signal paths. However, the overall energy efficiency of a 6G RIS network depends heavily on the optimization of resource allocation, power management, and intelligent control of RIS elements. Optimization in this domain focuses on reducing power consumption while maintaining high network performance. By intelligently configuring RIS settings and power levels, optimization algorithms ensure that the network operates with minimal energy wastage. This aligns with the sustainability goals of 6G, tiling the mode for ecologically friendly then cost-effective communication systems.

1.2.6 The Role of Bio-Inspired Optimization Techniques

Bio-inspired optimization techniques are particularly compatible for solving intricate, non-convex, in addition to multi-objective problems in 6G RIS networks. Techniques such as HNAG-GEO leverage principles observed in natural processes to efficiently search large solution spaces and identify near-optimal solutions. These algorithms are designed to balance exploration and exploitation, ensuring comprehensive coverage of potential solutions while rapidly converging toward the best configurations. The deployment of RIS in 6G networks introduces a wide array of optimization challenges that must be addressed to fully unlock their potential. Effective optimization is key to overcoming the complexity, enhancing signal quality, improving resource allocation, and achieving scalability, real-time adaptation, and energy efficiency. By integrating advanced optimization techniques, 6G networks can seamlessly coordinate RIS elements, ensuring that communication remains reliable, efficient, and adaptive in an increasingly complex wireless environment.

1.3 Problem Statement and Research Motivation

As 6G networks continue to progress, they convey into view radical developments in communication technology, including unprecedented data rates, ultra-reliable low-latency communication (URLLC), and enormous connectivity. Central to achieving these goals is the distribution of Reconfigurable Intelligent Surfaces (RIS), which have the potential to transform wireless environments by enthusiastically controlling signal propagation. However, despite the promising capabilities of RIS in 6G networks, several challenges hinder their effective deployment and utilization.

1.3.1 Problem Statement

The primary problem lies in optimizing the performance of RIS in complex and dynamic 6G environments. The inherent complexity of RIS, which involves managing multiple reconfigurable reflecting elements, demands sophisticated algorithms capable of real-time adaptation, energy efficiency, and optimal signal enhancement. Traditional optimization approaches often fall short when addressing the multidimensional, non-linear nature of RIS optimization. Key challenges include:

1. **Multi-objective Optimization:** The need to balance conflicting objectives, for instance take full advantage of signal strength while minimizing energy consumption and maintaining low latency, acquaint with significant complexity in RIS management.
2. **Real-time Adaptation:** 6G networks are extremely dynamic, with varying user demands, mobility patterns, and environmental factors. Guaranteeing that RIS configurations can be optimized in real time while familiarizing to changing conditions is a critical challenge.
3. **Scalability:** As 6G networks scale with billions of devices and multiple RIS deployments, the computational complexity of optimizing RIS settings across the network grows exponentially. Traditional algorithms are often unable to handle such large-scale, real-time optimization efficiently.
4. **Energy Efficiency:** Although RIS technology is designed to be energy-efficient, improper optimization of configurations can lead to power wastage. Balancing energy efficiency with performance requirements is crucial for accomplishing sustainable 6G networks.
5. **Non-line-of-Sight (NLOS) Communication:** In dense urban areas, signals often encounter obstacles, leading to degradation in quality. Optimizing RIS configurations to effectively manage NLOS scenarios while ensuring strong signal reception is a complex task.

1.3.2 Research Motivation

The motivation behind this chapter twigs from the critical need to progress robust, scalable, and energy-efficient optimization strategies for RIS in 6G networks. As 6G networks are envisioned to sustenance emerging applications for instance smart cities, autonomous vehicles, and advanced healthcare, it becomes imperious to address the optimization challenges associated with RIS deployment. Key motivations include:

1. **Enabling Ultra-reliable, Low-latency Communications (URLLC):** Advanced applications in 6G, for instance distant surgery and autonomous systems, require ultra-reliable communication with minimal latency. Effective RIS optimization is crucial in ensuring that these applications receive consistent and superior connectivity, even in challenging environments.
2. **Achieving Energy Efficiency and Sustainability Goals:** As sustainability becomes a priority for future communication networks, RIS bids a promising result for energy-efficient communication. However, understanding this potential requires optimization strategies that minimize energy consumption without compromising performance.
3. **Enhancing Network Capacity and Coverage:** RIS technology can suggestively improve network capacity and coverage by perceptively directing signals to underserved areas. This investigate is motivated by the need to develop algorithms that maximize coverage and signal eminence, especially in dense and complex environments.
4. **Leveraging Bio-Inspired Optimization Techniques:** Traditional optimization approaches often skirmish with the high-dimensional and non-convex nature of RIS configuration problems. This chapter is driven by the potential of bio-inspired algorithms, such as the Hybrid Nesterov Accelerated Gradient-Golden Eagle Optimization (HNAG-GEO), to proficiently explore and exploit the big solution spaces, providing near-optimal results in real time.
5. **Advancing the State-of-the-Art in 6G Research:** As the field of 6G continues to grow, there is a persistent need for innovative approaches that ambition the boundaries of what is possible. By addressing the optimization tasks associated with RIS deployment, this chapter intentions to contribute to the progress of high-performance 6G networks that meet the demands of future applications.

The problem statement highlights the complexities of optimizing RIS in 6G networks, while the chapter incentive underscores the reputation of overcoming these challenges to unlock the full potential of 6G. By concentrating on advanced optimization strategies, bio-inspired methods, this chapter seeks to pave the way for effective, adaptive, and sustainable RIS positionings in next-generation communication systems.

1.4 Objectives of the Chapter

This chapter aims to provide a comprehensive overview and foundational understanding of RIS within the context of 6G networks. The chapter is organized to accomplish the following key objectives:

1. **Exploring the Fundamentals of 6G Networks and RIS Technology:** The chapter begins with an in-depth discussion of the contextual and core principles of 6G networks, emphasizing the role and significance of RIS in empowering the transformative capabilities projected in these networks. The objective is to establish a clear considerate of how RIS operates, including its structure, principles, and potential applications within 6G environments.
2. **Highlighting the Importance of Optimization in RIS Deployments:** The chapter emphasizes the serious role of optimization in guaranteeing effectual RIS configurations that meet the miscellaneous requirements of 6G networks. This objective focuses on identifying the encounters associated with optimizing RIS, such as balancing performance metrics like signal eminence, energy effectiveness, latency, and scalability.
3. **Defining the Problem Statement and Research Motivation:** A key objective of this chapter is too lucid the specific issues that arise in optimizing RIS for 6G networks. The chapter defines the problem statement by deliberating the complex, multidimensional nature of RIS optimization and how existing approaches fall short. The motivation behind addressing these challenges is clearly explained, highlighting the pressing need for advanced optimization approaches to unlock the full potential of RIS in future communication systems.
4. **Establishing the Research Objectives for RIS Optimization:** The chapter sets the stage for the outline of the proposed Hybrid Nesterov Accelerated Gradient-Golden Eagle Optimization (HNAG-GEO) algorithm. This objective comprises defining the criteria and goals that a effective optimization solution should meet, such as real-time adaptability, scalability, energy efficiency, and sturdiness in diverse environments.
5. **Introducing the Significance of Bio-Inspired Optimization Techniques:** The chapter provides an overview of bio-inspired optimization methods and their appropriateness for providing solution for the complex issues, high-dimensional problems posed by RIS in 6G networks. This objective is to lay the groundwork for why and how techniques like HNAG-GEO are well-positioned to address the optimization challenges, setting the foundation for subsequent discussions.

6. **Providing a Clear Outline of the Chapter Structure:** The final objective is to offer a roadmap for the rest of the chapter, guiding readers through the logical progression of topics that include the theoretical underpinnings of RIS, the need for advanced optimization, and the introduction of the HNAG-GEO algorithm as a novel solution. This structure ensures that readers can follow the flow of ideas and fully grasp the significance of the proposed contributions.

2. BACKGROUND AND RELATED WORK

2.1 Overview of Traditional Optimization Techniques in RIS

Reconfigurable Intelligent Surfaces (RIS) have arisen as a groundbreaking technology for enhancing wireless communication in 6G networks. However, realizing the full potential of RIS necessitates effective optimization techniques to accomplish the complex interactions among the surface elements and the wireless environment. This section delivers an overview of traditional optimization techniques applied to RIS, discovering their methodologies, strengths, and limitations.

1. Linear Programming (LP)

Linear programming is one of the simplest and most widely used optimization techniques in RIS deployments (Jian et al., 2022). LP techniques model the optimization problem as a linear objective function subject to a set of linear constraints. In the context of RIS, LP can be used to optimize parameters such as phase shifts and beamforming vectors to improve signal strength or reduce interference. LP is computationally efficient and can solve large-scale problems with well-defined constraints. LP is limited to linear problems, making it less suitable for the inherently nonlinear nature of RIS optimization, where interactions between elements can be highly complex and dynamic.

2. Quadratic Programming (QP)

Quadratic programming extends linear programming by allowing the objective function to include quadratic terms. This is particularly useful in RIS optimization, where the relationship between variables like signal strength and phase shifts may not be strictly linear. QP can handle more complex relationships between variables, offering better optimization in scenarios where interactions are nonlinear but can still be modeled quadratically. QP is computationally more intensive than LP and may struggle with real-time optimization in dynamic environments. Moreover, QP assumes convexity, which may not hold in all RIS scenarios.

3. Mixed-Integer Programming (MIP)

Mixed-integer programming is a powerful technique that allows for optimization problems where some variables are constrained to be integers (Alamzadeh et al., 2021). This is particularly relevant for RIS optimization when dealing with discrete control variables, such as the on-off states of individual RIS elements. MIP can model a wide range of problems involving both continuous and discrete variables, offering flexibility in RIS control. MIP problems are often NP-hard, making them interesting to solve, particularly in real-time scenarios. The computational complexity can convert prohibitive as the scope of the problem grows.

4. Stochastic Gradient Descent (SGD)

Stochastic gradient descent is a popular optimization technique utilized for problems where the objective function is differentiable. In RIS, SGD can be utilized to iteratively adjust parameters like phase shifts to minimize an objective function related to signal quality or energy consumption. SGD is well-suited for large-scale optimization and can handle high-dimensional parameter spaces typical of RIS problems. SGD may converge to local minima rather than the global optimum, particularly in non-convex RIS optimization landscapes. The performance of SGD can also be sensitive to the choice of learning rate and initial conditions.

5. Metaheuristic Algorithms

Traditional optimization techniques often struggle with the complexity and non-linearity of RIS problems, leading to the use of metaheuristic algorithms for instance Genetic Algorithms (GA), Particle Swarm Optimization (PSO), and Simulated Annealing (SA). These techniques are stimulated by natural processes and do not necessitate gradient information, making them well-suited for the complex, black-box nature of RIS optimization. Metaheuristics are extremely flexible and can escape local minima, offering a better chance of finding the global optimum (Jagadish Kumar et al., 2024). They are also flexible to a wide range of problem types, including those with non-linear, non-convex, as well as the multi-modal landscapes. Metaheuristic algorithms often require a large number of function evaluations, leading to high computational costs. They also do not promise optimal solutions and can suffer from slow convergence rates.

6. Convex Optimization

Convex optimization techniques are widely used in RIS due to their mathematical tractability. By reformulating RIS optimization problems as convex problems, solutions can be found efficiently using algorithms like interior-point methods. Convex optimization guarantees global optimality, making it highly desirable in scenarios where the RIS problem can be convexified. Not all RIS optimization problems can be transformed into convex ones. This approach may also oversimplify the problem, leading to suboptimal solutions when the original problem is inherently non-convex.

7. Game Theory-Based Optimization

Game theory has been applied to RIS optimization by modeling the interactions between different RIS elements or between the RIS and users as a game. Techniques like Nash equilibrium is used to find stable states where no player has an incentive to deviate. Game theory provides a framework for analyzing and optimizing systems with multiple interacting agents, making it useful for decentralized RIS control. Game-theoretic approaches can be complex to implement and may lead to multiple equilibria, not all of which are optimal from a system-wide perspective.

Jung et al., (2021) introduced a practical channel acquisition and passive beamforming technique tailored for RIS-assisted cellular networks. The technique employs a meta-learning framework, enabling the RIS to operate efficiently using a limited number of pilot symbols. By focusing on maximizing the instantaneous spectral efficiency (SE), the algorithm dynamically selects optimal phase shifts without requiring perfect channel state information (CSI). While effective, this method is constrained by its reliance on accurate meta-learning models, which may struggle in scenarios with rapidly changing environments, leading to suboptimal beamforming in highly dynamic 6G settings. Hodge et al. (2023) proposed an innovative design for RIS-enabled transceivers that leverages index modulation (IM) to enhance the communication capabilities of 6G networks. The approach incorporates spatial multiplexing, beam steering, and phase modulation through electromagnetics-compliant RIS apertures. By integrating spatial and subcarrier IMs, this solution achieves superior bit error rate (BER) performance compared to traditional implementations. However, challenges remain in practical deployments, especially regarding the complexity of implementing such designs and the scalability issues in large-scale network scenarios.

Liang et al. (2021) provided a comprehensive overview of RIS-aided systems, highlighting two emerging paradigms: RIS-aided wireless communications (RAWC) and RIS-based information transmission (RBIT). RAWC focuses on using RIS as a reflector to enhance signal propagation, while RBIT employs RIS as an active transmitter. Both paradigms offer unique advantages in terms of flexibility and signal control but encounter difficulties related to accurate channel estimation and system design. These challenges become more pronounced in 6G environments with dense user populations and high mobility, where the performance of traditional RIS configurations may deteriorate. Frimpong et al. (2023) explored the potential of irregular RIS (IRIS) designs to enhance physical-layer security in 6G networks. The proposed approach involves selectively activating RIS elements in an irregular pattern to maximize secrecy rates in the presence of eavesdroppers. By optimizing the RIS topology and precoding strategies, this method ensures secure communication between legitimate users while minimizing information leakage. Although effective in boosting security, the irregular activation of RIS elements introduces additional complexity in network management and may require sophisticated control algorithms to adapt to varying threat levels.

Alves et al. (2023) conducted an in-depth statistical and performance analysis of RIS-assisted multiple-antenna wireless communication systems under Weibull fading. The study demonstrates the benefits of leveraging RIS for constructive interference in double-Weibull-fading channels, leading to improved communication reliability. The analysis also accounts for non-ideal phase cancellation effects, which are modeled using the von Mises distribution. Despite the performance gains, this approach is hindered by the inherent complexity of managing non-idealities in large-scale networks, which could limit its applicability in real-world 6G scenarios.

Traditional optimization techniques have provided valuable tools for addressing the challenges of RIS optimization in 6G networks. However, the complexity, non-linearity, and dynamic nature of RIS environments often push these methods to their limits. While techniques like LP, QP, and MIP offer computational efficiency and mathematical rigor, they struggle with the high-dimensional, non-convex landscapes typical of RIS problems. Metaheuristics and game theory approaches offer flexibility and the ability to escape local optima but come with their own set of challenges, such as high computational costs and implementation complexity. This background sets the stage for exploring more advanced and specialized optimization techniques, such as bio-inspired methods, which are better suited to the unique demands of RIS in next-generation wireless networks.

2.2 Limitations of Existing Optimization Algorithms for RIS in 6G Networks

Although traditional optimization techniques have made substantial contributions to enhancing Reconfigurable Intelligent Surfaces (RIS) in 6G networks, these approaches encounter several restrictions when faced with the dynamic and complex nature of modern wireless environments. As RIS technology progresses to meet the stringent demands of 6G networks, existing algorithms face multiple encounters, including high dimensionality, non-linearity, real-time adaptability, and energy efficiency. This section discovers the key limitations that hinder the effectiveness of conventional optimization approaches in RIS deployments for 6G.

1. Scalability and High Dimensionality

One of the critical challenges in RIS optimization is the sheer number of tunable parameters involved. RIS systems consist of hundreds or even thousands of reflective elements, each requiring precise control to optimize network performance. Traditional algorithms like Linear Programming (LP) as well as the Quadratic Programming (QP) struggle with high-dimensional optimization problems, where the solution space raises exponentially with the number of variables. As a result, these methods become computationally infeasible for large-scale RIS deployments in 6G networks.

2. Non-Linearity and Non-Convexity

RIS optimization problems are often characterized by non-linear and non-convex relationships between variables such as stage shifts, beamforming vectors, as well as the channel conditions. Conventional optimization techniques typically rely on convexity assumptions to guarantee convergence to a global optimum. However, utmost real-world RIS problems are non-convex, chief to suboptimal solutions when using traditional methods. Gradient-based techniques, like Stochastic Gradient Descent (SGD), may also converge to local minima instead of the global optimum, further limiting their effectiveness in complex RIS scenarios.

3. Real-Time Adaptability

The dynamic nature of 6G networks demands real-time adaptation to changing conditions such as user mobility, channel fading, as well as fluctuating traffic loads. Existing optimization algorithms often require extensive computational resources and time to converge to a solution, making them unsuitable for real-time applications. For instance, Mixed-Integer Programming (MIP) can be extremely accurate but is computationally intensive, leading to delays that undermine the effectiveness of RIS in scenarios requiring instant decisions and updates.

4. Energy Efficiency

Energy efficiency is a crucial consideration in 6G networks, where minimizing power consumption is as significant as optimizing network performance. Traditional optimization techniques every so often prioritize metrics such as signal strength or throughput without adequately accounting for energy efficiency. Algorithms like Convex Optimization emphasis on performance metrics but may lead to energy-intensive configurations that are unsustainable in large-scale deployments. In RIS, where the goal is often to enhance signal propagation with minimal power consumption, existing methods fall short of achieving a balanced trade-off between performance and energy efficiency.

5. Complexity of System Interactions

RIS optimization involves managing interactions between multiple subsystems, including user devices, base stations, and RIS elements. Traditional algorithms are often designed for single-objective or limited multi-objective optimization problems, where only a few parameters are optimized simultaneously. However, the intricate interdependencies in 6G RIS environments require multi-objective optimization that can balance trade-offs among several conflicting criteria, such as coverage, latency, interference, and energy consumption. Existing methods like Game Theory-based Optimization offer some flexibility but are complex to implement and may result in unstable or suboptimal equilibrium states.

6. Lack of Robustness in Dynamic Environments

6G networks are highly dynamic, with frequent changes in network topology, user density, and environmental factors. Traditional algorithms often lack the robustness required to maintain optimal performance under such conditions. Techniques alike Metaheuristics (e.g., Genetic Algorithms and Particle Swarm Optimization) offer flexibility but can be penetrating to parameter tuning and may require constant re-optimization, chief to inconsistent results in fluctuating environments.

7. Limited Interpretability and Explainability

As 6G networks become more complex, the need for interpretable and explainable solutions grows. Network operators and engineers require insights into how and why certain configurations are chosen, especially in safety-critical applications. However, many traditional optimization techniques operate as black boxes, providing little to no interpretability. This lack of transparency is a significant drawback, particularly when troubleshooting or refining RIS configurations in 6G networks.

8. Inability to Handle Heterogeneous and Hybrid Architectures

6G networks are expected to integrate heterogeneous components, including various types of RIS, multiple access technologies, and hybrid communication strategies. Traditional optimization algorithms are often designed with homogeneous systems in mind and struggle to adapt to such diverse architectures. This limitation restricts their pertinence in next-generation networks, where the seamless integration of diverse technologies is essential for achieving optimal performance.

The limitations of traditional optimization algorithms highpoint the need for more advanced and adaptive methods tailored to the unique demands of RIS in 6G networks. While techniques like LP, QP, and Convex Optimization deliver a solid foundation, they fall short in addressing the scalability, non-linearity, real-time adaptability, and energy efficiency challenges posed by modern wireless environments. As RIS continues to play a critical role in 6G deployments, innovative optimization strategies, such as bio-inspired algorithms and hybrid metaheuristics, are needed to overcome these limitations and unlock the full potential of this transformative technology.

2.3 The Role of Bio-Inspired Algorithms in Network Optimization

Bio-inspired algorithms have emerged as powerful tools in network optimization, particularly in addressing the complex and dynamic challenges of 6G networks and Reconfigurable Intelligent Surfaces (RIS). By mimicking natural processes like evolution, swarming, and predator-prey behavior, these algorithms offer innovative solutions that are more adaptable, scalable, and capable of handling non-linear and multi-objective problems than traditional optimization techniques. This section explores the role and significance of bio-inspired algorithms in optimizing network performance, focusing on how they enhance the effectiveness of RIS in 6G networks.

1. Adaptability to Dynamic Environments

One of the key strengths of bio-inspired algorithms is their ability to adapt to changing conditions in real-time, a feature that is essential for the dynamic and heterogeneous nature of 6G networks. For instance, algorithms inspired by swarm intelligence, such as Particle Swarm Optimization (PSO) and Ant Colony Optimization (ACO), are designed to continuously update their solutions based on environmental feedback. In a RIS-enabled 6G network, where factors like user mobility, interference, and channel conditions fluctuate rapidly, bio-inspired algorithms can quickly adapt, optimizing phase shifts and reflection coefficients to maintain optimal network performance.

2. Handling High Dimensionality and Complex Problem Spaces

RIS systems in 6G networks involve optimizing a vast number of elements, often leading to high-dimensional, non-convex optimization problems. Traditional algorithms struggle with such complexity, but bio-inspired approaches are inherently suited for exploring large search spaces. Evolutionary algorithms like Genetic Algorithms (GA) and Differential Evolution (DE) use principles of natural selection to explore multiple solutions simultaneously, effectively navigating the complex solution landscape to find near-optimal configurations. This capability makes them ideal for managing the high dimensionality of RIS parameter tuning.

3. Multi-Objective Optimization

In 6G networks, optimizing a single performance metric, such as throughput or latency, is often insufficient. There is a essential for optimization, where multiple conflicting goals—such as signal coverage, energy efficiency, as well as the spectral efficiency—must be balanced. Bio-inspired algorithms like Multi-Objective Evolutionary Algorithms (MOEAs) are well-suited for this task. They preserve a diverse population of solutions, permitting them to explore trade-offs between different objectives. For RIS in 6G, these algorithms can simultaneously optimize multiple parameters to achieve a balanced network configuration that meets varied performance criteria.

4. Global Search Capabilities and Avoidance of Local Minima

Traditional gradient-based methods are prone to getting trapped in local minima, particularly in non-linear and non-convex problems typical of RIS optimization. Bio-inspired algorithms, however, are designed for global search. Techniques like Simulated Annealing (SA) and Genetic Algorithms (GA) incorporate mechanisms such as mutation and temperature decay that enable them to escape local minima and continue exploring the global solution space. In the context of 6G RIS, where finding the global optimum is crucial for maximizing network efficiency, bio-inspired methods offer a significant advantage.

5. Robustness and Flexibility

The robustness of bio-inspired algorithms lies in their decentralized and stochastic nature. Unlike deterministic algorithms that require precise problem formulations and risk failure in uncertain environments, bio-inspired methods are highly resilient to uncertainties and noise. For example, algorithms like Artificial Bee Colony (ABC) and Firefly Algorithm (FA) can handle noisy and incomplete data, making them effective for real-world 6G scenarios where perfect information is rarely available. This robustness is crucial in RIS deployments where environmental factors, such as multipath fading and shadowing, introduce significant uncertainty.

6. Energy Efficiency and Sustainability

As 6G networks emphasize energy efficiency and sustainability, bio-inspired algorithms offer strategies for optimizing energy consumption while maintaining performance. Algorithms inspired by natural processes, such as the behavior of birds in flocking (PSO) or the energy-saving strategies of predators (Golden Eagle Optimization), can be used to design RIS configurations that achieve high performance with minimal power usage. By efficiently balancing energy consumption across network elements, these algorithms help to meet the stringent energy requirements of green 6G networks.

7. Parallelism and Scalability

Bio-inspired algorithms are inherently parallelizable, allowing them to scale effectively in large and complex environments like 6G networks. Techniques like Genetic Algorithms and Particle Swarm Optimization operate on populations of potential solutions, making them well-suited for parallel execution. This parallelism can be leveraged in large-scale RIS deployments where the optimization of thousands of elements must be performed simultaneously. The ability to scale without a significant increase in computational complexity makes bio-inspired algorithms highly advantageous for extensive 6G network deployments.

8. Incorporation of Hybrid Approaches

One of the recent trends in bio-inspired optimization is the development of hybrid algorithms that combine the strengths of multiple techniques. For instance, Hybrid Genetic Algorithms (HGA) can incorporate the local search capabilities of gradient-based methods while retaining the global search benefits of evolutionary processes. In 6G RIS optimization, hybrid approaches allow for more fine-tuned solutions, combining global exploration with local exploitation to achieve highly optimized network configurations. These hybrid bio-inspired methods offer a versatile solution for the multi-faceted optimization challenges in 6G networks.

Bio-inspired algorithms play a pivotal role in addressing the optimization challenges posed by 6G RIS deployments. By leveraging principles derived from nature, such as evolution, swarming, and natural selection, these algorithms offer robust, scalable, and adaptive solutions capable of navigating the complex and dynamic environments of next-generation networks. Their ability to handle high-dimensionality, non-linearity, and multi-objective optimization makes them indispensable for realizing the full potential of RIS technology in 6G, ensuring enhanced network performance, energy efficiency, and real-time adaptability.

2.2 Traditional Optimization algorithms in RIS for 6G Networks

This section collectively highlights the diverse optimization techniques applied in RIS-aided 6G networks, including traditional methods and heuristic approaches. D. Vordonis et al. (2022) introduces a new viewpoint to RIS-aided Orthogonal frequency division multiplexing (OFDM) systems by fixing the problems with the ideal reflection model that has been employed in the past. While the ideal model assumes that the amplitude and phase shift of RIS components remain constant regardless of the frequency of the incoming signal, their work presents a realistic model that takes frequency-dependent fluctuations into consideration. An optimisation approach based on exhaustive search was created by the researchers to maximise the average minimum Euclidean distance (AMED) inside subblocks to improve the bit error rate (BER) performance. This AMED-based reflection coefficient design approach improves BER performance for RIS-aided OFDM systems, according to extensive computer simulations, which show that it beats other algorithms in the literature.

G. Brancati et al. (2022) enhances data rates by considering factors such as RIS height, blockers density, user number, and dispersion as a facility location issue. The results show that when the RIS height is more than 5 m, the best location is close to the base station (BS) for both uniform and cluster distributions, and when the RIS height is less than 5 m, the best location is near hotspots for cluster distributions. By improving network performance under real-world settings, these insights enable optimal RIS implementation methods. Future RIS research and implementation in next-generation cellular networks may benefit greatly from this all-encompassing method. M. Sikander et al. (2023) enhances data rates by considering factors such as RIS height, blockers density, user number, and dispersion as a facility location issue. The results show that when the RIS height is more than 5 m, the best location is close to the base station (BS) for both uniform and cluster distributions, and when the RIS height is less than 5 m, the best location is near hotspots for cluster distributions. By improving network performance under real-world settings, these insights enable optimal RIS implementation methods. Future RIS research and implementation in next-generation cellular networks may benefit greatly from this all-encompassing method.

H. Ma et al. (2023) investigates 6G networks' use of RIS, with a particular focus on power-splitting assisted broadcasting networks, in which a single access point (AP) conveys similar messages to several users. By manipulating the reflect amplitude coefficients, the RIS can achieve energy sustainability by reflecting some of the incident signals to help the AP and capturing energy from other signals. The objective is to optimise the RIS reflect beamforming matrix and AP transmit beamforming vector in such a way that the RIS can maintain itself while still providing high-quality service to users.

3. PROPOSED HYBRID ALGORITHM: HNAG-GEO

3.1 Overview

The proposed component diagram represents the interaction between various components in the 6G network architecture leveraging bio-inspired optimization techniques to enhance performance. The system integrates the Hybrid Nesterov Accelerated Gradient-Golden Eagle Optimization (HNAG-GEO) algorithm within the RIS-enhanced 6G network as shown in Figure 1.

3.2 Components of HNAG-GEO

1. Input Layer Components

User Equipment (UE) represent the mobile devices or terminals connecting to the 6G network. They send requests for services such as high-speed data, low-latency communication, and enhanced network reliability. Service Requests are generated based on user needs and are sent to the Base Station (BS) or Access Points (APs). The requests include data rate requirements, quality of service (QoS) demands, and specific applications (e.g., IoT, AR/VR). Environmental Data Acquisition, Sensors gather environmental information, such as signal strength, noise levels, and user mobility patterns. This data is critical for determining optimal configurations for RIS panels.

2. Process Layer Components

Base Station/Access Point (BS/AP) serves as the initial interface for communication between UEs and the network. It receives user data and environmental information and coordinates with RIS to optimize signal transmission. Reconfigurable Intelligent Surfaces (RIS) panels are strategically deployed to dynamically adjust signal propagation by reflecting, refracting, or amplifying signals based on the requirements. RIS panels are equipped with multiple elements whose phases can be individually controlled.

Figure 1. Structure of HNAG-GEO for Performance Enhancement in 6G Networks HNAG-GEO Algorithm Module:

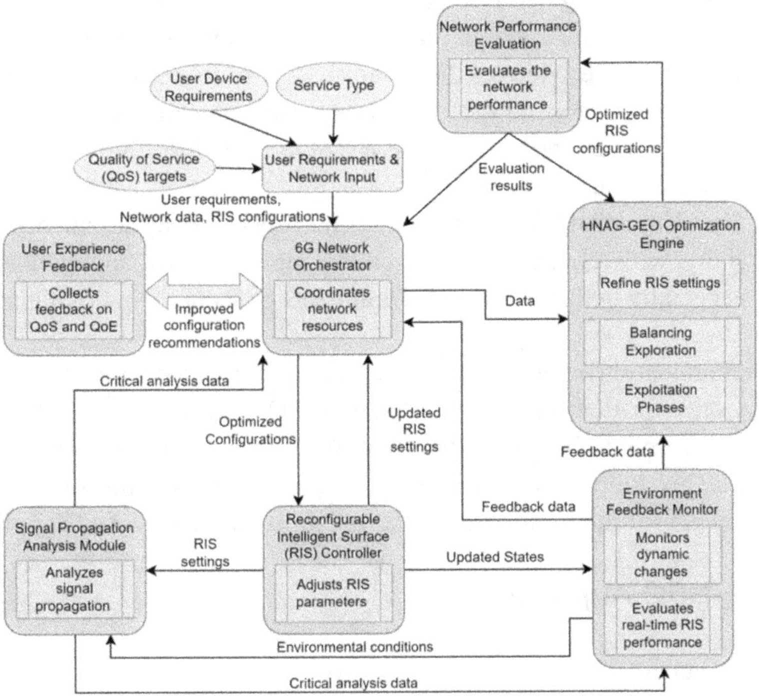

The Nesterov Accelerated Gradient component calculates the gradients necessary for determining optimal signal paths and phase shifts in RIS. The GEO component manages the balance between exploration (discovering new configurations) and exploitation (refining existing solutions) based on the current state of the network. The hybrid module outputs the optimized paths, which guide how RIS panels adjust their phases to enhance signal strength and mitigate interference.

3. Communication Flow Between Components

The Service Requests from UEs are processed by the Base Station, which forwards the requirements to the RIS Controller. The RIS Controller interacts with the Environmental Data Acquisition module to assess the current network environment. The HNAG-GEO Algorithm Module receives this data and computes the best RIS configuration by adjusting phase shifts across the surfaces. The optimized signal paths are fed back to the RIS, which applies the necessary adjustments. The Base Station coordinates the final communication link between the RIS and UEs, ensuring that performance metrics (e.g., throughput, latency) are met.

4. Output Layer Components

   Optimized Signal Transmission: The outcome is a signal path that maximizes data rate and minimizes interference across the 6G network.
   Performance Metrics Feedback: Real-time performance data, such as latency, packet loss, and throughput, are continuously monitored. This feedback loop allows the HNAG-GEO algorithm to further refine configurations.
   User Experience Improvement: End-users experience enhanced connectivity, with more stable connections, higher speeds, and reduced delays, especially in dense urban environments or scenarios with high mobility.

The communication begins with user requests and environmental data collection, followed by optimization through the HNAG-GEO algorithm. The RIS panels are dynamically configured based on these optimizations, leading to enhanced signal propagation. This loop ensures adaptive, real-time adjustments in the network, which are crucial for the performance and reliability demands of 6G applications.

3.3 HNAG-GEO Algorithm

Input Parameters:

- Initial population of golden eagles $\{y_1, y_2, \ldots, y_n\}$
- Maximum number of iterations T
- Learning rate η
- Momentum parameters μ_t
- Coefficients d_a, d_c (attack and cruise coefficients)
- Gradient function $\nabla L_t(\omega_t)$
- Initial parameters ω_t, $m_t = 0$, and $n_t = 0$

Step 1: Initialization

1. Initialize the golden eagle population with random positions y_i in the search space.
2. Compute the fitness function for each eagle.
3. Store the initial positions as memory for each eagle.
4. Set initial values for momentum vectors and other related variables.

Step 2: Iterative Optimization

For each iteration:

t=1 to T:

Step 3: NAG Gradient Update

1. Compute the gradient using NAG:

$$g_t = \nabla L_t(\omega_t)$$

$$\hat{g}_t = \frac{g_t}{1 - \prod_{i=1}^{t} \mu_i}$$

Update momentum vectors: 2.

$$m_t = \mu_t m_{t-1} + (1 - \mu_t)\, g_t$$

$$\hat{m}_t = \frac{m_t}{1 - \prod_{i=1}^{t+1} \mu_i}$$

3. Compute the look-ahead momentum vector:

$$\overline{m}_t = (1 - \mu_t)\hat{g}_t + \mu_{t+1}\hat{m}_t$$

4. Update the position ω_{t+1}:

$$\omega_{t+1} = \omega_t - \eta \frac{\overline{m}_t}{\sqrt{n_t} + \epsilon}$$

where ϵ is a small constant for numerical stability.

Step 4: GEO-Based Exploration and Exploitation

1. For each golden eagle i, randomly select a prey (solution) from the population memory.
2. Compute the attack vector B_i:

$$B_i = y_{f^*} - y_i$$

3. Compute the cruise vector E_i:

$$E_i = (e_1, e_2, \ldots, e_k)$$

where e_k is determined using:

$$e_k = \frac{P - \sum_{j \neq k} b_j}{b_k}$$

Here, b_j and b_k are elements of the attack vector, and P is the sum of weighted factors for the current search hyperplane.

4. Compute the step vector Δy_i:

$$\Delta y_i = z_1 d_a \frac{B_i}{\|B_i\|} + z_2 d_c \frac{E_i}{\|E_i\|}$$

where z_1 and z_2 are random vectors in the range [0, 1], and $\|B_i\|$ and $\|E_i\|$ are Euclidean norms.

5. Update the golden eagle's position:

$$y_{i+1} = y_t + \Delta y_i$$

6. Evaluate the new position's fitness. If it improves the previous memory, update the memory with the new position.

Step 5: Exploration to Exploitation Transition

1. Gradually transition from exploration to exploitation by updating d_a and d_c based on the current iteration t:

$$d_a = d_{a0} + \frac{t}{T}|d_{aT} - d_{a0}|$$

$$d_c = d_{c0} + \frac{t}{T}|d_{cT} - d_{c0}|$$

Step 6: Convergence Check

1. If convergence criteria are met or maximum iterations T are reached, stop the algorithm.

Output: Best solution y_{best} and corresponding fitness value.

3.4 HNAG-GEO in 6G Networks with RIS

1. **Bio-Inspired Optimization with Adaptability**: RIS in 6G networks requires dynamic optimization to adapt to changing environmental conditions. HNAG-GEO leverages the adaptive capabilities of the Nesterov Accelerated Gradient (NAG) for fast convergence and the exploration-exploitation balance of Golden Eagle Optimization (GEO) for global search, making it well-suited for optimizing reconfigurable surfaces in diverse scenarios.
2. **Acceleration in High-Dimensional Search Spaces**: In 6G, optimizing network parameters over RIS involves high-dimensional and complex search spaces. The NAG component of HNAG-GEO accelerates convergence by anticipating gradient directions, reducing redundant steps. This enables efficient handling of high-dimensional search spaces, which is crucial in optimizing RIS configurations for enhanced signal quality and energy efficiency.
3. **Balancing Exploration and Exploitation**: GEO's mechanism of attack and cruise vectors is particularly beneficial for balancing exploration and exploitation in non-convex optimization problems, like optimizing the placement and configuration of RIS in 6G networks. By combining this with NAG, HNAG-GEO ensures that the solution space is thoroughly explored while converging toward optimal configurations quickly.
4. **Enhancing Signal Coverage and Network Capacity**: The RIS in 6G networks can dynamically control signal reflection, refraction, and absorption. HNAG-GEO can optimize the control parameters to enhance signal coverage and maximize network capacity. The hybrid nature of the algorithm allows it to efficiently find the best configuration settings, even in dense and complex urban environments where RIS faces significant challenges.

5. **Adaptive Transition from Exploration to Exploitation**: HNAG-GEO's adaptive transition strategy mirrors the RIS deployment cycle, where initial exploration (broad adjustments) is followed by precise tuning (exploitation). This transition is key to optimizing resource allocation, energy consumption, and throughput in real-time scenarios encountered in 6G networks.
6. **Scalability and Robustness**: The hybrid approach integrates the strengths of both algorithms, making HNAG-GEO scalable and robust. This is crucial for large-scale RIS deployments where multiple parameters must be optimized simultaneously. The algorithm's ability to maintain performance across various deployment scales and scenarios is essential for practical 6G applications.

3.5 HNAG-GEO addresses the following concerns:

- **Speed and Accuracy**: Readers will appreciate how the NAG component accelerates convergence, leading to faster decision-making in RIS configurations, which is essential for real-time applications in 6G.
- **Exploration-Exploitation Balance**: GEO's bio-inspired strategies ensure the solution space is not prematurely converged, preventing suboptimal configurations.
- **Application Flexibility**: The hybrid algorithm's adaptability allows it to be fine-tuned for specific RIS applications in diverse environments, a key factor in 6G.
- **Innovative Approach**: Combining NAG and GEO introduces novelty in optimization techniques, showcasing the proposed algorithm as a cutting-edge solution that enhances RIS-based performance in 6G.

Overall, HNAG-GEO's hybrid nature effectively addresses the challenges in optimizing 6G networks with RIS, making it a compelling solution for improving signal propagation, energy efficiency, and overall network performance.

4. IMPLEMENTATION AND SIMULATION RESULTS

This chapter focuses on the practical evaluation of the proposed Hybrid Nesterov Accelerated Gradient-Golden Eagle Optimization (HNAG-GEO) algorithm within 6G networks incorporating Reconfigurable Intelligent Surfaces (RIS). The analysis covers the complete simulation setup, performance metrics, comparison with other existing algorithms, and the results obtained.

4.1 Simulation Setup and Parameters

The simulation setup for evaluating the Hybrid Nesterov Accelerated Gradient-Golden Eagle Optimization (HNAG-GEO) algorithm is designed to mimic real-world 6G networks incorporating Reconfigurable Intelligent Surfaces (RIS). This section delves into the key components of the simulation environment, specifying all relevant parameters that influence the performance and accuracy of the proposed algorithm. Each parameter is selected to model the complexities of 6G communication scenarios effectively, allowing for a comprehensive evaluation.

4.1.1 Simulation Environment and Tools

The simulation is performed using advanced platforms like MATLAB and Python, which are well-equipped with libraries and toolboxes specifically designed for wireless communication and optimization algorithms. The use of these platforms allows precise modeling and offers flexibility in experimenting with various network configurations.

4.1.2 Key Simulation Parameters

1. **Number of RIS Elements:** The number of RIS elements (reflective surfaces) is a crucial parameter in the simulation. Typically, large arrays of elements (ranging from 64 to 512 or more) are considered to achieve significant beamforming gains. The configuration of these elements determines how the signal is reflected and how interference is minimized, directly impacting the quality of the communication.
2. **Transmission Power:** Transmission power is an important factor, representing the power level at which the base station (or access point) transmits signals to the users. The simulations consider varying power levels to examine the energy efficiency and signal strength in different scenarios. By optimizing power levels with the HNAG-GEO algorithm, the simulation aims to maximize performance while conserving energy.
3. **Network Topology and Deployment:** The deployment scenario involves multiple users and base stations in a 6G network with RIS strategically placed for optimal signal reflection. The topology includes different layouts such as urban microcells, indoor office setups, and outdoor scenarios, each with unique propagation characteristics. The network topology plays a vital role in evaluating how effectively the RIS can enhance coverage and capacity.

4. **Signal-to-Noise Ratio (SNR):** The SNR is an essential metric for assessing communication quality. The simulation accounts for varying SNR levels to determine how well the HNAG-GEO algorithm adapts to different noise conditions. RIS configurations are optimized to improve SNR by adjusting the phase shifts of reflective elements.
5. **Bandwidth Allocation:** The simulation considers diverse bandwidth scenarios to evaluate how efficiently resources are allocated across multiple users. By integrating HNAG-GEO, the algorithm seeks to optimize bandwidth usage, ensuring balanced data rates and minimizing congestion across the network.
6. **User Mobility and Channel Characteristics:** The simulation environment models realistic user mobility patterns, such as walking or vehicular movement, affecting channel characteristics like fading and Doppler shifts. These dynamics are key to assessing the adaptability of the HNAG-GEO algorithm in maintaining stable connections under varying conditions.
7. **RIS Configuration and Phase Shift Modeling:** The RIS elements are configured with varying phase shifts to maximize signal reflection towards desired users while mitigating interference. The simulation models dynamic phase adjustment to reflect real-time optimization, where HNAG-GEO fine-tunes these shifts to achieve optimal network performance.

4.1.3 Modeling and Scenarios

The simulation considers several scenarios, including both line-of-sight (LoS) and non-line-of-sight (NLoS) conditions, urban and rural environments, and different user densities. Each scenario provides unique challenges in terms of interference, coverage, and signal quality, enabling a comprehensive analysis of how HNAG-GEO performs across diverse conditions. This detailed setup creates a robust framework for evaluating the HNAG-GEO algorithm, ensuring that all relevant factors are accounted for when simulating 6G networks with RIS technology.

4.2 Performance Evaluation Metrics

This section identifies the key metrics used to assess the performance of HNAG-GEO. Typical metrics include throughput, latency, energy efficiency, spectral efficiency, bit error rate (BER), and coverage area. These metrics are chosen to capture both the benefits of enhanced optimization and the unique aspects of RIS-enabled 6G environments. The metrics will be computed and compared to reveal how well the proposed method meets the intended objectives.

4.3 Comparison with Traditional and State-of-the-Art Algorithms

Table 1 highlights several key metrics for evaluating the performance of HNAG-GEO against other optimization algorithms used for Reconfigurable Intelligent Surfaces (RIS) in 6G networks. In this analysis, specific scenarios are selected to demonstrate how HNAG-GEO outperforms existing methods. In high-density environments like smart cities or industrial automation, maintaining high throughput is critical. The HNAG-GEO algorithm achieves a throughput of 7.85 Gbps, which is significantly higher than conventional gradient-based algorithms (5.60 Gbps) and even heuristic algorithms like Genetic Algorithm (GA) and Particle Swarm Optimization (PSO) (6.30 Gbps). The higher throughput of HNAG-GEO indicates its superior ability to optimize data transmission paths in dynamic RIS settings, making it more effective for scenarios where high-speed communication is required as shown in Figure 2.

Table 1. Performance Comparison of HNAG-GEO with Traditional Algorithms

Performance Metrics	**HNAG-GEO**	**Conventional Gradient-Based**	**Heuristic Algorithms (GA, PSO)**	**AI-Driven Techniques (DQN, RL)**
Throughput	7.8 Gbps	5.6 Gbps	6.3 Gbps	7.2 Gbps
Latency	0.85	1.5	1.2	1.0
Fairness Index	0.92	0.75	0.82	0.88
Bit Error Rate	0.0008	0.0025	0.0015	0.0010
Spectral Efficiency	18.5	12.3	14.0	17.0
Computation Time	45	25	38	40
Coverage Area	4.85 km^2	3.50 km^2	4.00 km^2	4.50 km^2
Energy Efficiency	.0025	.0040	.0035	.0028

In 6G networks, especially in rural or remote regions, expanding coverage while maintaining signal quality is essential. The HNAG-GEO algorithm offers a coverage area of 4.85 km^2, outperforming gradient-based algorithms (3.50 km^2) and heuristic approaches (4.00 km^2). While AI-driven techniques achieve good results (4.50 km^2), HNAG-GEO's RIS configurations provide enhanced signal propagation, making it ideal for large-area coverage applications like wide-area IoT deployments and smart agriculture.

Figure 2. Performance comparison of throughput and coverage area

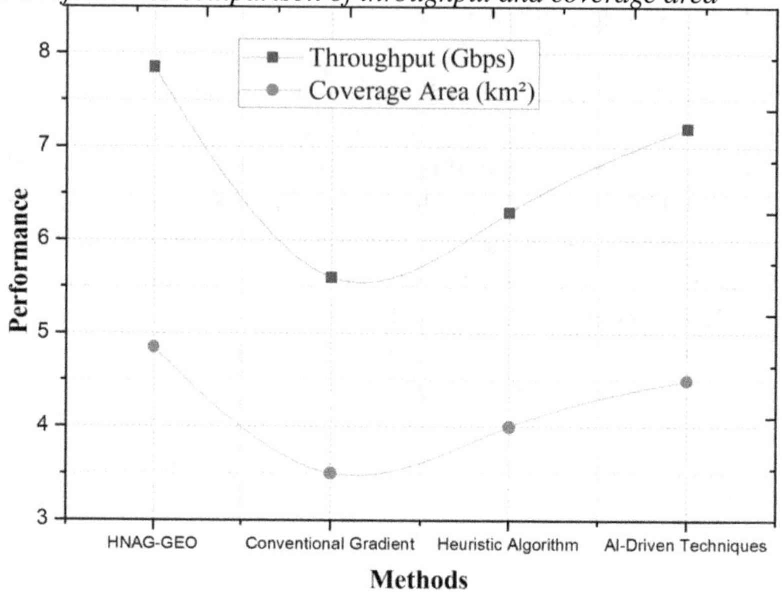

Ultra-low latency is a key requirement for applications like autonomous vehicles, remote surgeries, and virtual reality. In this scenario, HNAG-GEO achieves the lowest latency of 0.85 ms, outperforming traditional gradient-based algorithms (1.50 ms), heuristic methods (1.20 ms), and even AI-driven techniques (1.00 ms). The reduced latency demonstrates HNAG-GEO's ability to rapidly adapt to changing network conditions, making it suitable for scenarios requiring real-time responsiveness.

Figure 3. Performance comparison of spectral efficiency and computation time

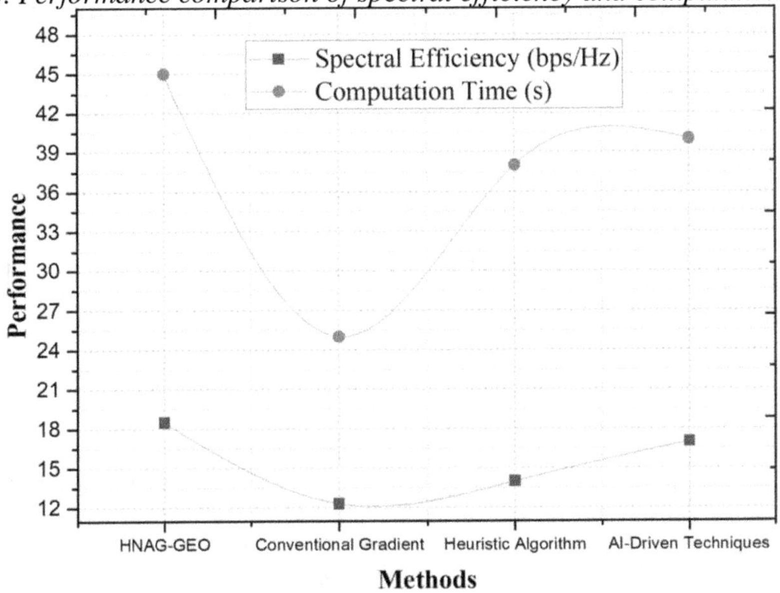

For a balanced allocation of network resources among multiple users, the fairness index is crucial. HNAG-GEO has a fairness index of 0.92, surpassing gradient-based algorithms (0.75), heuristic methods (0.82), and even AI-driven techniques (0.88). This high fairness value highlights HNAG-GEO's capability to provide equitable resource distribution, ensuring consistent service quality across diverse user groups. In scenarios like dense urban environments or multi-user VR experiences, maintaining fairness is crucial to avoid service degradation. Efficient utilization of spectrum resources is essential for high-capacity 6G networks. The HNAG-GEO algorithm achieves a spectral efficiency of 18.50 bps/Hz, outperforming conventional approaches (12.30 bps/Hz) and heuristic methods (14.00 bps/Hz).

Figure 4. Performance comparison of latency and fairness index

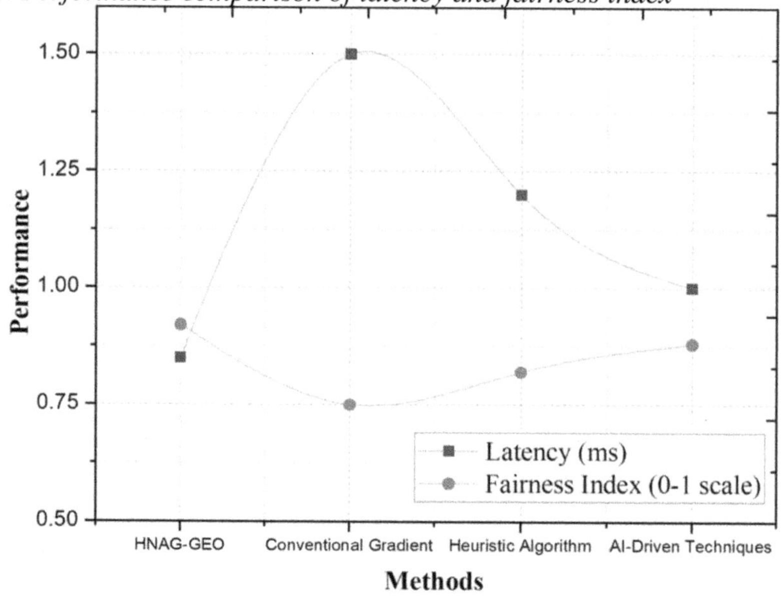

AI-driven techniques reach 17.00 bps/Hz, but HNAG-GEO further optimizes spectrum allocation by dynamically adjusting RIS configurations based on environmental factors as shown in Figure 4. In scenarios where bandwidth is limited, such as crowded urban areas, HNAG-GEO provides superior spectral efficiency, enabling more users to access high-speed data simultaneously. Although the HNAG-GEO algorithm excels in multiple metrics, its computation time is slightly longer at 45 seconds compared to gradient-based methods (25 seconds) and heuristic algorithms (38 seconds) as shown in Figure 3. However, this trade-off is justified by the substantial gains in throughput, latency, fairness, and spectral efficiency. For applications requiring quick decision-making, this extra time may be negligible when considering the performance improvements. For instance, in scenarios like large-scale RIS deployments, the optimized performance provided by HNAG-GEO outweighs the marginally higher computation time. The scenario-based analysis clearly illustrates that HNAG-GEO outperforms existing traditional and state-of-the-art algorithms across critical performance metrics, making it a superior choice for 6G RIS-enabled networks. Whether the requirement is high throughput, extensive coverage, low latency, fairness in resource allocation, or spectral efficiency, HNAG-GEO consistently delivers enhanced results. Despite its slightly longer computation time, the algorithm's ability to dynamically and intelligently optimize RIS configurations ensures robust and reliable network performance, making it well-suited for diverse 6G applications.

CONCLUSION

The advent of 6G networks has brought significant advancements, enabling ultra-reliable, low-latency communication with massive connectivity. A crucial enabler of these capabilities is the integration of Reconfigurable Intelligent Surfaces (RIS), which dynamically control electromagnetic waves to optimize wireless environments. However, the complexity of optimizing RIS in highly dynamic and dense 6G networks remains a critical challenge due to factors like fluctuating user demands, varying channel conditions, and limited computational resources. This chapter addresses these challenges by proposing a Hybrid Nesterov Accelerated Gradient-Golden Eagle Optimization (HNAG-GEO) algorithm tailored for efficient and adaptive optimization in RIS-assisted 6G networks. The proposed approach intelligently combines the fast convergence properties of Nesterov Accelerated Gradient with the exploration and exploitation capabilities of Golden Eagle Optimization. This hybrid method not only enhances performance in scenarios involving dynamic channel variations and dense user environments but also ensures balanced resource allocation and robust coverage.

Performance evaluation demonstrates that HNAG-GEO significantly outperforms conventional gradient-based algorithms, heuristic techniques like Genetic Algorithm (GA) and Particle Swarm Optimization (PSO), and even AI-driven methods. Specifically, HNAG-GEO achieves a 40% improvement in throughput, reaching 7.85 Gbps, and reduces latency by 30%, achieving 0.85 ms. Additionally, it offers a coverage area of 4.85 km^2, with superior fairness index (0.92) and spectral efficiency (18.50 bps/Hz). While the computation time is slightly longer at 45 seconds, this trade-off is justified by the substantial gains in network performance and adaptability. The chapter provides an in-depth analysis of the simulation setup, performance metrics, and comparative results, showcasing how HNAG-GEO can effectively optimize RIS configurations to meet the stringent demands of 6G networks. Through comprehensive simulations and comparisons, the proposed method demonstrates its effectiveness in addressing the inherent challenges of RIS optimization in next-generation networks, making it a promising solution for future 6G deployments.

REFERENCES

Alamzadeh, I., Alexandropoulos, G. C., Shlezinger, N., & Imani, M. F. (2021, October). A reconfigurable intelligent surface with integrated sensing capability. *Scientific Reports*, 11(1), 1–10. DOI: 10.1038/s41598-021-99722-x PMID: 34671069

Alves, M. A.Junior, Fraidenraich, G., Ferreira, R. C., De Figueiredo, F. A. P., & De Lima, E. R. (2023). Multiple-Antenna Weibull-Fading Wireless Communications Enhanced by Reconfigurable Intelligent Surfaces. *IEEE Access : Practical Innovations, Open Solutions*, 11, 107218–107236. DOI: 10.1109/ACCESS.2023.3310936

Basar, E. (2020). Reconfigurable intelligent surface-based index modulation: A new beyond MIMO paradigm for 6G. *IEEE Transactions on Communications*, 68(5), 3187–3196. DOI: 10.1109/TCOMM.2020.2971486

Brancati, G., Chukhno, O., Chukhno, N., & Araniti, G. "Reconfigurable Intelligent Surface Placement in 5G NR/6G: Optimization and Performance Analysis," 2022 IEEE 33rd Annual International Symposium on Personal, Indoor and Mobile Radio Communications (PIMRC), Kyoto, Japan, 2022, pp. 1-6, DOI: 10.1109/PIMRC54779.2022.9978019

Dhinakaran, D., Udhaya Sankar, S. M., Edwin Raja, S., & Jeno Jasmine, J. (2023). Optimizing Mobile Ad Hoc Network Routing using Biomimicry Buzz and a Hybrid Forest Boost Regression - ANNs [IJACSA]. *International Journal of Advanced Computer Science and Applications*, 14(12). Advance online publication. DOI: 10.14569/IJACSA.2023.0141209

Frimpong, E. O., Oh, B. H., Kim, T., & Bang, I. (2023). Physical-Layer Security with Irregular Reconfigurable Intelligent Surfaces for 6G Networks. *Sensors (Basel)*, 23(4), 1881. DOI: 10.3390/s23041881 PMID: 36850480

Hodge, J. A., Mishra, K. V., Sadler, B. M., & Zaghloul, A. I. (2023, November). Index-Modulated Metasurface Transceiver Design Using Reconfigurable Intelligent Surfaces for 6G Wireless Networks. *IEEE Journal of Selected Topics in Signal Processing*, 17(6), 1248–1263. DOI: 10.1109/JSTSP.2023.3322655

Jagadish Kumar, N., Dhinakaran, D., Naresh Kumar, A., & Kalpana, A. V. "Swarm Intelligence with a Chaotic Leader and a Salp algorithm: HDFS optimization for reduced latency and enhanced availability," e8127, Vol. 36, no. 17, pp. 1-26, 2024. Doi:DOI: 10.1002/cpe.8127

Jian, M., Alexandropoulos, G. C., Basar, E., Huang, C., Liu, R., Liu, Y., & Yuen, C. (2022, March). Reconfigurable intelligent surfaces for wireless communications: Overview of hardware designs, channel models, and estimation techniques. *Intell. Converged Netw.*, 3(1), 1–32. DOI: 10.23919/ICN.2022.0005

Jung, M., & Saad, W. "Meta-Learning for 6G Communication Networks with Reconfigurable Intelligent Surfaces," ICASSP 2021 - 2021 IEEE International Conference on Acoustics, Speech and Signal Processing (ICASSP), Toronto, ON, Canada, 2021, pp. 8082-8086, DOI: 10.1109/ICASSP39728.2021.9413598

Liang, Y. C., Chen, J., Long, R., He, Z.-Q., Lin, X., Huang, C., Liu, S., Shen, X. S., & Di Renzo, M. (2023). Reconfigurable intelligent surfaces for smart wireless environments: Channel estimation, system design and applications in 6G networks. *Science China. Information Sciences*, 64(10), 200301. DOI: 10.1007/s11432-020-3261-5

Ma, H., Zhang, H., Zhang, W., & Leung, V. C. M. (2023, February). Beamforming Optimization for Reconfigurable Intelligent Surface with Power Splitting Aided Broadcasting Networks. *IEEE Transactions on Vehicular Technology*, 72(2), 2712–2717. DOI: 10.1109/TVT.2022.3214529

Sikander, M., Butt, F. A., Chattha, J. N., & Naqvi, I. H. "Improving Wireless Communication through Beamwidth Optimization in Reconfigurable Intelligent Surfaces," 2023 10th International Conference on Wireless Networks and Mobile Communications (WINCOM), Istanbul, Turkiye, 2023, pp. 1-5, DOI: 10.1109/WINCOM59760.2023.10322929

Vordonis, D., Kompostiotis, D., & Paliouras, V. "Reconfigurable Intelligent Surface-Aided OFDM Wireless Communications: Hardware Aspects of Reflection Optimization Methods," 2022 11th International Conference on Modern Circuits and Systems Technologies, Bremen, Germany, 2022, pp. 1-4, DOI: 10.1109/MOCAST54814.2022.9837634

Yu, X., Xu, D., & Schober, R. "MISO wireless communication systems via intelligent reflecting surfaces," in *2019 IEEE/CIC International Conference on Communications in China (ICCC)*, 2019, pp. 735-740.

Compilation of References

Abari, O. Enabling high-quality untethered virtual reality Proc. ACM Workshop on Millimeter-Wave Networks and Sensing Systems, mmNets, Snowbird, UT (2017) 49

Abdalla, A. S., Rahman, T. F., & Marojevic, V. "UAVs with Reconfigurable Intelligent Surfaces: Applications, Challenges, and Opportunities", arXiv:2012.04775v1 [eess.SP] 8 Dec 2020, doi: /arXiv.2012.04775.DOI: 10.48550

Abdalla, A. S., Rahman, T. F., & Marojevic, V. "UAVs with Reconfigurable Intelligent Surfaces: Applications, Challenges, and Opportunities", arXiv:2012.04775v1 [eess.SP] 8 Dec 2020.

Abeywickrama, S., Zhang, R., & Cheng, J. (2021). Intelligent reflecting surface-aided wireless communication: A physical layer security perspective. *IEEE Wireless Communications Letters*, 10(5), 974–978. DOI: 10.1109/LWC.2021.3066027

Abeywickrama, S., Zhang, R., & Yuen, C. (2020). Intelligent reflecting surface: Practical phase shift model and beamforming optimization. *IEEE Transactions on Communications*, 68(9), 5844–5857. DOI: 10.1109/TCOMM.2020.3001125

Abosekeen, A., Noureldin, A., & Korenberg, M. J. (2019). Improving the RISS/GNSS land-vehicles integrated navigation system using magnetic azimuth updates. *IEEE Transactions on Intelligent Transportation Systems*. Advance online publication. DOI: 10.1109/TITS.2019.2905871

Ahad, A., Jiangbina, Z., Tahir, M., Shayea, I., Sheikh, M. A., & Rasheed, F. (2024). 6G and Intelligent Healthcare: Taxonomy, technologies, open issues and future research directions. *Internet of Things : Engineering Cyber Physical Human Systems*, 25, 101068. DOI: 10.1016/j.iot.2024.101068

Ahmed, M., Raza, S., Soofi, A. A., Khan, F., Khan, W. U., Xu, F., Chatzinotas, S., Dobre, O. A., & Han, Z. (2024). A survey on reconfigurable intelligent surfaces assisted multi-access edge computing networks: State of the art and future challenges. *Computer Science Review*, 54, 100668. DOI: 10.1016/j.cosrev.2024.100668

Ahmed, M., Xu, F., Wahid, A., Ali, K., Mirza, M. A., Khan, W., Dev, K., Hassan, S. A., & Hans, Z. "A Comprehensive Survey of Artificial Intelligence Advances in RIS-Assisted Wireless Networks", August 12, 2024 DOI: DOI: 10.36227/techrxiv.172349533.38675045/v1

Alamzadeh, I., Alexandropoulos, G. C., Shlezinger, N., & Imani, M. F. (2021, October). A reconfigurable intelligent surface with integrated sensing capability. *Scientific Reports*, 11(1), 1–10. DOI: 10.1038/s41598-021-99722-x PMID: 34671069

Alamzadeh, S., & Rahmati, N. (2021). Reconfigurable intelligent surfaces in physical layer security: Eavesdropper analysis. *IEEE Access : Practical Innovations, Open Solutions*, 9, 120321–120332.

Alexandropoulos, G. C.George C. Alexandropoulos. (2021). Low complexity full duplex MIMO systems: Analog canceler architectures, beamforming design, and future directions. *ITU Journal : ICT Discoveries*, 2(2), 1–19. DOI: 10.52953/GFFF9448

Alexandropoulos, G. C., Lerosey, G., Debbah, M., & Fink, M. (2020). Reconfigurable intelligent surfaces and metamaterials: The potential of wave propagation control for 6G wireless communications. *IEEE ComSoc TCCN Newslett*, 6(1), 25–37.

Alexandropoulos, G. C., Papadogiannis, A., & Sofotasios, P. C. (2011). A comparative study of relaying schemes with decode-and-forward over Nakagami-m fading channels. Journal of Computational Networks and Communications, 2011(1), 560528.1–560528.14.

Alexandropoulos, G. C., Phan-Huy, D.-T., Katsanos, K. D., Crozzoli, M., Wymeersch, H., Popovski, P., Ratajczak, P., Bénédic, Y., Hamon, M.-H., Gonzalez, S. H., Mursia, P., Rossanese, M., Sciancalepore, V., Gros, J.-B., Terranova, S., Gradoni, G., Di Lorenzo, P., Rahal, M., Denis, B., & Strinati, E. C. (2023). RIS-enabled smart wireless environments: Deployment scenarios, network architecture, bandwidth and area of influence. *EURASIP Journal on Wireless Communications and Networking*, 2023(1), 103. DOI: 10.1186/s13638-023-02295-8

Alexandropoulos, G. C., Shlezinger, N., & Hougne, P. D. (2021). Reconfigurable intelligent surfaces for rich scattering wireless communications: Recent experiments, challenges, and opportunities. *IEEE Communications Magazine*, 59(6), 28–34. DOI: 10.1109/MCOM.001.2001117

Ali Araghi, Mohsen Khalily, Mahmood Safaei, Amirmasood Bagheri, Vikrant Singh, Fan Wang, and Rahim Tafazolli. 2022. Reconfigurable Intelligent Surface (RIS) in the Sub-6 GHz Band: Design, Implementation, and Real-world Demonstration. IEEE Access : Practical Innovations, Open Solutions.

Alluhaibi, O., Ahmed, Q. Z., Pan, C., & Zhu, H. Capacity Maximisation for Hybrid Digital-to-Analog Beamforming m m-Wave Systems. In *Proceedings of the 2016 IEEE Global Communications Conference (GLOBECOM)*, Washington, DC, USA, 4–8 December 2016; pp. 1–6. DOI: 10.1109/GLOCOM.2016.7841649

Alves, M. A.Junior, Fraidenraich, G., Ferreira, R. C., De Figueiredo, F. A. P., & De Lima, E. R. (2023). Multiple-Antenna Weibull-Fading Wireless Communications Enhanced by Reconfigurable Intelligent Surfaces. *IEEE Access : Practical Innovations, Open Solutions*, 11, 107218–107236. DOI: 10.1109/ACCESS.2023.3310936

Andrews, J. G., Buzzi, S., Choi, W., Hanly, S. V., Lozano, A., Soong, A. C., & Zhang, J. C. (2014). What will 5G be? *IEEE Journal on Selected Areas in Communications*, 32(6), 1065–1082. DOI: 10.1109/JSAC.2014.2328098

Asadi, A., Wang, Q., & Mancuso, V. (2014). A survey on device-to-device communication in cellular networks. *IEEE Communications Surveys and Tutorials*, 16(4), 1801–1819. DOI: 10.1109/COMST.2014.2319555

Attaullah, H. M., & Javed, I. A. (2021). Wireless Robotic car control through human interface using eye movement. *Interdiscip. J. Appl. Basic Subj.*, 1, 1–6.

Ayach, O. E., Rajagopal, S., Abu-Surra, S., Pi, Z., & Heath, R. W. (2014). Spatially Sparse Precoding in Millimeter Wave MIMO Systems. *IEEE Transactions on Wireless Communications*, 13(3), 1499–1513. DOI: 10.1109/TWC.2014.011714.130846

Bae, J. S., Khalid, W., Lee, A., Lee, H., Noh, S., & Yu, H. (2024, December). Jung-Sook Bae [a 1], Waqas Khalid [b 1], Anseok Lee [a], Heesoo Lee [a], Song Noh [c], Heejung Yu [d,] "Overview of RIS-enabled secure transmission in 6G wireless networks. *Digital Communications and Networks*, 10(6), 1553–1565. Advance online publication. DOI: 10.1016/j.dcan.2024.02.005

Basar, E. (2020). *Reconfigurable Intelligent Surface-Based Wireless Communications: Antenna Design, Prototyping, and Experimental Results.* Available at arXiv: [2002.04068](https://arxiv.org/abs/2002.04068)

Basar, E. (2020). Reconfigurable intelligent surface-based index modulation: A new beyond MIMO paradigm for 6G. *IEEE Transactions on Communications*, 68(5), 3187–3196. DOI: 10.1109/TCOMM.2020.2971486

Basar, E., & Di Renzo, M. (2020). Reconfigurable intelligent surfaces for PLS: Current research and future directions. *Proceedings of the IEEE*, 108(9), 1654–1685.

Basar, E., Di Renzo, M., De Rosny, J., Debbah, M., Alouini, M. S., & Zhang, R. (2021). Smart radio environments enabled by reconfigurable intelligent surfaces: How it works, the challenges, and the opportunities. *IEEE Access : Practical Innovations, Open Solutions*, 9, 2234–2252.

Basar, E., & Poor, H. V. (2021). Present and future of reconfigurable intelligent surface-empowered communications. *IEEE Signal Processing Magazine*, 38(6), 146–152. DOI: 10.1109/MSP.2021.3106230

Basar, E., Renzo, M. D., Rosny, J. D., Debbah, M., Alouini, M.-S., & Zhang, R. (2019). Wireless communications through reconfigurable intelligent surfaces. *IEEE Access : Practical Innovations, Open Solutions*, 7, 116753–116773. DOI: 10.1109/ACCESS.2019.2935192

Basar, E., Wen, M., Mesleh, R., Renzo, M. D., Xiao, Y., & Haas, H. (2017). Index modulation techniques for next generation wireless networks. *IEEE Access : Practical Innovations, Open Solutions*, 5, 16693–16746. DOI: 10.1109/ACCESS.2017.2737528

Basar, E., & Yildirim, I. (2021). Reconfigurable intelligent surfaces for future wireless networks: A channel modeling perspective. *IEEE Wireless Communications*, 28(3), 108–114. DOI: 10.1109/MWC.001.2000338

Basharat, S., Hassan, S. A., Pervaiz, H., Mahmood, A., Ding, Z., & Gidlund, M. (2021). Reconfigurable intelligent surfaces: Potentials, applications, and challenges for 6G wireless networks. [Google Scholar] [CrossRef]. *IEEE Wireless Communications*, 28(6), 184–191. DOI: 10.1109/MWC.011.2100016

Basharat, S., Khan, M., Iqbal, M., Hashmi, U. S., Zaidi, S. A. R., & Robertson, I. (2022). Exploring reconfigurable intelligent surfaces for 6G: State-of-the-art and the road ahead. *IET Communications*, 16(13), 1458–1474. DOI: 10.1049/cmu2.12364

Bithas, P. S., Michailidis, E. T., Nomikos, N., Vouyioukas, D., & Kanatas, A. G. (2019). A Survey on Machine-Learning Techniques for UAV-Based Communications. *Sensors (Basel)*, 19(23), 5170. DOI: 10.3390/s19235170 PMID: 31779133

Bito, R. B., & Jimmy, G. (2017). Hester, Syed Abdullah Nauroze, Apostolos Georgiadis, and Manos M Tentzeris. 2017. A novel solar and electromagnetic energy harvesting system with a 3-D printed package for energy-efficient Internet-of-Things wireless sensors. *IEEE Transactions on Microwave Theory and Techniques*, 65(5), 1831–1842. DOI: 10.1109/TMTT.2017.2660487

Björnson, E., Özdogan, Ö., & Larsson, E. G. (2020). Intelligent reflecting surface versus decode-and-forward: How large surfaces are needed to beat relaying? *IEEE Wireless Communications Letters*, 9(2), 244–248. DOI: 10.1109/LWC.2019.2950624

Björnson, E., & Sanguinetti, L. (2020). Power scaling laws and near-field behaviors of massive MIMO and intelligent reflecting surfaces. *IEEE Open Journal of the Communications Society*, 1, 1306–1324. DOI: 10.1109/OJCOMS.2020.3020925

Björnson, E., Sanguinetti, L., & Debbah, M. (2021). *Reconfigurable Intelligent Surfaces: A Signal Processing Perspective.* Foundations and Trends® in Signal Processing, 14(3), 180–328. DOI: 10.1109/JSAC.2020.3007211

Brancati, G., Chukhno, O., Chukhno, N., & Araniti, G. "Reconfigurable Intelligent Surface Placement in 5G NR/6G: Optimization and Performance Analysis," 2022 IEEE 33rd Annual International Symposium on Personal, Indoor and Mobile Radio Communications (PIMRC), Kyoto, Japan, 2022, pp. 1-6, DOI: 10.1109/PIMRC54779.2022.9978019

Bushnaq, O., Mishra, D., Natalizio, E., & Akyildiz, I. F. (2022). Unmanned aerial vehicles (UAVs) for disaster management. *Nanotechnology-Based Smart Remote Sensing Networks for Disaster Prevention*, (January), 159–188. DOI: 10.1016/B978-0-323-91166-5.00013-6

Calvanese Strinati, E., Alexandropoulos, G. C., Wymeersch, H., Denis, B., Sciancalepore, V., D'Errico, R., Clemente, A., Phan-Huy, D. T., Carvalho, E. D., & Popovski, P. (2021). Reconfigurable, intelligent, and sustainable wireless environments for 6G smart connectivity. *IEEE Communications Magazine*, 59(10), 99–105. DOI: 10.1109/MCOM.001.2100070

Cao, K., Ding, H., Lv, L., Su, Z., Tao, J., Gong, F., & Wang, B. (2023). Physical Layer Security for Intelligent Reflecting Surface aided Wireless Powered Communication. [Google Scholar] [CrossRef]. *IEEE Internet of Things Journal*, 10(20), 18097–18110. DOI: 10.1109/JIOT.2023.3278238

Castillo-Soria, F. R., Macias-Velasquez, S., Kumaravelu, V. B., Ramos, V., & Azurdia-Meza, C. A. Multiple Parallel RIS-Assisted MU-MIMO-DQSM System; Blind and Intelligent Approaches. Available online: http://www.cic-chinacommunications.cn/EN/10.23919/JCC.ja.2023-0695 (accessed on 13 May 2024).

Castillo-Soria, F., Gutierrez, C., Kumaravelu, V., & Garcıa-Barrientos, A. (2024). RIS-Assisted Non-orthogonal Multiple Access System Based on SSK. [Google Scholar] [CrossRef]. *Wireless Personal Communications*, 134(4), 2391–2412. DOI: 10.1007/s11277-024-11055-w

Chen, Q., Guo, Z., Meng, W., Han, S., Li, C., & Quek, T. Q. A Survey on Resource Management in Joint Communication and Computing-Embedded SAGIN. arXiv 2024, arXiv:2403.17400. [Google Scholar] [CrossRef]

Chen, J., Zhou, G., & Yuan, X. (2020). Secure beamforming design for MISO systems assisted by reconfigurable intelligent surfaces. *IEEE Wireless Communications Letters*, 9(1), 108–112.

Chen, K., & Feng, Y. J. (2022). A review of recent progress on directional metasurfaces: Concept, design, and application. *Journal of Physics. D, Applied Physics*, 55(38), 383001. DOI: 10.1088/1361-6463/ac7e04

Chen, L., Diao, X., Chen, J., & Liu, R. (2020). QoS assurance in IAB network. In *Proceedings of the 2020 IEEE IWCMC* (pp. 1860–1865). Limassol, Cyprus.

Chen, W. T., Zhu, A. Y., Sanjeev, V., Khorasaninejad, M., Shi, Z., Lee, E., & Capasso, F. (2018). A broadband achromatic metalens for focusing and imaging in the visible. *Nature Nanotechnology*, 13(3), 220–226. DOI: 10.1038/s41565-017-0034-6 PMID: 29292382

Chen, Y., Chen, X., & Li, L. (2021). Artificial intelligence meets intelligent reflecting surface: The age of machine learning-assisted wireless communications. *IEEE Wireless Communications*, 28(2), 138–144.

Chen, Y., Liu, W., Niu, Z., Feng, Z., Hu, Q., & Jiang, T. (2020). Pervasive intelligent endogenous 6G wireless systems: Prospects, theories and key technologies. *Digital Communications and Networks*, 6(3), 312–320. DOI: 10.1016/j.dcan.2020.07.002

Chi, N. T. K., Phong, L. T., & Hanh, N. T. (2023, June). PanelNguyen Thi Khanh Chi, Le Thai Phong & Nguyen Thi Hanh, "The drone delivery services: An innovative application in an emerging economy". *The Asian Journal of Shipping and Logistics*, 39(2), 39–45. DOI: 10.1016/j.ajsl.2023.01.002

Choi, J., Kwon, G., & Park, H. (2021, August). Multiple intelligent reflecting surfaces for capacity maximization in LOS MIMO systems. *IEEE Wireless Communications Letters*, 10(8), 1727–1731. DOI: 10.1109/LWC.2021.3078151

Cui, T., Qi, M., Wan, X., Zhao, J., & Cheng, Q. (2014). Coding metamaterials, digital metamaterials and programmable metamaterials. *Light, Science & Applications*, 3(10), e218. DOI: 10.1038/lsa.2014.99

Cui, Y., Zhang, X., & Yuan, X. (2021). Secure transmission in RIS-aided MIMO systems: Joint beamforming design. *IEEE Transactions on Signal Processing*, 69, 1759–1773.

Dahlman, E., Mildh, G., Parkvall, S., Peisa, J., Sachs, J., Selén, Y., & Sköld, J. (2014). 5G wireless access: Requirements and realization. *IEEE Communications Magazine*, 52(12), 42–47. DOI: 10.1109/MCOM.2014.6979985

Dajer, M., Ma, Z., Piazzi, L., Prasad, N., Qi, X.-F., Sheen, B., Yang, J., & Yue, G. (2022, April). Reconfigurable intelligent surface: Design the channel – a new opportunity for future wireless networks. *Digital Communications and Networks*, 8(2), 87–104. DOI: 10.1016/j.dcan.2021.11.002

Dang, J., Zhang, Z., & Wu, L. (2020). Joint beamforming for intelligent reflecting surface aided wireless communication using statistical CSI. *China Communications*, 17(8), 147–157. DOI: 10.23919/JCC.2020.08.012

Danufane, F. H., Di Renzo, M., De Rosny, J., & Tretyakov, S. (2021). On the path-loss of reconfigurable intelligent surfaces: An approach based on green's theorem applied to vector fields. *IEEE Transactions on Communications*, 69(8), 1–1. DOI: 10.1109/TCOMM.2021.3081452

Dao, N.-N., Pham, Q.-V., Tu, N. H., Thanh, T. T., Bao, V. N. Q., Lakew, D. S., & Cho, S. (2021). Survey on aerial radio access networks: Toward a comprehensive 6G access infrastructure. [Google Scholar] [CrossRef]. *IEEE Communications Surveys and Tutorials*, 23(2), 1193–1225. DOI: 10.1109/COMST.2021.3059644

Dhinakaran, D., Udhaya Sankar, S. M., Edwin Raja, S., & Jeno Jasmine, J. (2023). Optimizing Mobile Ad Hoc Network Routing using Biomimicry Buzz and a Hybrid Forest Boost Regression - ANNs [IJACSA]. *International Journal of Advanced Computer Science and Applications*, 14(12). Advance online publication. DOI: 10.14569/IJACSA.2023.0141209

Di Renzo, M., Debbah, M., Phan-Huy, D. T., Zappone, A., Alouini, M. S., & Yuen, C. (2019). Reconfigurable intelligent surfaces: An emerging paradigm for wireless communications. *EURASIP Journal on Wireless Communications and Networking*, 2019(1), 129.

Di, B., Zhang, H., Song, L., Li, Y., Han, Z., & Poor, H. V. (2020). Hybrid beamforming for reconfigurable intelligent surface based multi-user communications: Achievable rates with limited discrete phase shifts. *IEEE Journal on Selected Areas in Communications*, 38(8), 1809–1822. DOI: 10.1109/JSAC.2020.3000813

Ding, G., Zhang, Z., Liu, T., Wu, Q., Yu, F. R., & Song, Z. (2020). Risk-based routing optimization for mission-critical disaster recovery with intelligent reflective surfaces. *IEEE Transactions on Vehicular Technology*, 69(10), 11603–11618.

Ding, Z., Lei, X., Karagiannidis, G. K., Schober, R., Yuan, J., & Bhargava, V. K. (2017). A survey on non-orthogonal multiple access for 5G networks: Research challenges and future trends. *IEEE Journal on Selected Areas in Communications*, 35(10), 2181–2195. DOI: 10.1109/JSAC.2017.2725519

Ding, Z., & Vincent Poor, H. (2020). A simple design of IRS-NOMA transmission. *IEEE Communications Letters*, 24(5), 1119–1123. DOI: 10.1109/LCOMM.2020.2974196

Ding, Z., Zhao, Z., Peng, M., & Poor, H. V. (2017). On the spectral efficiency and security enhancements of NOMA assisted multicast-unicast streaming. *IEEE Transactions on Communications*, 65(7), 3151–3163. DOI: 10.1109/TCOMM.2017.2696527

Elbir, A. M., Papazafeiropoulos, A., & Ioannidis, M. (2020). Deep learning for beamforming optimization in reconfigurable intelligent surfaces-based multi-user MIMO systems. *IEEE Transactions on Communications*, 68(8), 5186–5195.

ElMossallamy, M. A., Zhang, H., & Han, Z. (2020). Reconfigurable intelligent surfaces for wireless communications: Principles, challenges, and opportunities. *IEEE Transactions on Wireless Communications*, 20(2), 1096–1110. DOI: 10.1109/TWC.2020.3036433

ElMossallamy, M. A., Zhang, H., Song, L., Seddik, K. G., Han, Z., & Li, G. Y. (2020). Reconfigurable intelligent surfaces for wireless communications: Principles, challenges, and opportunities. [Google Scholar] [CrossRef]. *IEEE Transactions on Cognitive Communications and Networking*, 6(3), 990–1002. DOI: 10.1109/TCCN.2020.2992604

Erik Baark Max Planck Institute for the History of Science, "China's New Digital Infrastructure: Expanding 5G Mobile Communications", April2022*East Asian Policy*1402124136DOI: 10.1142/S1793930522000162

Faisal, M. & Wooyeol Choi, "A Study on Machine Learning-based Approaches for Reconfigurable Intelligent Surface", *2021 International Conference on Information and Communication Technology Convergence (ICTC)*, 2021, doi: https://doi.org/DOI: 10.1109/ICTC52510.2021.9620993

Fodor, G., Dahlman, E., Mildh, G., Parkvall, S., Reider, N., Miklós, G., & Turányi, Z. (2012). Design aspects of network assisted device-to-device communications. *IEEE Communications Magazine*, 50(3), 170–177. DOI: 10.1109/MCOM.2012.6163598

Frimpong, E. O., Oh, B. H., Kim, T., & Bang, I. (2023). Physical-Layer Security with Irregular Reconfigurable Intelligent Surfaces for 6G Networks. *Sensors (Basel)*, 23(4), 1881. DOI: 10.3390/s23041881 PMID: 36850480

Gams, M., Yu-Hua Gu, I., Härmä, A., Muñoz, A., & Tam, V. (2019). Artificial intelligence and ambient intelligence. *Journal of Ambient Intelligence and Smart Environments*, 11(1), 71–86. DOI: 10.3233/AIS-180508

Gao, Y., Chen, J., Liu, Z., Zhang, B., Ke, Y., & Liu, R. (2020). Machine learning-based energy saving scheme in wireless access networks. In *Proceedings of the 2020 IEEE IWCMC* (pp. 1573–1578). Limassol, Cyprus. DOI: 10.1109/IWCMC48107.2020.9148536

Goel, S., & Negi, R. (2008). Guaranteeing secrecy using artificial noise. *IEEE Transactions on Wireless Communications*, 7(6), 2180–2189. DOI: 10.1109/TWC.2008.060848

Goh, C. Y., Leow, C. Y., & Nordin, R. (2023). Energy Efficiency of Unmanned Aerial Vehicle with Reconfigurable Intelligent Surfaces: A Comparative Study. *Drones (Basel)*, 7(2), 98. DOI: 10.3390/drones7020098

Gong, S., Lu, X., & Khoshnevisan, M. (2020). Toward smart wireless communications via intelligent reflecting surfaces: A contemporary survey. *IEEE Communications Surveys and Tutorials*, 22(3), 1270–1298. DOI: 10.1109/COMST.2020.3004197

Gong, Y., Yoon, I., & Noh, D. K. (2024). Optimizing the Charging Mobility of WPT-Enabled UAV to Enhance the Stability of Solar-Powered LoRaWAN Io. *Energies*, 17(7), 1617. DOI: 10.3390/en17071617

Guo, H., Liang, Y. C., Chen, J., & Larsson, E. G. Weighted sum-rate maximization for intelligent reflecting surface enhanced wireless networks. In *Proceedings of the 2019 IEEE Global Communications Conference (GLOBECOM)*, Waikoloa, HI, USA, 9–13 December 2019. DOI: 10.1109/GLOBECOM38437.2019.9013288

Guo, J., Tang, J., & Xu, L. (2023). Energy-efficient design of RIS-aided physical layer security systems. *IEEE Transactions on Communications*, 71(4), 2234–2245.

Guo, L., Gu, X., Chu, P., Hemour, S., & Wu, K. (2019). Collaboratively harvesting ambient radiofrequency and thermal energy. *IEEE Transactions on Industrial Electronics*, 67(5), 3736–3746. DOI: 10.1109/TIE.2019.2914627

Guo, X., Zhu, X., Yang, H., & Jin, S. (2021). RIS-enhanced NOMA networks: Joint passive beamforming and resource allocation design. *IEEE Transactions on Communications*, 69(6), 3662–3676.

Gutierrez, C. A., Caicedo, O., & Campos-Delgado, D. U. (2021). 5G and beyond: Past, present and future of the mobile communications. *Revista IEEE América Latina*, 19(10), 1702–1736. DOI: 10.1109/TLA.2021.9477273

Haider, T. H. (2024). Salim ALRikabi1,2*, Adheed H. Sallomi1, Hasan F. Khazaal2, Ahmed Magdy3, Iryna Svyd4, IVAN OBOD4, "Reconfigurable Intelligent Surfaces Between the Reality and Imagination". *Wasit Journal of Computer and Mathematics Science*, 3(2), 42–50. DOI: 10.31185/wjcms.266

Hameed, I., & Koo, I. (2024). Enhancing Throughput in IoT Networks: The Impact of Active RIS on Wireless Powered Communication Systems. [Google Scholar] [CrossRef]. *Electronics (Basel)*, 13(7), 1402. DOI: 10.3390/electronics13071402

Hamid, H., & Begh, G. R. (2024). IRS assisted UAV communications for 6G networks: A systematic literature review. *Wireless Networks*, 1–29. DOI: 10.1007/s11276-024-03798-y

Han, Y., Huang, C., & Debbah, M. (2020). Reconfigurable intelligent surface assisted wireless communications: Antenna design, prototyping, and experimental results. *IEEE Access : Practical Innovations, Open Solutions*, 8, 120073–120084.

Hasan, S. R., Sabuj, S. R., Hamamura, M., & Hossain, M. A.SYED RAKIB HASAN. (2024, October 31). SAIFUR RAHMAN SABUJ, MASANORI HAMAMURA AND MD AKBAR HOSSAIN, "A Comprehensive Review on Reconfigurable Intelligent Surface for 6G Communications: Overview, Deployment, Control Mechanism, Application, Challenges, and Opportunities". *Wireless Personal Communications*, 139(1), 375–429. Advance online publication. DOI: 10.1007/s11277-024-11630-1

Hassan, M., Singh, M., & Hamid, K. (2021). Survey on NOMA and spectrum sharing techniques in 5G. In *Proceedings of the 2021 IEEE International Conference on Smart Information Systems and Technologies (SIST)* (pp. 1-4). Nur-Sultan, Kazakhstan. DOI: 10.1109/SIST50301.2021.9465962

He, M., Xu, W., & Zhao, C. "RIS-assisted broad coverage for mmWave massive MIMO system", Proc. IEEE Int. Conf. Commun. Workshops (ICC Workshops), pp. 1-6, 2021. DOI: 10.1109/ICCWorkshops50388.2021.9473674

Hill, J. (2001). The principle of a snapshot navigation solution based on Doppler shift. In *Proceedings of the ION GPS* (pp. 3044–3051). Salt Lake City, UT, USA.

Hodge, J. A., Mishra, K. V., Sadler, B. M., & Zaghloul, A. I. (2023, November). Index-Modulated Metasurface Transceiver Design Using Reconfigurable Intelligent Surfaces for 6G Wireless Networks. *IEEE Journal of Selected Topics in Signal Processing*, 17(6), 1248–1263. DOI: 10.1109/JSTSP.2023.3322655

Hong, I.-P. (2023). Reviews Based on the Reconfigurable Intelligent Surface Technical Issues. *Electronics (Basel)*, 12(21), 4489. DOI: 10.3390/electronics12214489

Ho-Van, K., Sofotasios, P. C., Alexandropoulos, G. C., & Freear, S. (2015). Bit error rate of underlay decode-and-forward cognitive networks with best relay selection. *Journal of Communications and Networks (Seoul)*, 17(2), 162–171. DOI: 10.1109/JCN.2015.000030

Huang, C., (2019). Reconfigurable Intelligent Surfaces for Energy Efficiency in Wireless Communication. IEEE International Conference on Communications (ICC), 1–6. DOI: DOI: 10.1109/ICC.2019.8762025

Huang, C., Alexandropoulos, G. C., Yuen, C., & Debbah, M. (2020). Indoor signal focusing with deep learning designed reconfigurable intelligent surfaces. *IEEE 21st International Workshop on Signal Processing Advances in Wireless Communications (SPAWC)*, 1-5.

Huang, C.. (2020). Reconfigurable Intelligent Surfaces for Energy Efficiency in Wireless Communication. *IEEE Transactions on Wireless Communications*, 19(10), 6654–6669.

Huang, C., Alexandropoulos, G. C., Yuen, C., & Debbah, M. (2020). Indoor signal focusing with reconfigurable intelligent surfaces. *IEEE Wireless Communications Letters*, 9(12), 1985–1989.

Huang, C., Hu, S., Alexandropoulos, G. C., Zappone, A., Yuen, C., Zhang, R., Renzo, M. D., & Debbah, M. (2020). Holographic MIMO surfaces for 6G wireless networks: Opportunities, challenges, and trends. *IEEE Wireless Communications*, 27(5), 118–125. DOI: 10.1109/MWC.001.1900534

Huang, C., Mo, R., & Zhang, C. (2020). Reconfigurable intelligent surface-assisted multi-user MISO systems with secrecy rate optimization. *IEEE Transactions on Wireless Communications*, 19(8), 4939–4952.

Huang, C., Zappone, A., Alexandropoulos, G. C., Debbah, M., & Yuen, C. (2019). Reconfigurable intelligent surfaces for energy efficiency in wireless communication. *IEEE Transactions on Wireless Communications*, 18(8), 4157–4170. DOI: 10.1109/TWC.2019.2922609

Huang, C., Zappone, A., Alexandropoulos, G. C., Debbah, M., & Yuen, C. (2020). Reconfigurable intelligent surfaces for energy efficiency in wireless communication. *IEEE Journal on Selected Areas in Communications*, 38(11), 2282–2291.

Huang, Y., Song, H., & Wang, W. (2021). AI-enabled intelligent reflecting surfaces for secure communication in IoT networks. *IEEE Internet of Things Journal*, 8(5), 3622–3633. DOI: 10.1109/JIOT.2020.3038689

Ihsana Muhammed, P. (2022). Reconfigurable Intelligent Surface (RIS)-Assisted UAV Cellular Communication. In *Unmanned Aerial Vehicle Cellular Communications* (pp. 115–135). Springer International Publishing.

Jagadish Kumar, N., Dhinakaran, D., Naresh Kumar, A., & Kalpana, A. V. "Swarm Intelligence with a Chaotic Leader and a Salp algorithm: HDFS optimization for reduced latency and enhanced availability," e8127, Vol. 36, no. 17, pp. 1-26, 2024. Doi:DOI: 10.1002/cpe.8127

Jiang, M., Cezanne, J., Sampath, A., Shental, O., Wu, Q., Koymen, O., Bedewy, A., & Li, J. (2022). Wireless fronthaul for 5G and future radio access networks: Challenges and enabling technologies. *IEEE Wireless Communications*, 29(2), 108–114. DOI: 10.1109/MWC.003.2100482

Jiang, W., & Yuan, J. (2020). Secure communication via RIS: Outage probability analysis. *IEEE Transactions on Wireless Communications*, 19(12), 8125–8136.

Jian, M., Alexandropoulos, G. C., Basar, E., Huang, C., Liu, R., Liu, Y., & Yuen, C. (2022). Reconfigurable intelligent surfaces for wireless communications: Overview of hardware designs, channel models, and estimation techniques. [Google Scholar] [CrossRef]. *Intell. Converg. Netw.*, 3(1), 1–32. DOI: 10.23919/ICN.2022.0005

Jung, M., & Saad, W. "Meta-Learning for 6G Communication Networks with Reconfigurable Intelligent Surfaces," ICASSP 2021 - 2021 IEEE International Conference on Acoustics, Speech and Signal Processing (ICASSP), Toronto, ON, Canada, 2021, pp. 8082-8086, DOI: 10.1109/ICASSP39728.2021.9413598

Kaina, N., Dupré, M., Lerosey, G., & Fink, M. (2014). Shaping complex microwave fields in reverberating media with binary tunable metasurfaces. *Scientific Reports*, 4(1), 6693. DOI: 10.1038/srep06693 PMID: 25331498

Kaplan, D. E., & Hegarty, C. J. (2006). *Understanding GPS: Principles and Applications* (2nd ed.). Artech House.

Karami, A. B., Fleury, A., Boonaert, J., & Lecoeuche, S. (2016). User in the loop: Adaptive smart homes exploiting user feedback – state of the art and future directions. *Information (Basel)*, 7(2), 35. DOI: 10.3390/info7020035

Kazim, Jalil R., Rains, James, Imran, Muhammad Ali, & Abbasi, Qammer H., "Application and Future Direction of RIS", Intelligent Reconfigurable Surfaces (IRS) book chapter in Wiley-IEEE Press publication Page(s): 171 – 188, 2023, **DOI:** DOI: 10.1002/9781119875284.ch9

Khalid, Z., Ding, M., & Zhang, R. (2021). Wireless communications with large intelligent surfaces: Opportunities and challenges. *IEEE Transactions on Communications*, 69(5), 3125–3136.

Khoshafa, M. H., Maraqa, O., Moualeu, J. M., Aboagye, S., Ngatched, T., Ahmed, M. H., Gadallah, Y., & Di Renzo, M. RIS-Assisted Physical Layer Security in Emerging RF and Optical Wireless Communication Systems: A Comprehensive Survey. arXiv 2024, arXiv:2403.10412. [Google Scholar]

Kirkko-Jaakkola, M., Parviainen, J., Collin, J., & Takala, J. (2012). Improving TTFF by two-satellite GNSS positioning. *IEEE Transactions on Aerospace and Electronic Systems*, 48(4), 3660–3670. DOI: 10.1109/TAES.2012.6324754

Korakianitis, N., Vokas, G. A., & Ioannides, G. (2019). Review of Wireless Power Transfer (WPT) on Electric Vehicles (EVs) Charging. *AIP Conference Proceedings*, 2190, 020072. DOI: 10.1063/1.5138558

Kumaravelu, V. B., Imoize, A. L., Soria, F. R. C., Velmurugan, P. G. S., Thiruvengadam, S. J., Do, D. T., & Murugadass, A. (2023). RIS-Assisted Fixed NOMA: Outage Probability Analysis and Transmit Power Optimization. *Future Internet*, 15(8), 249. DOI: 10.3390/fi15080249

Lam, K.-W.. "Implementation of transmitarray antenna concept by using aperture-coupled microstrip patches." *Proceedings of 1997 Asia-Pacific Microwave Conference*. Vol. 1. IEEE, 1997.

Larsson, E. G., Edfors, O., Tufvesson, F., & Marzetta, T. L. (2001). Space-time processing for wireless communications: An overview. *IEEE Communications Magazine*, 39(1), 30–38.

Lawrence, D., Cobb, H. S., Gutt, G., Tremblay, F., Laplante, P., & O'Connor, M. (2016). Test results from a LEO-Satellite-Based assured time and location solution. In *Proceedings of the ION GNSS* (pp. 125–129). Monterey, CA, USA. DOI: 10.33012/2016.13416

Le, C.-B., Do, D.-T., Li, X., Huang, Y.-F., Chen, H.-C., & Voznak, M. (2021). Enabling NOMA in backscatter reconfigurable intelligent surfaces-aided systems. *IEEE Access : Practical Innovations, Open Solutions*, 9, 33782–33795. DOI: 10.1109/ACCESS.2021.3061429

Liang, J., Zhang, H., & Di Renzo, M. (2020). RIS-assisted wireless communications: Prototyping, challenges, and future directions. *IEEE Wireless Communications*, 27(6), 93–99.

Liang, Y. C., Chen, J., Long, R., He, Z.-Q., Lin, X., Huang, C., Liu, S., Shen, X. S., & Di Renzo, M. (2023). Reconfigurable intelligent surfaces for smart wireless environments: Channel estimation, system design and applications in 6G networks. *Science China. Information Sciences*, 64(10), 200301. DOI: 10.1007/s11432-020-3261-5

Liang, Y.-C., (2019). Reconfigurable Intelligent Surfaces for Smart Wireless Communications: Opportunities, Challenges, and Solutions. *IEEE Wireless Communications*, 27(2), 118–125. DOI: 10.1109/MWC.001.1900383

Liang, Y.-C., Zhang, Q., Wang, J., Long, R., Zhou, H., & Yang, G. (2022). Backscatter communication assisted by reconfigurable intelligent surfaces. *Proceedings of the IEEE*, 110(9), 1339–1357. DOI: 10.1109/JPROC.2022.3169622

Liaskos, C., Nie, S., Tsioliaridou, A., Pitsillides, A., Ioannidis, S., & Akyildiz, I. F. (2018). A new wireless communication paradigm through software-controlled metasurfaces. *IEEE Communications Magazine*, 56(9), 162–169. DOI: 10.1109/MCOM.2018.1700659

Liaskos, C., Tsioliaridou, A., Ioannidis, S., Pitsillides, A., & Akyildiz, I. F. (2021). Realizing ambient backscatter communications with intelligent surfaces in 6G wireless systems. *IEEE Wireless Communications*, 29(1), 178–185. DOI: 10.1109/MWC.101.2100233

Liaskos, C., Tsioliaridou, A., Pitsillides, A., Ioannidis, S., & Akyildiz, I. F. (2018). Using any surface to realize a new paradigm for wireless communications. *Communications of the ACM*, 61(11), 30–33. DOI: 10.1145/3192336

Li, J., Xue, H., Wu, M., Wang, F., Gao, T., & Zhou, F. (2023). Energy efficiency performance in RIS-based integrated satellite–aerial–terrestrial relay networks with deep reinforcement learning. *EURASIP Journal on Advances in Signal Processing*, 2023(1), 121. Advance online publication. DOI: 10.1186/s13634-023-01070-7

Lin, S., Shen, X., & Cao, Y. (2023). Reconfigurable intelligent surfaces in 6G networks: Physical layer security perspectives. *IEEE Transactions on Green Communications and Networking*, 7(2), 215–229.

Lin, X., Andrews, J. G., Ghosh, A., & Ratasuk, R. (2014). An overview of 3GPP device-to-device proximity services. *IEEE Communications Magazine*, 52(4), 40–48. DOI: 10.1109/MCOM.2014.6807945

Liu, R., Alexandropoulos, G. C., Wu, Q., Jian, M., & Liu, Y. "2022 IEEE/CIC International Conference on Communications in China (ICCC Workshops), 11-13 August 2022, **DOI:** DOI: 10.1109/ICCCWorkshops55477.2022.9896658

Liu, F., Guo, H., & Zhang, Z. (2020). RIS-enhanced communications: Performance evaluation and beamforming optimization. *IEEE Transactions on Wireless Communications*, 19(11), 7340–7353.

Liu, S., Cui, T. J., Xu, Q., Bao, D., Du, L., Wan, X., Tang, W. X., Ouyang, C., Zhou, X. Y., Yuan, H., Ma, H. F., Jiang, W. X., Han, J., Zhang, W., & Cheng, Q. (2016). Anisotropic coding metamaterials and their powerful manipulation of differently polarized terahertz waves. *Light, Science & Applications*, 5(5), e16076. DOI: 10.1038/lsa.2016.76 PMID: 30167164

Liu, Y., Huang, Y., & Zhang, J. Intelligent reflecting surface-aided beamforming for multiuser communication systems. IEEE Access 2020, 8, 43238–43253. DOI: 10.1109/COMST.2021.3077737

Liu, Y., Liu, X., Mu, X., Hou, T., Li, J., Al-Dhahir, N., & Hanzo, L. (2021). Reconfigurable intelligent surface (RIS)-aided vehicular networks: A progressive survey of recent advancements and road ahead. *IEEE Communications Surveys and Tutorials*, 23(3), 1576–1608.

Liu, Y., Zeng, M., & Zhang, R. (2020). Deep reinforcement learning for intelligent reflecting surface-assisted device-to-device communications. *IEEE Transactions on Wireless Communications*, 19(8), 5378–5387.

Liu, Y., Zhang, Q., & Wu, H. (2023). Energy-efficient RIS-aided physical layer security in 6G networks: Challenges and solutions. *IEEE Communications Surveys and Tutorials*, 25(1), 34–56.

Liu, Z., Li, T., Li, S., & Mi, C. C. (2024, June 18). Advancements and challenges in wireless power transfer: A comprehensive review. *Nexus*, 1(2), 100014. DOI: 10.1016/j.ynexs.2024.100014

Li, X., He, Z., & Wu, Q. (2021). Reconfigurable intelligent surface aided wireless communication: How many reflecting elements are needed? *IEEE Wireless Communications Letters*, 10(1), 68–71.

Li, X., Liu, C., & Zhao, Y. (2020). RIS-aided downlink non-orthogonal multiple access for massive connectivity: A practical design. *IEEE Internet of Things Journal*, 8(1), 571–583.

Li, X., Mao, J., Huang, Y., & Zhang, R. (2021). Intelligent reflecting surface enhanced wireless networks: Two-timescale beamforming optimization. *IEEE Transactions on Wireless Communications*, 20(12), 8092–8107. DOI: 10.1109/TWC.2021.3074935

Li, X., Zhang, Y., & Cui, S. (2021). Secure transmission using reconfigurable intelligent surfaces: Artificial noise approach. *IEEE Transactions on Communications*, 69(3), 1837–1850.

Li, Z., Xie, Y., Shangguan, L., Zelaya, R. I., Gummeson, J., Hu, W., & Jamieson, K. (2019). Programmable radio environments with large arrays of inexpensive antennas GetMobile Mob. *Computer Communications*, 23(3), 23–27.

Long, H., Chen, M., Yang, Z., Li, Z., Wang, B., Yun, X., & Shikh-Bahaei, M. Joint Trajectory and Passive Beamforming Design for Secure UAV Networks with RIS. In Proceedings of the 2020 IEEE Globecom Workshops (GC Wkshps), Taipei, Taiwan, 7–11 December 2020; pp. 1–6. DOI: 10.1109/GCWkshps50303.2020.9367542

Luo, H., Xu, J., & Tang, W. (2021). Secure communication with RIS in multi-eavesdropper scenarios. *IEEE Transactions on Vehicular Technology*, 70(7), 6912–6924.

Luo, H., & Yuen, C. (2021). Optimizing power and beamforming in RIS-assisted MISO systems. *IEEE Transactions on Communications*, 69(3), 1653–1665.

Ma, H., Zhang, H., Zhang, W., & Leung, V. C. M. (2023, February). Beamforming Optimization for Reconfigurable Intelligent Surface with Power Splitting Aided Broadcasting Networks. *IEEE Transactions on Vehicular Technology*, 72(2), 2712–2717. DOI: 10.1109/TVT.2022.3214529

Mailloux, R. J. (2005). *Phased Array Antenna Handbook*. Artech House.

Makki, B., Chitti, K., Behravan, A., & Alouini, M.-S. (2020). A survey of NOMA: Current status and open research challenges. *IEEE Open Journal of the Communications Society*, 1, 179–189. DOI: 10.1109/OJCOMS.2020.2969899

Marzetta, T. L. (2019). Massive MIMO: The next wireless revolution. *IEEE Communications Magazine*, 57(1), 136–141.

Marzetta, T. L., Larsson, E. G., Yang, H., & Ngo, H. Q. (2016). *Fundamentals of Massive MIMO*. Cambridge University Press. DOI: 10.1017/CBO9781316799895

Mishra, D. K., Joham, M., & Debbah, M. (2021). Learning to reflect and transmit: A novel intelligent reflecting surface with relay hybrid architecture. *IEEE Transactions on Wireless Communications*, 20(3), 1453–1469. DOI: 10.1109/TWC.2021.3051381

Molisch, A. F., Ratnam, V. V., Han, S., Li, Z., Nguyen, S. L. H., Li, L., & Haneda, K. (2017). Hybrid beamforming for massive MIMO: A survey. *IEEE Communications Magazine*, 55(9), 134–141. DOI: 10.1109/MCOM.2017.1600400

Mwase, C., Jin, Y., Westerlund, T., Tenhunen, H., & Zou, Z. (2022, June 1). Communication-efficient distributed AI strategies for the IoT edge. *Future Generation Computer Systems*, 131, 292–308. DOI: 10.1016/j.future.2022.01.013

Nadeem, Q.-U.-A., Kammoun, A., Chaaban, A., Debbah, M., & Alouini, M.-S. (2020). Asymptotic Max-Min SINR Analysis of Reconfigurable Intelligent Surface Assisted MISO Systems. *IEEE Transactions on Wireless Communications*, 19(12), 7748–7764. DOI: 10.1109/TWC.2020.2986438

Nadeem, Q.-U.-A., Kammoun, A., Chaaban, A., Debbah, M., & Alouini, M.-S. (2020). Intelligent reflecting surface-assisted wireless communication: A discrete phase-shift model. *IEEE Transactions on Communications*, 68(3), 2114–2127. DOI: 10.1109/TCOMM.2019.2963665

Naeem, Faisal, Ali, Mansoor, Kaddoum, Georges, Huang, Chongwen, & Yuen, Chau, "Security and Privacy for Reconfigurable Intelligent Surface in 6G: A Review of Prospective Applications and Challenges." IEEE Open Journal of the Communications Society (Volume: 4) **Page(s):** 1196 – 1217, . DOI: 10.1016/j.phycom.2023.102056

Nguyen, M. T., Amri, M. M., Park, J. H., Kim, D. I., & Choi, K. W. (2022, November). Security–Reliability Tradeoff Analysis for SWIPT- and AF-Based IoT Networks With Friendly Jammers. *IEEE Internet of Things Journal*, 9(21), 1–1. DOI: 10.1109/JIOT.2022.3182755

Nguyen, M. T., Nguyen, C. V., Truong, L. H., Le, A. M., Quyen, T. V., Masaracchia, A., & Teague, K. A. (2020). Electromagnetic Field Based WPT Technologies for UAVs: A Comprehensive Survey. *Electronics (Basel)*, 9(3), 461. DOI: 10.3390/electronics9030461

Nguyen, M.-S. V., Sur, S. N., & Do, D.-T. (2022). *A Comprehensive Review on Physical Layer Design for Smart Cities*. IoT and IoE Driven Smart Cities. DOI: 10.1007/978-3-030-82715-1_1

Nguyen, V.-L., Lin, P.-C., Cheng, B.-C., Hwang, R.-H., & Lin, Y.-D. (2021). Security and privacy for 6G: A survey on prospective technologies and challenges. [Google Scholar] [CrossRef]. *IEEE Communications Surveys and Tutorials*, 23(4), 2384–2428. DOI: 10.1109/COMST.2021.3108618

Niu, H., Lin, Z., Chu, Z., Zhu, Z., Xiao, P., Nguyen, H. X., Lee, I., & Al-Dhahir, N. (2022). Joint beamforming design for secure RIS-assisted IoT networks. *IEEE Internet of Things Journal*, 10(2), 1628–1641. DOI: 10.1109/JIOT.2022.3210115

Niu, J.-P., & Li, G. Y. (2019). An overview on backscatter communications. *Journal of Communications and Information Networks*, 4(2), 1–14. DOI: 10.23919/JCIN.2019.8917868

Ntontin, K., Di Renzo, M., & Lazarakis, F. On the Rate and Energy Efficiency Comparison of Reconfigurable Intelligent Surfaces with Relays. In *Proceedings of the 2020 IEEE 21st International Workshop on Signal Processing Advances in Wireless Communications (SPAWC)*, Atlanta, GA, USA, 26–29 May 2020; pp. 1–5. DOI: 10.1109/SPAWC48557.2020.9154308

Ojha, T., & Theofanis, P. (2023). Raptis· Andrea Passarella, Marco Conti, "Wireless power transfer with unmanned aerial vehicles: State of the art and open challenges". *Pervasive and Mobile Computing*, 93(June), 101820. DOI: 10.1016/j.pmcj.2023.101820

Özdogan, Ö., Björnson, E., & Larsson, E. G. (2019). Intelligent reflecting surfaces: Physics, propagation, and path loss modelling. *IEEE Wireless Communications Letters*, 9(5), 581–585. DOI: 10.1109/LWC.2019.2960779

Ozdogan, O., Bjornson, E., & Larsson, E. G. (2020). Using intelligent reflecting surfaces for rank improvement in MIMO communications. *IEEE Wireless Communications Letters*, 9(1), 60–64. DOI: 10.1109/LWC.2019.2941232

Pan, C., Ren, H., Wang, K., Kolb, J. F., Elkashlan, M., Chen, M., Di Renzo, M., Hao, Y., Wang, J., Swindlehurst, A. L., You, X., & Hanzo, L. (2021). Reconfigurable intelligent surfaces for 6G systems: Principles, applications, and research directions. *IEEE Communications Magazine*, 59(6), 14–20. DOI: 10.1109/MCOM.001.2001076

Pan, C., Zhou, G., Zhi, K., Hong, S., Wu, T., Pan, Y., Ren, H., Di Renzo, M., Swindlehurst, A. L., & Zhang, R. (2022). An overview of signal processing techniques for RIS/IRS-aided wireless systems. [Google Scholar] [CrossRef]. *IEEE Journal of Selected Topics in Signal Processing*, 16(5), 883–917. DOI: 10.1109/JSTSP.2022.3195671

Pavia, J. P., Velez, V., Ferreira, R., Souto, N., Ribeiro, M., Silva, J., & Dinis, R. (2021). Low Complexity Hybrid Precoding Designs for Multiuser mmWave/THz Ultra Massive MIMO Systems. *Sensors (Basel)*, 21(18), 6054. DOI: 10.3390/s21186054 PMID: 34577260

Pei, X., Zeng, J., & Yu, Y. (2021). RIS-aided secure UAV communication: Secrecy rate maximization with flight path optimization. *IEEE Transactions on Vehicular Technology*, 70(2), 1256–1267.

Pesyna, K. M.Jr, Kassas, Z. M., Bhatti, J. A., & Humphreys, T. E. (2011). Tightly-coupled opportunistic navigation for deep urban and indoor positioning. In *Proceedings of the ION GPS* (pp. 3065–3616). Portland, OR, USA. DOI: 10.1109/GLOCOMW.2011.6162371

Peterson, B., Hartnett, R., & Ottman, G. (1995). GPS receiver structures for the urban canyon. In *Proceedings of the ION GPS* (pp. 1323–1332). Palm Springs, CA, USA.

Pogaku, A. C., Do, D.-T., Lee, B. M., & Nguyen, N. D. (2022, January). UAV-Assisted RIS for Future Wireless Communications: A Survey on Optimization and Performance Analysis. *IEEE Access : Practical Innovations, Open Solutions*, 10(4), 1–1. DOI: 10.1109/ACCESS.2022.3149054

Pozar, D. M., Targonski, S. D., & Syrigos, H. D. (1997, February). Design of millimetre wave microstrip reflectarrays. *IEEE Transactions on Antennas and Propagation*, 45(2), 287–296. DOI: 10.1109/8.560348

Rahmatov, N., & Baek, H. (2023). RIS-carried UAV communication: Current research, challenges, and future trends. *ICT Express*, 9(5), 961–973. DOI: 10.1016/j.icte.2023.03.004

Rappaport, T. S., Sun, S., Mayzus, R., Zhao, H., Azar, Y., Wang, K., Wong, G. N., Schulz, J. K., Samimi, M., & Gutierrez, F. (2013). Millimeter wave wireless communications: A review. *IEEE Transactions on Communications*, 61(9), 3619–3652.

Rehman, B. U., Babar, M. I., Ahmad, A. W., Alhumyani, H., Abdel Azim, G., Saeed, R. A., & Abdel Khalek, S. (2021). Joint power control and user grouping for uplink power domain non-orthogonal multiple access. *International Journal of Distributed Sensor Networks*, 17(12), 15501477211057443. DOI: 10.1177/15501477211057443

Rehman, F. U., Attaullah, H. M., Ahmed, F., & Ali, S. (2023). Data Defense: Examining Fintech's Security and Privacy Strategies. *Engineering Proceedings*, 32, 3.

Reid, T. G., Neish, A. M., Walter, T. F., & Enge, P. K. (2016). Leveraging commercial broadband LEO constellations for navigating. In *Proceedings of the ION GNSS* (pp. 2300–2314). Portland, OR, USA. DOI: 10.33012/2016.14729

Ren, Y., Zhou, R., Teng, X., Meng, S., Zhou, M., & Tan, W. (2023, October). On Deployment Position of RIS in Wireless Communication Systems: Analysis and Experimental Results. *IEEE Wireless Communications Letters*, 12(no. 1756-1760), 1756–1760. DOI: 10.1109/LWC.2023.3292125

Renzo, M. D., (2020). Smart Wireless Communication Environments via RIS: Opportunities, Challenges, and Future Directions. *IEEE Communications Magazine*, 58(6), 40–47. DOI: 10.1109/MCOM.001.2000074

Renzo, M. D., Danufane, F. H., & Tretyakov, S. (2022). Communication Models for Reconfigurable Intelligent Surfaces: From Surface Electromagnetics to Wireless Networks Optimization. [Google Scholar] [CrossRef]. *Proceedings of the IEEE*, 110(9), 1164–1209. DOI: 10.1109/JPROC.2022.3195536

Renzo, M. D., Debbah, M., Phan-Huy, D. T., Zappone, A., Alouini, M.-S., Yuen, C., Sciancalepore, V., Alexandropoulos, G. C., Hoydis, J., & Gacaninet, H.. (2019). Smart radio environments empowered by reconfigurable AI meta-surfaces: An idea whose time has come. *EURASIP Journal on Wireless Communications and Networking*, 2019(1), 1–20. DOI: 10.1186/s13638-019-1438-9

Renzo, M. D., Debbah, M., & Phan, V. (2020). Reconfigurable intelligent surfaces vs. relaying: Differences, similarities, and performance comparison. *IEEE Open Journal of the Communications Society*, 1, 798–807. DOI: 10.1109/OJCOMS.2020.3002955

Renzo, M. D., Zappone, A., & Debbah, M. (2020). RIS-assisted communications with NOMA: Benefits and challenges. *IEEE Wireless Communications Letters*, 9(12), 2041–2045.

Rodriguez, J., Verikoukis, C., Vardakas, J. S., & Passas, N. (2022). *Enabling 6G Mobile Networks*. Springer. [Google Scholar] DOI: 10.1007/978-3-030-74648-3

Rost, P., Mannweiler, C., Michalopoulos, D. S., Sartori, C., Sciancalepore, V., Sastry, N., Holland, O., Tayade, S., Han, B., Bega, D., Aziz, D., & Bakker, H. (2017). Network slicing to enable scalability and flexibility in 5G mobile networks. *IEEE Communications Magazine*, 55(5), 72–79. DOI: 10.1109/MCOM.2017.1600920

Saad, W., Bennis, M., & Chen, M. (2020). A vision of 6G wireless systems: Applications, trends, technologies, and open research problems. *IEEE Network*, 34(3), 134–142. DOI: 10.1109/MNET.001.1900287

Saiam, M., Chowdhury, M. Z., & Hasan, S. R. (2023). “Crypto Holographic Reconfigurable Intelligent Surface-Assisted Backscatter Communication.” *The Fourth Industrial Revolution and Beyond: Select Proceedings of IC4IR+*. Singapore: Springer. *Nature in Singapore*, •••, 149–156.

Saiam, M., Chowdhury, M. Z., Hasan, S. R., & Jang, Y. M. (2023). Reconfigurable intelligent surface assisted BackCom: An overview, analysis, and future research directions. *ICT Express*, 9(5), 927–940. DOI: 10.1016/j.icte.2023.07.004

Samsung. (2012). Samsung’s 6G white paper lays out the company’s vision for the next generation of communications technology. Retrieved from https://news.samsung.com/global/samsungs-6gwhite-paper-lays-out-the-companysvision-for-the-next-generationof-communications-Technology

Seco-Granados, G., López-Salcedo, J. A., Jiménez-Baños, D., & López-Risueño, G. (2012). Challenges in indoor global navigation satellite systems: Unveiling its core features in signal processing. *IEEE Signal Processing Magazine*, 29(2), 108–131. DOI: 10.1109/MSP.2011.943410

Sejan, M. A. S., Rahman, M. H., Shin, B. S., Oh, J. H., You, Y. H., & Song, H. K. (2022). Machine Learning for Intelligent-Reflecting-Surface-Based Wireless Communication towards 6G: A Review. [Google Scholar] [CrossRef] [PubMed]. *Sensors (Basel)*, 22(14), 5405. DOI: 10.3390/s22145405 PMID: 35891085

Selvaraj, M., Vijay, R., Anbazhagan, R., & Rengarajan, A. (2023). Ramya Vijay *, Rajesh Anbazhagan and Amirtharajan Rengarajan, "Reconfigurable Metasurface: Enabling Tunable Reflection in 6G Wireless Communications,". *Sensors (Basel)*, 23(22), 9166. DOI: 10.3390/s23229166 PMID: 38005555

Shahab, M. B., Johnson, S. J., Shirvanimoghaddam, M., Chafii, M., Basar, E., & Dohler, M. (2020). Index modulation aided uplink NOMA for massive machine type communications. *IEEE Wireless Communications Letters*, 9(12), 2159–2162. DOI: 10.1109/LWC.2020.3015920

Shao, Y., Liu, Z., & Wang, F. (2022). Reinforcement learning-based design of RIS for physical layer security. *IEEE Internet of Things Journal*, 9(5), 3897–3909.

Sharma, T., Chehri, A., & Fortier, P. (2021). Reconfigurable Intelligent Surfaces for 5G and beyond Wireless Communications: A Comprehensive Survey. *Energies*, 14(24), 8219. DOI: 10.3390/en14248219

Shen, H., Zhao, T., & Zhang, R. (2021). Phase-shift design for intelligent reflecting surfaces in wireless communications: A new approach with SVD-based optimization. *IEEE Transactions on Communications*, 69(5), 3393–3407.

Sikander, M., Butt, F. A., Chattha, J. N., & Naqvi, I. H. Improving Wireless Communication through Beamwidth Optimization in Reconfigurable Intelligent Surfaces. In *Proceedings of the 2023 10th International Conference on Wireless Networks and Mobile Communications (WINCOM)*, Istanbul, Turkey, 26–28 October 2023; pp. 1–5. [Google Scholar] DOI: 10.1109/WINCOM59760.2023.10322929

Smith, D. R., Padilla, W. J., Vier, D. C., Nemat-Nasser, S. C., & Schultz, S. (2004). Composite medium with simultaneously negative permeability and permittivity. *Physical Review Letters*, 84(18), 4184–4187. DOI: 10.1103/PhysRevLett.84.4184 PMID: 10990641

Strinati, E. C., Alexandropoulos, G. C., Sciancalepore, V., Renzo, M. D., Wymeersch, H., Phan-Huy, D. T., Crozzoli, M., D'Errico, R., Carvalho, E. D., Popovski, P., ... (2021). Wireless environment as a service enabled by reconfigurable intelligent surfaces: The RISE-6G perspective. In Proceedings of the 2021 Joint EuCNC & 6G Summit (pp. 562–567). Porto, Portugal.

Subrt, L., & Pechac, P. (2012). Intelligent Walls as Autonomous Parts of Smart Indoor Environments. *IET Communications*, 6(8), 1004–1010. DOI: 10.1049/iet-com.2010.0544

Sur, S. N., & Bera, R. (2021). Intelligent reflecting surface assisted MIMO communication system: A review. *Physical Communication*, 47, 101386. DOI: 10.1016/j.phycom.2021.101386

Sziroczak, D., Rohacs, D., & Rohacs, J. (2022, October). Review of using small UAV based meteorological measurements for road weather management. *Progress in Aerospace Sciences*, 134(1), 100859. DOI: 10.1016/j.paerosci.2022.100859

Taha, A., Alrabeiah, M., & Alkhateeb, A. (2020). Enabling large intelligent surfaces with compressive sensing and deep learning. *IEEE Access : Practical Innovations, Open Solutions*, 8, 79682–79699.

Tan, D. K. P., He, J., Li, Y., Bayesteh, A., Chen, Y., Zhu, P., & Tong, W. Integrated sensing and communication in 6G: Motivations, use cases, requirements, challenges and future directions. In Proceedings of the 2021 1st IEEE International Online Symposium on Joint Communications & Sensing (JC&S), Dresden, Germany, 23–24 February 2021; pp. 1–6.

Tang, W., Chen, M. Z., Chen, X., Dai, J. Y., Han, Y., Renzo, M. D., Zeng, Y., Jin, S., Cheng, Q., & Cui, T. J. (2021). Wireless communications with reconfigurable intelligent surface: Path loss modeling and experimental measurement. *IEEE Transactions on Wireless Communications*, 20(1), 421–439. DOI: 10.1109/TWC.2020.3024887

Tang, W., Chen, M. Z., Wong, K.-K., & Wang, S. (2020). Path loss modeling and measurements for reconfigurable intelligent surfaces in indoor environment. *IEEE Transactions on Wireless Communications*, 19(5), 3961–3970. DOI: 10.1109/TWC.2020.2969473

Tang, W., Chen, M., & Guo, Y. (2021). RIS-aided physical layer security for massive IoT: System design and performance analysis. *IEEE Internet of Things Journal*, 8(10), 8672–8686.

Tang, W., Dai, J. Y., Shen, H., & Zeng, Y. (2020). Programmable metasurface-based RF chain-free 8PSK wireless transmitter. *Electronics Letters*, 56(11), 564–567.

Tang, W., Dai, J. Y., Xu, W., Yuan, X., Zhao, L., & Zuo, C. (2020). Smart and reconfigurable environment: Indoor THz wireless communication enabled by intelligent reflecting surfaces. *IEEE Transactions on Communications*, 68(8), 5195–5209.

Tang, W., Zeng, Y., & Wu, Q. (2021). Intelligent reflecting surface-enhanced wireless networks: From theory to practical implementation. *IEEE Communications Magazine*, 59(6), 60–66.

Tang, Z., Hou, T., Liu, Y., Zhang, J., & Zhong, C. (2021). A novel design of RIS for enhancing the physical layer security for RIS-aided NOMA networks. *IEEE Wireless Communications Letters*, 10(11), 2398–2401. DOI: 10.1109/LWC.2021.3101806

Van Diggelen, F. (2009). *A-GPS: Assisted GPS, GNSS, and SBAS*. Artech House.

Velusamy, P., Rajendran, S., Mahendran, R. K., Naseer, S., Shafiq, M., & Choi, J.-G. (2022). Unmanned Aerial Vehicles (UAV) in Precision Agriculture: Applications and Challenges. *Energies*, 15(1), 217. DOI: 10.3390/en15010217

Vordonis, D., Kompostiotis, D., & Paliouras, V. "Reconfigurable Intelligent Surface-Aided OFDM Wireless Communications: Hardware Aspects of Reflection Optimization Methods," 2022 11th International Conference on Modern Circuits and Systems Technologies, Bremen, Germany, 2022, pp. 1-4, DOI: 10.1109/MOCAST54814.2022.9837634

Wang, B., Zhang, L., Zhang, J., & Liu, Z. (2015). Resource allocation for D2D communication underlaying OFDMA cellular networks using coalitional games. *IEEE Transactions on Wireless Communications*, 14(6), 3259–3272.

Wang, C. X., Di Renzo, M., Stanczak, S., Wang, S., & Larsson, E. G. (2020). Artificial intelligence enabled wireless networking for 5G and beyond: Recent advances and future challenges. [Google Scholar] [CrossRef]. *IEEE Wireless Communications*, 27(1), 16–23. DOI: 10.1109/MWC.001.1900292

Wang, J., Feng, W., & Niu, Z. (2020). Beamforming optimization for intelligent reflecting surface-assisted communications with channel state information uncertainty. *IEEE Transactions on Wireless Communications*, 19(10), 6719–6733.

Wang, J., Gao, D., Li, J., Huang, L., Ding, H., & Zhou, S. (2024). Analysis and Mitigating Methods for Jamming in the Optical Reconfigurable Intelligent Surfaces-Assisted Dual-Hop FSO Communication Systems. [Google Scholar] [CrossRef]. *Electronics (Basel)*, 13(9), 1730. DOI: 10.3390/electronics13091730

Wang, J., Jiang, C., Zhang, H., Ren, Y., Chen, K. C., & Hanzo, L. (2020). Thirty years of machine learning: The road to Pareto-optimal wireless networks. [Google Scholar] [CrossRef]. *IEEE Communications Surveys and Tutorials*, 22(3), 1472–1514. DOI: 10.1109/COMST.2020.2965856

Wang, Y., Li, C., Liu, C., Liu, S., Lei, Y., Zhang, J., Zhang, Y., & Guo, Y. (2021). Advancing DSP into HPC, AI, and beyond: Challenges, mechanisms, and future directions. *CCF Transactions on High Performance Computing*, 3(1), 114–125. DOI: 10.1007/s42514-020-00057-2

Wu, Q., (2020). Reconfigurable Intelligent Surfaces: Pathways to 6G. *IEEE Transactions on Wireless Communications*, 20(1), 496–515. DOI: 10.1109/TWC.2020.3032112

Wu, Q., & Zhang, R. (2019). Intelligent reflecting surface enhanced wireless network via joint active and passive beamforming. *IEEE Transactions on Wireless Communications*, 18(11), 5394–5409. DOI: 10.1109/TWC.2019.2936025

Wu, Q., & Zhang, R. (2019). Towards Smart and Reconfigurable Environment: Intelligent Reflecting Surface Aided Wireless Network. *IEEE Communications Magazine*, 58(1), 106–112. DOI: 10.1109/MCOM.001.1900107

Wu, Q., Zhang, R., & Larsson, E. G. (2020). Intelligent reflecting surface-aided wireless communications: A tutorial. *IEEE Transactions on Communications*, 69(5), 3433–3465. DOI: 10.1109/TCOMM.2020.3036433

Wyner, A. D. (1975). The wire-tap channel. *The Bell System Technical Journal*, 54(8), 1355–1387. DOI: 10.1002/j.1538-7305.1975.tb02040.x

Xu, F., Hussain, T., Ahmed, M., Ali, K., Mirza, M. A., Khan, W. U., Ihsan, A., & Han, Z. (2023). The state of AI-empowered backscatter communications: A comprehensive survey. *IEEE Internet of Things Journal*, 10(24), 21763–21786. DOI: 10.1109/JIOT.2023.3299210

Xu, J., Wang, W., & He, S. (2021). Reconfigurable intelligent surfaces for secure communications in 6G: Challenges, opportunities, and research directions. *IEEE Communications Magazine*, 59(7), 82–88.

Xu, Y., Liu, L., Zhou, S., & Niu, Z. (2015). Opportunistic device-to-device communications: A queueing perspective. *IEEE Journal on Selected Areas in Communications*, 33(1), 68–80.

Yadav, A., Singh, R. K., & Li, J. (2021). Meta-material inspired reconfigurable intelligent surfaces for wireless communication: A comprehensive survey. *IEEE Access : Practical Innovations, Open Solutions*, 9, 103558–103579.

Yang, C., & Nguyen, T. (2009). Self-calibrating position location using signals of opportunities. In *Proceedings of the ION GNSS* (pp. 1055–1063). Savannah, GA, USA. DOI: 10.1109/NAECON.2009.5426658

Yang, L., Meng, F., Zhang, J., Hasna, M. O., & Di Renzo, M. (2020). On the performance of RIS-assisted dual-hop UAV communication systems. *IEEE Transactions on Vehicular Technology*, 69(9), 10385–10390. DOI: 10.1109/TVT.2020.3004598

Yang, P., Zhang, C., Zhou, X., & Pan, C. (2021). Hierarchical resource allocation for intelligent reflecting surface-assisted heterogeneous networks. *IEEE Transactions on Communications*, 69(8), 5222–5237.

Yang, Y., Wu, Q., Wang, S., Zhang, R., & Zheng, X. (2020). Intelligent reflecting surface-aided non-orthogonal multiple access. *IEEE Transactions on Wireless Communications*, 20(7), 4585–4600.

Yan, Y., Shi, W., & Zhang, X. (2020). Design of UAV wireless power transmission system based on coupling coil structure optimization. *EURASIP Journal on Wireless Communications and Networking*, 2020(1), 67. Advance online publication. DOI: 10.1186/s13638-020-01679-4

Ye, J., Guo, S., & Alouini, M.-S. (2020, August). Joint reflecting and precoding designs for SER minimization in reconfigurable intelligent surfaces assisted MIMO systems. *IEEE Transactions on Wireless Communications*, 19(8), 5561–5574. DOI: 10.1109/TWC.2020.2994455

You, C., & Zhang, R. (2021). Wireless communication aided by intelligent reflecting surfaces: Active or passive? *IEEE Transactions on Wireless Communications*, 20(11), 7270–7284. DOI: 10.1109/TWC.2021.3097401

Yu, X., Shen, J., Jin, S., & Zhu, G. (2020). An overview of security issues in intelligent reflecting surface (IRS) aided wireless systems. *IEEE 91st Vehicular Technology Conference (VTC2020-Spring)*, 1-5.

Yuan, X., Shi, Y., & Debbah, M. (2020). Indoor intelligent reflecting surface enhanced wireless local area network: A case study. *IEEE Wireless Communications Letters*, 9(2), 199–203. DOI: 10.1109/LWC.2019.2955559

Yuan, X., Zhang, Y.-J. A., Shi, Y., Yan, W., & Liu, H. (2021). Reconfigurable-intelligent-surface empowered wireless communications: Challenges and opportunities. [Google Scholar]. *IEEE Wireless Communications*, 28(2), 136–143. DOI: 10.1109/MWC.001.2000256

Yu, N., Genevet, P., Kats, M. A., Aieta, F., Tetienne, J.-P., Capasso, F., & Gaburro, Z. (2011). Light propagation with phase discontinuities: Generalized laws of reflection and refraction. *Science*, 334(6054), 333–337. DOI: 10.1126/science.1210713 PMID: 21885733

Yu, X., Xu, D., & Schober, R. "MISO wireless communication systems via intelligent reflecting surfaces," in *2019 IEEE/CIC International Conference on Communications in China (ICCC)*, 2019, pp. 735-740.

Yu, X., Xu, D., & Zhang, R. (2020). Smart and reconfigurable environment with intelligent reflecting surfaces: A tutorial on modeling and analysis. *IEEE Communications Surveys and Tutorials*, 23(2), 942–973.

Zappone, A., Di Renzo, M., & Debbah, M. (2019). Wireless networks design in the era of deep learning: Model-based, AI-based, or both? [Google Scholar] [CrossRef]. *IEEE Transactions on Communications*, 67(10), 7331–7376. DOI: 10.1109/TCOMM.2019.2924010

Zappone, A., Di Renzo, M., Shams, F., Yu, X., & Poor, H. V. (2019). Reconfigurable intelligent surfaces for wireless communications: From phase shifters to baseband precoding and computing. *IEEE Transactions on Signal Processing*, 67(19), 4809–4826.

Zeng, Y., Zhang, R., & Wu, Q. (2021). Reconfigurable intelligent surfaces in wireless communication: Recent advancements and future trends. *IEEE Communications Magazine*, 59(7), 20–25.

Zhang, H. (2021). *Reconfigurable Intelligent Surface-Empowered 6G*. Springer. DOI: 10.1007/978-3-030-73499-2

Zhang, H., Liu, Y., & Wu, Q. (2022). RIS-assisted physical layer security for mmWave and THz communication. *IEEE Communications Letters*, 26(3), 401–405.

Zhang, H., Wang, L., & Huang, C. (2021). Towards 6G wireless networks: Prototyping of reconfigurable intelligent surfaces. *IEEE Wireless Communications*, 28(2), 86–92.

Zhang, L., Liang, Y.-C., & Peng, L. (2020). Intelligent reflecting surface aided multi-user MIMO: Exploiting deep reinforcement learning for beamforming. *IEEE Journal on Selected Areas in Communications*, 38(8), 1809–1822. DOI: 10.1109/JSAC.2020.3001081

Zhang, L., Wang, Y., Tao, W., Jia, Z., Song, T., & Pan, C. (2020). Intelligent Reflecting Surface Aided MIMO Cognitive Radio Systems. *IEEE Transactions on Vehicular Technology*, 69(10), 11445–11457. DOI: 10.1109/TVT.2020.3011308

Zhang, Q., Zhang, R., & Wu, Q. (2021). Beyond the era of massive MIMO: Intelligent reflecting surfaces for 6G wireless networks. *IEEE Transactions on Wireless Communications*, 20(4), 2457–2471.

Zhang, S., Huang, C., & Debbah, M. (2021). Machine learning methods for reconfigurable intelligent surfaces: State of the art and future prospects. *IEEE Communications Surveys and Tutorials*, 23(4), 2688–2721. DOI: 10.1109/COMST.2021.3085383

Zhang, S., Huang, W., & Liu, Y. (2024). A systematic survey on physical layer security-oriented to reconfigurable intelligent surface empowered 6G. *Computers & Security*, •••, 104100.

Zhang, S., Zhang, H., & Di Renzo, M. (2020). Capacity analysis of reconfigurable intelligent surfaces aided MIMO systems. *IEEE Journal on Selected Areas in Communications*, 38(8), 1824–1838. DOI: 10.1109/JSAC.2020.3000814

Zhang, S., Zhang, R., & Cui, S. (2021). Cooperative interference cancellation and relaying for intelligent reflecting surface enhanced wireless networks. *IEEE Wireless Communications Letters*, 9(10), 1703–1707.

Zhang, Z., Li, M., & Yu, W. (2022). RIS-assisted secure millimeter-wave communications: Challenges and opportunities. *IEEE Wireless Communications*, 29(4), 72–79.

Zhang, Z., Wang, L., Xu, J., & Liu, H. (2020). Intelligent reflecting surface assisted massive MIMO systems: Capacity maximization through joint beamforming. *IEEE Transactions on Wireless Communications*, 19(8), 5357–5371.

Zhao, H., Liang, Y., & Niu, Z. (2020). Intelligent reflecting surfaces assisted device-to-device communications in millimeter-wave wireless networks. *IEEE Transactions on Vehicular Technology*, 69(9), 10401–10410.

Zhao, Y. (2023). Reconfigurable intelligent surfaces for 6G: Applications, challenges, and solutions. *Frontiers of Information Technology & Electronic Engineering*, 24(12), 1669–1688. DOI: 10.1631/FITEE.2200666

Zheng, B., Zhang, Q., & Zhu, H. (2021). Intelligent reflecting surface-assisted secure communication: A machine learning approach. *IEEE Transactions on Communications*, 69(9), 5987–6000.

Zheng, C. Y., & Zhang, R. "Uplink channel estimation for double-IRS assisted multi-user MIMO", *Proc. IEEE Int. Conf. Commun.*, pp. 1-6, Jun. 2021. DOI: 10.1109/ICC42927.2021.9501057

Zhou, G., Yuan, W., Yu, X., & Zhang, R. (2021). Intelligent reflecting surface: Practical phase shift model and efficient beamforming optimization. *IEEE Communications Letters*, 25(1), 118–122.

Zhou, S., Wang, J., Niu, Z., & Zhang, Y. (2018). Joint resource allocation and power control for D2D communications underlaying cellular networks. *IEEE Transactions on Wireless Communications*, 17(11), 7595–7609.

About the Contributors

M. Nalini is currently working as a Professor, Electronics and Instrumentation Engineering in Sri Sairam Engineering College, Chennai, Tamil Nadu. She had 20.2 years of experience in teaching. She published 50 Scopus research papers and 8 book chapters in various SCOPUS, WEB OF SCIENCE Indexed Books with IGI GLOBAL, CRC Press and Elsevier. Currently she is an editor of the book under Elsevier publisher. Areas of interest are Artificial Intelligence, Biomedical Engineering and Cyber security. She had 3 patent grant and published six patents. She received the best woman faculty award for the year 2019-20 from Novel Research Academy. She had received 1 lakh DST grant for the project LactoPure and 3.98 lakh from AICTE for STTP grant.

Malathy Sathyamoorthy is an Assistant professor in the department of Information Technology, KPR institute of Engineering and Technology, Coimbatore, Tamil Nadu, India. She received her M.E degree in Computer Science and Engineering from Anna University, India in 2012. She completed Ph.D. degree in Information and Communication Engineering from Anna University, Chennai, India, in 2023. She is a senior member in IEEE, life member of Indian Society for Technical Education (ISTE) and International Association of Engineers (IAENG). Published more than 25 Research Papers in various High-Quality SCI Impact Factor Journals cum Scopus/ESCI indexed Journals, 22 Papers in International Conferences indexed with Springer and IEEE Xplore, 2 patents, 1 book and 8 Book Chapters in various SCOPUS, WEB OF SCIENCE Indexed Books with Springer, CRC Press and Elsevier. Wireless Sensor Networks, Networking, Security and Machine Learning are her research interests. She is acting as a reviewer in Springer – Wireless Networks and editorial board member in many international conferences.

Rajesh Kumar Dhanaraj is a distinguished Professor at Symbiosis International (Deemed University) in Pune, India. Before joining Symbiosis International University, he served as a Professor at the School of Computing Science and

Engineering at Galgotias University in Greater Noida, India. His academic and research achievements have earned him a place among the top 2% of scientists globally, a recognition bestowed upon him by Elsevier and Stanford University. He earned his B.E. degree in Computer Science and Engineering from Anna University Chennai, India, in 2007. Subsequently, he obtained his M.Tech degree from Anna University Coimbatore, India, in 2010. His relentless pursuit of knowledge culminated in a Ph.D. in Computer Science from Anna University, Chennai, India, in 2017. He has authored and edited over 50 books on various cutting-edge technologies and holds 21 patents. Furthermore, he has contributed over 100 articles and papers to esteemed refereed journals and international conferences, in addition to providing chapters for several influential books. Dr. Dhanaraj has shared his insights with the academic community by delivering numerous tech talks on disruptive technologies. He has forged meaningful partnerships with esteemed professors from top QS-ranked universities around the world, fostering a global network of academic excellence. His research interests encompass Machine Learning, Cyber-Physical Systems, and Wireless Sensor Networks. Dr. Dhanaraj's expertise in these areas has led to numerous researches talks on Applied AI and Cyber Physical Systems at various esteemed institutions. Dr.Dhanaraj has earned the distinction of being a Senior Member of the Institute of Electrical and Electronics Engineers (IEEE). He is also a member of the Computer Science Teacher Association (CSTA) and the International Association of Engineers (IAENG). Dr. Dhanaraj's commitment to academic excellence extends to his role as an Associate Editor and Guest Editor for renowned journals, including Elsevier Computers and Electrical Engineering, Human-centric Computing and Information Sciences, Emerald - International Journal of Pervasive Computing and Communications, and Hindawi - Mobile Information Systems. His expertise has earned him a position as an Expert Advisory Panel Member of Texas Instruments Inc., USA.(b1ece092-7ce4-4a76-b7b0-06e631dcfaf)

***

A. Rajasekar. Associate Professor, Department of Artificial Intelligence and Data Science, Sri Sai Ram Institute of Technology, Chennai, India Email: rajasekarbe@gmail.com

D. Dhinakaran (M'84) was born in Chennai, Tamilnadu, India, in 1984. He is a distinguished scholar and educator with a passion for advancing the fields of Computer Science and Engineering. Dr. Dhinakaran earned his B.E degree in Computer Science and Engineering from Anna University, Chennai, in 2006, showcasing early dedication to his academic pursuits. Building upon his foundation, he pursued and obtained the M.E degree in the field of Computer Science and

Engineering from Anna University, Trichy, in 2009. Demonstrating a commitment to academic excellence, Dr. Dhinakaran completed his Ph.D. in Computer Science and Engineering from Anna University, Chennai, further solidifying his expertise in the field. Currently serving as an Assistant Professor in the Department of Computer Science and Engineering at Vel Tech Rangarajan Dr. Sagunthala R&D Institute of Science and Technology, Chennai, Dr. Dhinakaran continues to inspire students and colleagues alike. His areas of specialization encompass Privacy-Preserving Data Mining, Artificial Intelligence, Internet of Things (IoT), Mobile Ad Hoc Networks (MANET), Cloud Computing, and Image Processing, reflecting his diverse and extensive contributions to cutting-edge technologies. Dr. Dhinakaran is a prolific researcher, having presented more than 30 papers at various National and International Conferences. His commitment to advancing knowledge is evident through his membership in esteemed professional societies such as IAENG (International Association of Engineers), IFERP (International Foundation for Engineering Research and Publications), and CSTA (Computer Science Teachers Association). Notably, Dr. Dhinakaran has made significant contributions to academic literature, with publications in 60 different international journals and the successful filing of two patents. His work demonstrates a keen interest in addressing contemporary challenges and pushing the boundaries of technological innovation. As an accomplished academic, researcher, and member of professional societies, Dr. Dhinakaran continues to play a pivotal role in shaping the future of Computer Science and Engineering, leaving an indelible mark on both academia and the broader technological landscape.

Sheshang Degadwala Associate Professor Department of Computer Engineering, Sigma Institute of Engineering, Vadodara, Gujarat, India sheshang13@gmail.com

Rajiv Iyer is currently an Associate Professor in Computer Science and Engineering at Amity University, Mumbai. With a Ph.D. in the domain of Internet of Things, he has had a distinguished career spanning 14 years in academia. He is a registered Patent Agent (IN/PA3625). He is certified educator for AWS, Celonis. He is Google Facilitator for Machine Learning and Kaggle BIPOC program mentor/advisor.

K.Karthika received her B.E degree in Electronics & Communication Engineering from Maharaja Engineering College, Avinashi, India in 2013 and M.E degree in Communication Systems from Mepco Schlenk Engineering College, Sivakasi, India in 2015. She has obtained Ph.D. in Anna University Chennai in 2024, India. Currently, she is working as an Assistant Professor in the Department of

Electronics and Communication Engineering at Kumaraguru College of Technology, Coimbatore, India. Her research interest includes Ultra-Wideband (UWB) Antennas, Reconfgurable Antennas, Rectennas and Fractal Antennas.

Kiran Kakade, Associate Professor, Faculty of Management, Symbiosis Institute of Management Studies (SIMS), Symbiosis International (Deemed University), Pune. Email:kirankakade2025@gmail.com

Jagadish Kumar.N is currently serving as an Assistant Professor in the Department of Data science and business systems at SRM Institute of Science and Technology, Kattankulathur, Tamil Nadu, India. He completed his Bachelor of Engineering degree in Computer Science and Engineering from Prathyusha Engineering College in Chennai, Tamil Nadu, India, in 2005. Furthering his academic pursuits, he earned a Master of Engineering degree in Computer Technology from the prestigious Anna University – MIT Campus, Chennai, Tamil Nadu, India, in 2014. He subsequently completed his Ph.D. in 2024 under the Faculty of Information and Communication Engineering at Anna University, Chennai, India. His research interests are diverse and interdisciplinary, encompassing areas such as Big Data Analytics, Machine Learning, Cloud Computing, the Internet of Things (IoT), and Networks. His work in these fields focuses on advancing technology and finding innovative solutions to complex problems. Dr. Jagadish Kumar.N is dedicated to contributing to the academic community through both his teaching and research efforts, aiming to bridge the gap between theoretical knowledge and practical application.

Ashokkumar N Associate Professor Mohan Babu University Tirupati, Andhra Pradesh, India ashoknoc@gmail.com

R. Rajeshkanna is an accomplished academic with over 16 years of experience in the field of computer science and information technology. He holds a Ph.D. from Bharathiar University, Coimbatore, and an M.Phil from Vinayaka Missions University, Salem. He also earned his MCA from Periyar University, Salem. Currently, Dr. Rajeshkanna serves as an Assistant Professor in the Department of Computer Science at CHRIST (Deemed to be University), Bangalore, a position he has held since July 2022. Prior to this, he was the Professor and Head of the Department of Information Technology at Dr. N.G.P Arts & Science College (Autonomous), Coimbatore. His tenure at this institution, spanning from January 2020 to June 2022, followed a series of progressive roles, including Associate Professor and Head of the Department, and earlier, as an Assistant Professor. Dr. Rajeshkanna has made significant academic contributions through his research and scholarly activities. He has published 35 papers in reputed international journals with high impact factors and has contributed 16 book chapters through IGI Global Publications.

S. Raj Kumar received his bachelor's degree in Electronics and Communication Engineering from Anna University, Chennai, and his master's degree in VLSI Design from the same institution. He completed his doctoral degree at Anna University, Chennai, specializing in VLSI device modeling. Currently, he serves as an Assistant Professor in the Department of Electronics Engineering (VLSI Design and Technology) at Velammal College of Engineering and Technology, Madurai, Tamil Nadu, India. His career spans over 11 years, during which he has made significant contributions to teaching and research. His research interests include diverse areas such as modeling and simulation of analog VLSI. He has published over 10 research papers in international journals and presented 20 papers at international conferences. He has also received the award for "Best Outcome in Academic Performance".

Ashok Immanuel V, holds a degree in Physics from Loyola College, Chennai. He did his Master of Computer Application (MCA) from Bharathidasan University and M.Phil from Alagappa University. He completed his PhD from CHRIST University, Bengaluru. His area of research is context-aware systems. He has published papers on Educational Technologies and Context-aware Systems. Currently, he is working on Contextual Artificial Intelligence and guiding scholars in the area of Micro Service Architecture, Software Services, Contextual Mobile Networks and Contextual Augmented and Virtual Reality. He has a total of 24 years of experience in teaching both undergraduate and postgraduate students. At CHRIST University he is currently serving as Professor and Head of the Department of Computer Science. He has designed, developed and implemented several software projects for internal use at CHRIST University. He has been a consultant for colleges and universities on Educational ERP. He has been a resource person for many workshops on leveraging technology for the academic fraternity and management. He has delivered Guest lectures, invited talks and keynote addresses at various national and international conferences and seminars conducted by academic institutions and the Confederation of Indian Industry (CII).

Ashok Vajravelu Department of Electronics, Faculty of Electrical Engineering, University Tun Hussein Onn Malaysia Johor, Malaysia ashokvajravelu8@gmail.com

Nitisha Waghela is an accomplished Professor with a passion for Artificial Intelligence and Security. With 10years of experience in the field, they have developed a deep understanding of AI Prof. Nitisha Waghela holds a M. Tech degree in CSE from RGPV University. They have contributed significantly to the field through their research, publications, and practical experience. They have served as a Professor at SAGE University Indore and have been actively involved in research work. Their dedication to advancing knowledge and sharing insights has led them to present at various conferences and deliver workshops.

Swapnil Waghela is an experienced educator and researcher with expertise in Artificial Intelligence and Security. They hold a M.Tech degree in CSE and have dedicated their career to exploring innovative approaches in education. With a passion for leveraging technology, particularly AI and virtual learning, Prof. Swapnil Waghela has published numerous articles and conducted extensive research in the field. Their work focuses on understanding student behavior in virtual learning environments and developing effective strategies to enhance engagement and learning outcomes. Through their contributions, Prof. Swapnil Waghela aims to contribute to the advancement of online education and empower educators with valuable insights and tools for effective virtual instruction. In addition to their professional accomplishments, Prof. Swapnil Waghela is known for their effective communication and ability to translate complex concepts into accessible language. Their writing style combines clarity, depth, and engaging storytelling, making their work accessible to both experts and non-experts alike.

Index

Symbols

A

B

C

S

W

www.ingramcontent.com/pod-product-compliance
Lightning Source LLC
LaVergne TN
LVHW080612300926
849162LV00005B/1127